A LOOK BACK
FOR
FUTURE GEOTECHNICS

A LOOK BACK
FOR
FUTURE GEOTECHNICS

Editors

Wu Shiming
Zhang Wohua
Richard D. Woods

A.A. BALKEMA/ROTTERDAM/BROOKFIELD/2001

ISBN 90 5809 218 6

A.A. Balkema, P.O Box 1675, 3000 BR Rotterdam, Netherlands
Fax: +31.10.4135947; E-mail: balkema@balkema.nl
Internet site: http:// www.balkema.nl

Distributed in USA and Canada by
A.A. Balkema Publishers, 2252, Ridge Road, Brookfield, Vermont 05036 USA
Fax: 802.276.3837; E-mail: Info@ashgate.com

Perface

Since modern soil mechanic took shape in the early-to-mid twentieth century, engineers and researchers throughout the world have made contributions to this new branch of civil engineering, now commonly referred to as Geotechnical engineering. During the past two decades in particular, owing to the acceleration of general developments in science and technology, the techniques and practice of Geotechnical Engineering have experienced enormous progress. Influential in this progress have been adaptations from related disciplines, modernized testing techniques, and the introduction of computer technology into daily practice as well as research in Geotechincal Engineering. Geotechincal Engineering has experienced enormous progress resulting in formation and maturing of many important sub-disciplines like, stability of earth masses, foundation engineering, ground improvement soil-structure interaction, soil dynamics and earthquake engineering, off-shore engineering, applications of numerical methods, constitutive modeling of soils, sophisticated laboratory and in-situ testing, and last, but by no means least geoenvironmental engineering.

To summarize the achievements in Geotechnical Engineering during the 20th century and encourage young colleagues to continue with improvements and innovations in the new century, Professors R.D Woods of the University of Michigan and Wu Shiming of Zhejiang University initiated the International Lecture Series on Geotechnical Engineering for the 21st Century. The Lecture Series turned out to be a great success with 12 renowned experts and scholars from around the world giving lectures at Zhejiang University during the period, early 1998 through April 1999. The publication of this book containing the written versions of those lectures is intended to inform technical personnel about many accomplishments in Geotechincal Engineering during the 20th Century, and provide some directions for research and development in the future. It is my hope that this book will provide inspiration and enlightenment for many new developments in the 21st Century.

December 10, 1999

Wu Shiming
Professor of Civil Engineering
Tongji University
Shanghai, P.R. China.

Contents

1

Biotechnical Slope Protection and Erosion Control

Donald H. Gray
The University of Michigan
Ann Arbor, Michigan,
USA

ABSTRACT

Biotechnical and soil bioengineering stabilization offer a cost-effective and attractive approach for stabilizing slopes against erosion and shallow mass movement. Biotechnical stabilization is characterized by the combined or integrated use of live vegetation with retaining structures and revetments. Vegetation can also be employed in combination with structural elements in ground cover systems. Rolled erosion control products (RECPs) are a good example of this combined approach. Soil bioengineering primarily entails the use of plants and live plant parts themselves. Plant roots and stems, serve as the main structural and mechanical elements in a slope protection system. These approaches capitalize on the advantages and benefits that vegetation confers for erosion control and slope protection and offset the limitations of using conventional slope plantings by themselves. Biotechnical stabilization has been used successfully in a variety of applications ranging from watershed restoration and streambank protection to stabilization of over-steepened cuts and embankment slopes along highways.

Key Words: Earth reinforcement, geotechnical engineering, soil stabilization, biotechnical stabilization, soil bioengineering, slope stability, vegetation.

1. INTRODUCTION

Biotechnical and soil bioengineering stabilization both entail the use of live materials—specifically vegetation. These approaches capitalize on the advantages and benefits that vegetation confers for erosion control and slope protection. The value of vegetation in civil engineering in general and the role of woody vegetation in particular for stabilizing slopes is now generally well recognized (Greenway, 1987; Coppin and Richards, 1990; Gray, 1994).

Biotechnical and soil bioengineering stabilization can be viewed as "mixed" construction approaches (see Table 1) that utilize both inert and live materials. While both approaches share this common attribute, they are characterized by subtle but important differences.

Table 1. Classification of slope protection and erosion control measures

	CATEGORY	EXAMPLES
	LIVE CONSTRUCTION Conventional Planting	* Grass seeding * Sodding * Transplants
MIXED CONSTRUCTION	Woody plants used primarily as reinforcements and barriers to soil movement	* Live staking * Live fascines * Brushlayering * Live slope grating
	Plant/structure associations	* Breast walls with slope face plantings * Revetments with slope face plantings · Tiered structures with bench plantings
	Woody plants grown in the frontal opening or interstices of retaining structures	* Planted cribwalls * Vegetated rock gabions * Vegetated geogrid walls * Vegetated breast walls
	Vegetation grown in the frontal opening or interstices of porous revetments and ground covers	* Joint plantings * Staked gabion mattresses * Vegetated concrete block revetments * Vegetated cellular grids * Turf reinforcement mats
	INERT CONSTRUCTION Conventional Structures	* Concrete gravity walls * Cylinder pile walls * Tie-back walls

Biotechnical stabilization describes the integrated or combined use of living vegetation and inert structural or mechanical components. The latter includes concrete, wood, stone, and geofabrics. Geofabrics refer to woven or non-woven geotextiles and geogrids made from either synthetic polymers or from natural materials such as jute and coir. Both biological and mechanical elements must function together in an integrated and complementary manner.

Soil bioengineering is a more specific term that refers primarily to the use of live plants and plant parts alone. Live cuttings and stems are purposely embedded and arranged in the ground or in earthen structures, where they serve as soil reinforcements, hydraulic drains, barriers to earth movement, and hydraulic pumps or wicks. Techniques such as live staking, live fascines, brushlayering, etc., fall into this category. Some soil bioengineering techniques also entail the use of live cuttings in combination with untreated wood logs, e.g., live slope gratings and live crib walls. Soil bioengineering treatments provide sufficient stability so that native vegetation can gain a foothold and eventually take over the role provided by inert structural components.

"Reinforced grass" a term coined by Hewlett et al. (1987) refers to a grass surface which has been artificially augmented with either soft armor or hard armor (e.g., articulated blocks). A typical soft-armor system consists of very porous, synthetic, three-dimensional mats (*turf reinforcement mats*) that are placed on the ground, filled with soil, and seeded. These mats improve the performance of the vegetation by tying (entangling) plant roots together and bridging over weak spots. Grassed surfaces reinforced in this manner resist the tractive force of high-velocity water flow better than a grassed surface by itself.

Live plants and variants of soil bioengineering have been used for centuries to control erosion problems on slopes and along river banks in different parts of the world—including mainland China. This paper briefly describes selection and design guidelines for biotechnical and soil bioengineering stabilization. The results of some applications or actual case studies are presented as well.

2. DESIGN AND SELECTION GUIDELINES

A wide variety of different soil bioengineering and biotechnical stabilization methods are available for both stream bank and upland slope protection. A brief review is provided here of design and selection guidelines; the reader is referred elsewhere (Gray and Sotir, 1996) for more details.

2.1 Biotechnical Stabilization

Biotechnical slope protection and erosion control entails the integrated or conjunctive use of plants and structures. Plants can be introduced and established in and around structural systems as noted in Table 1 under the category of "mixed construction" methods. Plants can be introduced, for example, on the benches of tiered retaining wall systems or alternatively they can be inserted and established in the interstices or frontal openings of porous revetments, cellular grids, and retaining structures. Open-front crib walls and gabion walls lend themselves well to the latter approach.

Vegetation can be introduced into virtually any type of porous revetment system. Joint planting, for example, which is a variant of a *live staking* (a soil bioengineering technique), consists of tamping or inserting live cuttings between bank armor stone or riprap. Live cuttings can also be introduced into gabion mattresses and articulated block revetments.

The structural components of biotechnical earth support and slope protection systems must be capable of resisting external forces causing sliding, overturning, and bearing capacity failure. In addition these structures must resist internal forces that cause shear, compression, and bending stresses.

2.2 Soil Bioengineering

Soil bioengineering methods can be used to prevent and control surficial erosion and shallow mass wasting. Different methods or combination of methods can be used on (1) natural hillslopes, (2) cut and fill slopes along roadways, (3) landfill covers, (4) spoil banks, and (5) streambanks. Some methods are better suited than others for particular site conditions and objectives. *Live fascines,* for example, provide good protection against erosion and are relatively easy to install on both cut and fill slopes. *Brushlayering,* on the other hand, provides better reinforcement and protection against shallow mass wasting but is more difficult to install on cut slopes. *Live crib walls* provide additional restraint at the base of slopes and also protect the toe.

Soil bioengineering methods can be used in combination with structural or conventional methods. Live stakes, for example, can be: (1) tamped into the ground to provide ancillary protection around a check dam in an eroding channel, (2) inserted through an erosion control netting on a steep slope to improve overall performance, or (3) tamped through openings in a rock revetment to increase lift-off resistance and interlocking of the armor units.

The range of unit costs for different soil bioengineering measures are listed in Table 2. The costs shown are the 'installed" costs of various soil bioengineering measures that were employed on actual projects. These unit costs can be used to obtain relative comparisons between different methods or a suitable inflation rate can be applied to obtain approximate current costs.

2.3 Reinforced Grass

A dense, herbaceous or grass cover comprises one of the best defenses against soil erosion. This is generally true provided the velocity of water flowing over the surface is not of sufficient duration and intensity to degrade the vegetative cover. In the latter case a more resistant cover must be employed or, alternatively, the vegetation can be reinforced or stabilized with two-dimensional, open-weave fabrics, meshes and nets or three-dimensional geosynthetic matting. These "soft armor" systems or biotechnical composites are composed of non-degradable elements which furnish temporary erosion protection, enhance vegetative establishment, and ultimately become intimately entangled with living plant tissue (roots) to extend the performance limits of vegetation. In addition to their reinforcing and anchoring function, some of these erosion control products also behave as mulches, e.g., coir netting, that enhance the establishment of vegetation.

Table 2. Unit costs for soil bioengineering measures (in '94 US $)

Method	Installed unit cost[1]
Live Staking	$1.50 → 3.50 per stake
Joint Planting	$2.00 → 9.00 per stake
Live Fascine	$5.00 → 9.00 per lineal foot ($16 → 30 per lineal meter)
Live Crib Wall	$10.00 → 25.00 per sq. ft of front face ($107 → 269 per sq. m of front face)
Brushlayer—Cut	$8.00 → 13.00 per lineal foot ($26 → 43 per lineal meter)
Brushlayer—Fill	$12.00 → 25.00 per lineal foot ($39 → 82 per lineal meter)
Vegetated Geogrid	$12.00 → 30.00 per lineal foot ($39 → 98 per lineal meter)
Live Slope Grating	$25.00 → 50.00 per sq. ft of front face ($269 → 537 per sq. m of front face)

[1]Installation includes 1) harvesting, 2) transportation, 3) storage, and 4) placement

This reinforced vegetation provides "permanent" protection against medium to high flows. If more aggressive, high velocity flows are encountered then "hard armor" systems must be used in conjunction with vegetation such as riprap, gabion mattresses, or articulated block revetments.

2.4 Rolled Erosion Control Products (RECPs)

Rolled Erosion Control Products are generally flexible nets or mats manufactured from both natural and synthetic materials that can be brought to a site, rolled out, and fastened down on a slope. They are manufactured in diverse forms and combinations of materials, have widely differing properties, and are designed for specific functions or site conditions. Materials used in their manufacture include wood excelsior, straw, coconut fiber, polyolefins, polyvinyl chloride (PVC), and nylon. Products manufactured from these materials include: *erosion control blankets (ECBs), turf reinforcement mats (TRMs),* and *geocellular containment systems (GCSs).* The combination of materials and prescribed structure used in their manufacture enable designers to incorporate some of the best attributes of long-fiber mulches with the tensile strength and reinforcing effect of dimensionally stable nets, meshes, and geotextiles.

These products may be used for either temporary or permanent erosion control, for aiding the establishment of a vegetative cover, for reinforcing and enhancing the effectiveness of mature grass cover under difficult or harsh site conditions, e.g., steep, highly erodible open slopes; channel bottoms and side slopes with moderately high velocities.

2.4.1 Erosion Control Blankets (ECBs)

Erosion control blankets are constructed from a variety of degradable organic and/or synthetic fibers that are woven, glued, or structurally bound with nettings or meshes. Erosion control blankets are manufactured typically of fibers such as straw, wood, excelsior, coconut, polypropylene, or a combination, stitched or glued to or between geosynthetic BOP netting or woven natural fiber netting.

ECBs must be placed in intimate contact with the ground surface for optimum effectiveness. Otherwise, erosion rills will form under the blanket. ECBs are normally fastened to the ground using staples and pins in a number and pattern recommended by the manufacturer according to slope and site conditions. Additionally, the placement of erosion control blankets and other rolled erosion control blankets into anchor trenches will enhance the stability of the installation.

Erosion control blankets are designed to assist in vegetation establishment and to provide temporary erosion protection. They are generally limited to areas where natural, unreinforced vegetation will eventually provide long-term stabilization and protection. They should be considered for use on sites requiring greater, more durable, and/or longer lasting protection than mulching alone or mulches with erosion control netting.

2.4.2 Turf Reinforcement Mats (TRMs)

Turf reinforcement mats consist of geosynthetic matting manufactured from various UV-stabilized synthetic fibers and filaments processed into permanent, high-strength, 3-D matrices. These mattings can be regarded as a type of permanent, "soft armor" alternative to rock riprap. They are designed for permanent and critical hydraulic applications such as drainage channels, where expected discharges result in velocities and tractive shear stresses that exceed the limits of mature, natural vegetation. A 3-D mat functions as an open, stable matrix for the entanglement of plant roots, stems, and soil which together form a coherent, living matrix.

2.4.3 Geocellular Containment Systems (GCSs)

Strictly speaking, a geocellular containment system is not a rolled, geosynthetic erosion control product. It consists of individual 3-dimensional cells up to 20 centimeters deep. The system is usually manufactured from polyethylene or polyester strips. When expanded into position, the cells have the appearance of large a honeycomb. The cells can be back-filled with soil, sand or gravel depending on the application. For conjunctive use with vegetation, the soil-back-filled cells are also seeded and fertilized.

2.4.4 Classification by Application/Performance

Ongoing efforts to classify rolled erosion control products according to application and performance are described by Lancaster and Austin (1994). They note that the various products span applications based on permanent vs. temporary control, and low- vs. high-velocity flows. They devised a classification system that separates the various products into the following categories: (1) Low-velocity degradable RECPs, (2) High-velocity degradable RECPs, and (3) Long Term, non-degradable RECPs, as shown in Table 3.

3. RECENT CASE STUDIES

3.1 Highway Cut Slope Repair

Both the California Department of Transportation and the US Forest Service have experimented with soil bioengineering measures to control erosion on cut and fill slopes in mountainous terrain along highways in the Sierra Nevada. In the project described herein a variety of soil bioengineering measures were used to repair recent slope failures along a highway cut in the eastern United States.

Roadway widening and encroachment during upgrading of a two-lane highway in Massachusetts, resulted in over-steepened slopes with marginal stability. Shallow surface failures occurred at the site in addition to deep-seated slumps in soil sections with little or no bedrock exposure. The site stratigraphy consisted of a silty, sandy residual soil with variable thickness overlying, a quartz-mica schist bedrock. A high groundwater table was present and ground water seeped from the slope face and fractures in the underlying bedrock.

The fractured nature of the bedrock, its uncertain depth and unfavorable attitude ruled out the use of structural retaining walls resting on the bedrock interface. A drained rock buttress at the toe of the cut would have satisfied mass stability requirements but would have left the top part of the cut exposed and vulnerable to seepage erosion. This limitation could have been overcome by running the rock all the way up the slope in the form of a rock blanket. This solution, however, was visually stark and environmentally incompatible. The solution (Gray and Sotir, 1996) that was eventually devised consisted of placing a 10-foot high rock buttress at the toe which in turn supported a drained, brushlayer fill above as shown in Fig. 1. A drainage course was placed behind brushlayer fill to intercept and divert seepage down into the rock toe. Live fascines were placed along the crest of the cut above the brushlayer fill.

The slope is now stable, well vegetated, and blends in with the natural surroundings. The brushlayers stabilized the cut and provided an opportunity for native vegetation to invade and establish on the slope as shown in Fig. 2.

Table 3. Performance classification of rolled erosion control products (RECPs) (adapted from Austin and Driver, 1995)

Required performance characteristics	Suitable types of RECPs	Properties and typical applications
LOW-VELOCITY DEGRADABLE	o single net, organic fiber ECBs o biodegradable natural fiber and photodegradable geosynthetic nets	o one- to two-season longevity o limited capacity in resisting damage and resisting erosion under severe conds. o slopes of moderate grade, length and runoff; channels where potential for damage during installation is minimal
HIGH-VELOCITY DEGRADABLE	o double net ECBs or high strength nettings and/or incr. quantities of organic fibers o dense, open-weave geotextiles or meshes (e.g., coir, jute, polyprop.)	o similar to low-velocity degradables in installation and function, but designed for more severe site conditions o heightened durability and longevity (1→5 years) o steeper slopes and high velocity channel linings where natural, unreinforced vegetation is expected to provide permanent stabilization
LONG-TERM NON-DEGRADABLE	o high strength geosynthetic mattings and cellular containment e.g., ECTRMs, TRMs, GCSs.	o provide immediate, high performance erosion protection followed by perm; reinforcement of established vegetation o steep slopes with very erodible soils; channel linings subjected to relatively high, long-duration flows.

Fig. 1. Brushlayer buttress fill used to repair and stabilize highway cut. View of slope three months after construction

Fig. 2. Brushlayer buttress fill two years later. The slope is stable and native vegetation has invaded site and become well established

As a result, the process of plant succession is well underway and, after three years, the project site has a pleasing, natural appearance.

3.2 Streambank Stabilization

The objective in this case was to repair and stabilize a high river bank slope adjacent a bayou in Houston, Texas, that had failed as a result of frequent flood discharges. The combination of natural flooding and controlled releases resulted in the abrupt rise and fall of water level in the bayou in addition to prolonged periods of high water. These hydrologic conditions, combined with sandy and silty soils with little cohesion, resulted in widespread erosion and streambank failure as shown in Fig. 3.

Successful treatment required redirecting the stream away from the foot of the bank, rebuilding and securing a toe berm, and stabilizing the slope face. Bank return flows during rapid drawdown after bankfull floods caused significant seepage erosion and slumping of the bank face. Accordingly, the interception and diversion of seepage water from the bank was also essential for successful stabilization. In addition to these functional considerations, visual and esthetic concerns dictated a treatment that blended with the natural riparian surroundings.

Fig. 3. Appearance of the streambank site before stabilization. Slope failures and erosion resulted in near vertical scarps and recession of the crest (from Gray and Sotir, 1996)

A multi-pronged approach was adopted to deal with the various site conditions and requirements (Gray and Sotir, 1996). A buttress fill with a grade of 0.5:1 (H:V) was reconstructed upon a foundation of wrapped concrete rubble installed in a 7-ft (2 m) deep toe-trench. The fill was constructed in 2-foot (61cm) lifts wrapped with a synthetic geogrid and burlap. Live brush layers long enough to extend from the undisturbed soil at the back and project several feet beyond the reconstructed fill face were placed on each wrapped soil layer.

Vertical strip or chimney drains were placed at the rear of the fill on 5-foot (1.5 m) centers. These drains intercepted and collected the bank seepage water and conducted it into a gravel drain beneath the fill. Live fascines were placed on the mid-slope bench of the reinforced fill. Live stakes were tamped into the ground along the edges and sides of the fill. A live boom (a rock groin capped with a prism of earth reinforced with criss-crossing fascines) was used to redirect the stream away from the bank. A view of the site immediately after construction is shown in Fig. 4.

The site has remained stable since construction, and the soil bioengineering installation has developed into a dense, riparian buffer. The site survived a major bankfull flood in the second year after construction. The supple stems of woody vegetation on the bank bent over during the flood and protected the bank against scour erosion.

Fig. 4. Biotechnically stabilized river bank after construction. Benched fill reinforced with vegetated geogrids rests atop rubble filled toe-tench (from Gray and Sotir, 1996)

4. FUTURE DEVELOPMENTS

Biotechnical stabilization is a dynamic and evolving field. New advances and developments continue to reshape the range of options and possibilities. Biotechnical groundcover systems, for example, were developed largely during the past decade. New ways of combining inert construction materials with living vegetation for slope protection and erosion control loom on the horizon (Gray and Sotir, 1996; Barker, 1994).

At the same time old uncertainties persist about the performance of living, constructed systems. The properties of plant materials and the characteristics of plant communities vary in time and space. It is not easy to ascertain much less predict the distribution and behavior of below ground root systems over time. This makes it difficult to analyze and quantify the contribution of vegetation to the stability of slopes.

There is no reason why biotechnical stabilization should concern itself solely with the conjunctive use of plant materials. Biological microorganisms, namely, fungal and bacterial colonies, are potential candidates for soil stabilization as well (Muir Wood et al., 1994). A coral reef is a quintessential biotechnical structure—the marine equivalent of a terrestrial hedgerow wall. A reef is built up with remains and excretions of microscopic marine organisms. The cementitious excretions of biological organisms can be thought of as type of "biological glue" that plays a role similar to that of plant roots.

Biotechnical and soil bioengineering methods greatly expand our range of options for stabilizing and protecting slopes against erosion and shallow mass movement. More to the point they allow us to carry out this mission in a cost-effective, visually attractive, and environmentally sensitive manner. Many soil bioengineering systems have been installed in the world that have met the test of time and that have ultimately resulted in stable slopes with a diverse habitat and natural appearance. Questions still remain with regard to quantification and performance evaluation; but answers to these questions will emerge as more experience is gained with this type of stabilization.

The use of biological organisms for biotechnical stabilization purposes, namely fungal and bacterial colonies, appears to be a promising new avenue of approach. As Muir Wood et al., 1994 have presciently noted, "... there is a continuous spectrum of biological influences on slope stability extending from microorganisms at one end to the effect of plant roots at the other." It remains for us to capitalize on these influences in a rational and productive manner.

REFERENCES

Barker, D.H. (1994). The way ahead: Continuing and future developments in vegetative slope engineering or ecoengineering. *Proceedings,* Intl. Conf. on Vegetation and Slopes, Institution of Civil Engineers, University Museum, Oxford, England, 29-30 September 1994.

Coppin, N.J. and Richards, I. (1990). Use of Vegetation in Civil Engineering, Butterworths: Sevenoaks, Kent (England).

Gray, D.H. (1994). Influence of Vegetation on the Stability of Slopes, *Proceedings,* Intl.

Conference on Vegetation and Slopes, Institution of Civil Engineers, University Museum, Oxford, England, 29-30 September, 1994, pp. 1-23.

Gray, D.H. and Leiser, A.T. (1982). Biotechnical Slope Protection and Erosion Control. Van Nostrand Reinhold: New York, N.Y.

Gray, D.H. and Sotir, R. (1992). Biotechnical stabilization of cut and fill slopes. *Proceedings,* ASCE-GT Specialty Conference on Slopes and Embankments, Berkeley, California, June 1992, Vol. 2, pp. 1395-1410

Gray, D.H. and Sotir, R. (1996). Biotechnical and Soil Bioengineering Slope Stabilization. John Wiley & Sons: New York, N.Y.

Greenway, D.R. (1987). Vegetation and Slope Stability, In: *Slope Stability,* edited by Anderson and Richards, John Wiley & Sons, N.Y.

Hewlett, H.W. et al. (1987). Design of reinforced grass waterways. CIRIA Report No. 116, Const. Indus. Res. & Info. Assoc., London, England, 118 pp.

Lancaster, T. and Austin, D.N. (1994). Classifying rolled erosion-control products. *Geotechnical Fabrics Report,* Vol. 12, No. 6, pp. 16-22.

Muir Wood, D., Meadows, A. and Murray, J.H.M. (1994). Effect of fungal and bacterial colonies on slope stability. In: *Vegetation and Slopes,* D.H. Barker (ed.), Institution of Civil Engineers, Proceedings of Intl. Conference held at the University Museum, Oxford, England, September 29-30, pp. 46-51.

2

Ground Behavior during Earthquake and its effect on Foundation Piles

Kenji Ishihara

*Department of Civil Engineering Science, University of Tokyo,
2641 Yamazaki, Noda-Shi, Chiba-ken, 278, Japan*

1. SOIL PROPERTY CHARACTERIZATION

1.1 Soil Conditions and its Strength during Earthquake

Introduction

The violent ground shaking during the Hyogoken-Nambu earthquake caused landfills in the port area to liquefy leading to widespread occurrence of ground deformation including settlement and lateral spreading. As a result of the ground distortion, various types of damage were incurred by infrastructure such as lateral shifting of quaywalls, breakage of gas and water mains. The landfills in the several affected areas were first reclaimed around 1953 by transporting soil from borrow areas at the foothill of the Rokko Mountains. Large amount of the soils was transported by a long-distance belt conveyer system to the sites of reclamation along the old shoreline. Several islands north of Rokko Island were constructed during this period. The second phase of land reclamation began in 1968 to construct two large man-made islands further offshore to the south. Port Island with an area of 436 hectare was constructed during the period from 1966 to 1980 by transporting soils from Suma which were carried in bottom dump-barges and placed under water at the site of reclamation. The 580 hectare Rokko Island was constructed between 1972 and 1990 by excavating soil materials from Suma and also from other sites in the Rokko Mountains. The grain size distribution of the materials used for the reclamation of Port and Rokko Islands are shown in Fig. 1.1

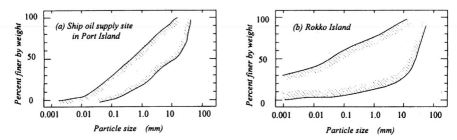

Fig. 1.1. Grain size distribution of the materials used for land reclamation:
(a) Ship oil supply site in Port Island; (b) Rokko Island

Laboratory Tests

Prior to the earthquake a series of cyclic triaxial tests was performed by Nagase et al. (1995) on undisturbed samples of Masado soils recovered from the bottom of a cut. The reclaimed site in the northern section of Port Island was excavated to provide an approach to the gate of a freeway tunnel which was under construction. Undisturbed samples were recovered in block and frozen in the field before they were brought to the laboratory. After carefully trimming into specimens 7.5 cm in diameter and 15 cm in height, they were placed in the triaxial chamber and consolidated to varying effective ambient pressures. In some tests, the consolidation pressure was reduced to produce a state of overconsolidation with the *OCR*-value of 2 and 4.

The grain size distribution curves of the samples tested are shown in Fig. 1.2 where it can be seen that the soil consists of about 5% fines, 55% gravel and 40% sand. The gravel is defined here as a material having a grain size larger than 2 mm, and the fines as the soil having a particle size smaller than 0.074

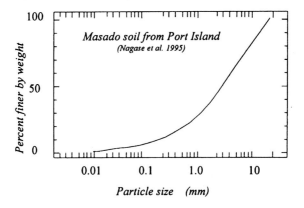

Fig. 1.2. Grain size distribution curve for the Masado soil tested

mm. It should be noted that the Masado soil prevailing in the coastal area of Kobe is well-graded, containing a fairly large proportion of gravel. Such a soil has generally been considered as a material practically "immune to liquefaction" and not been specified for detailed studies in the laboratory as well as in the field.

The results of the cyclic loading tests are presented in Fig. 1.3 in terms of the cyclic stress ratio plotted verses the number of cycles required to produce cyclic softening with development of 5% double-amplitude axial strain in the sample. Figure 1.3 indicates that the cyclic strength, defined as the cyclic stress ratio causing 5% D.A. axial strain in 20 cycles of load application, is about 0.18 for normally consolidated samples, which is a value commonly encountered in a loose to medium loose sand in a normally consolidated state. It can also be seen in Fig 1.3 that the cyclic strength increases significantly with increasing ratio of overconsolidation, taking on a value of 0.3 and 0.4, respectively, for the *OCR* of 2 and 4. The influence of *OCR* as stated above may be put in an empirical formula,

$$\left(\frac{\sigma_{dl}}{2\sigma_0'}\right)_{OCR} = (OCR)^n \cdot \left(\frac{\sigma_{dl}}{2\sigma_0'}\right)_N \tag{1-1}$$

where $(\sigma_{dl}/2\sigma_0')_N$ and $(\sigma_{dl}/2\sigma_0')_{OCR}$ denote respectively the cyclic strength of normally and overconsolidated specimens. The exponent n was shown to have a value of 0.5 by Ishihara and Takatsu (1978) based on cyclic triaxial tests on clean sand. Nagase et al. (1995) showed a value of $n = 0.40 - 0.45$ for the undisturbed samples obtained from the site in Port Island.

Another set of triaxial test results on undisturbed samples from Port Island

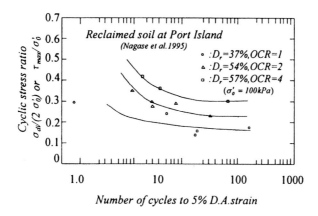

Fig. 1.3. Cyclic stress ratio verses number of cycles (Nagase et al., 1995)

was performed by Yasuda (1990). The samples were recovered by what is called triple-tube sampling and tested using the cyclic triaxial test apparatus. The outcome of these tests is presented in Fig. 1.6, together with data for the case of *OCR*-1 and 2 by Nagase et al. (1995) quoted from Fig. 1.3. It may be seen that the test data from Yasuda (1990) shows a cyclic strength of about 0.25 for 20 cycles of load application which is a value larger than the data obtained by Nagase et al. (1995) for the case of *OCR*-1.0.

Relative Density Interpretation

According to the method of JSSMFE (Japanese Society of Soil Mechanics and Foundation Engineering), it is stipulated that the maximum and minimum void ratio e_{max} and e_{min} be determined for the predominantly sand portion having particle size smaller than 2 mm while keeping its fines passing #200 mesh less than 5%. For the Masado soil, e_{max} and e_{min} were determined by Nagase et al. (1995) in accordance with these requirements, yielding values of e_{max} = 1.098 and e_{min} = 0.526. These values may well be taken as the limiting void ratios for the sand portion contained in the Masado soil.

The void ratio of the sand portion in the undisturbed samples can be determined without difficulty as follows. Suppose the gross void ratio of an undisturbed sample is assumed to be e = 0.48, for example. If the gravel fraction having particle size larger than 2 mm is 42% in weight, then the void ratio for the silt-sand portion e_s can be readily calculated as e_s = 0.48/(1−0.42)=0.828. The void ration thus determined for the silt-sand portion may be considered to represent the state of packing of the major constituent of soil which was responsible for the inducement of liquefaction during the Kobe earthquake. The limiting void ratios determined above by the method of JSSMFE are thus taken as the upper and lower bounds of the possible values of void ratio e_s for which the relative density of the silt-sand portion can be most logically defined. In the particular example stated above, the relative density is determined as 47% for the silt-sand portion. The relative density thus obtained is indicated in Fig. 1.3 for the specimens used in the cyclic triaxial tests on the undisturbed specimens from the site of Port Island. It may be seen that the cyclic strength as correlated with the relative density for the gravel containing silty sand is about the same as the correlation between these quantities that have been established for clean sand or silty sand. It should be noted therefore that if the relative density is determined for the silt-sand matrix by scalping the oversized gravel portion, it may be used as an index property to identify the liquefiability of a composite soil containing silt, sand and gravel. This concept is applicable only for the case of gravel content probably less than about 50% where gravel particles exist not in contact with each other in the matrix of silt and sand. For this condition, the presence of gravel may be assumed not to exert any influence on the overall resistance of the soil.

Back Calculation of Strength of Soil

As described in the report by Toki (1995), a set of records was monitored by four

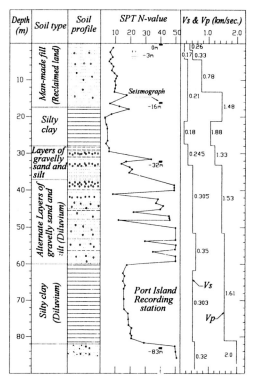

| Depth (m) | Soil type | Soil profile | SPT N-value | Vs & Vp (km/sec.) |

Fig. 1.4. Soil profile at the site of vertical array of seismograph (Toki, 1995)

accelerometers in a vertical array installed at a site in the northern part of Port Island. The soil profile and the time histories for two horizontal components obtained by Toki (1995) are reproduced in Figs. 1.4 and 1.5. The man-made fill to the depth of 18 m is composed basically of the same Masado soil as that at the tunnel site where undisturbed samples for the laboratory testing were recovered. In view of the low SPT N-value, about 5 to 10, the cyclic softening due to liquefaction appears to have taken place mainly in this near-surface deposit. In fact, the second thrust of acceleration in the south direction was shown to decline significantly at the surface (341 cm/sec^2) and at a depth of 16 m (340 cm/sec^2). A similar decrease in acceleration was also observed in the E-W component. It is, therefore, highly likely that the cyclic softening due to liquefaction occurred during one to two cycles of seismic load application, In addition, it may be reasonable to assume that the peak acceleration at the onset of liquefaction was $a = a_{max} = 340$ cm/sec^2. The cyclic stress ratio τ / σ_v at any instant of seismic shaking is estimated by the equation:

$$\frac{\tau}{\sigma_v'} = \frac{a}{g}(1 - 0.015Z)\frac{\sigma_v}{\sigma_v'}$$

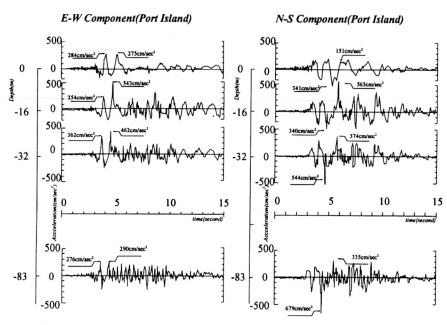

Fig. 1.5. Accelaration recorded at the site on the site on Port Island (Toki, 1995)

where a denotes the acceleration and σ_v, and σ_v, are the total and effective overburden pressures acting on a soil element located at a depth of Z (in meter). The depth of the ground water table at this location was estimated to be approximately 3 m. Introducing the value of $a = a_{max} = 340$ cm/sec^2 and $g = 980$ cm/ sec^2, the maximum stress ratio that must have been applied to soil elements at varying depths can be calculated. The maximum stress ratio thus computed from field observations was found to increase with increasing depth, to magnitudes of about 0.5 to 0.7 at shallow depths. This cyclic stress ratio is plotted in Fig. 1.6 versus the number of cycles which are assumed to be one to two as mentioned above. It should be noted that the cyclic stress ratio plotted is expressed in terms of the maximum stress divided by the mean confining stress σ_0' instead of the effective overburden stress σ_v'. The conversion from σ_v' to σ_0' was made simply through the use of the following relationship:

$$\sigma_0' = \frac{1 + 2K_0}{3} \sigma_v'$$

(1-3)

where K_0 is the coefficient of earth pressure at rest. This conversion represents the computed data in a way that permits direct comparison with the laboratory test data. While an estimate of an exact K_0-value is somewhat difficult, an assumed value between 0.5 and 0.75 is considered reasonable. Based on these values, the

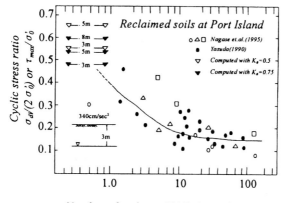

Number of cycles to 5% D.A. axial strain

Fig. 1.6. Summary plot of cyclic strength from laboratory tests and from computation

cyclic stress ratio τ_{max}/σ'_0 was calculated and is shown in Fig. 1.6.

Comparing the two sets of data, one from the laboratory and the other from the computation based on field measurements, it can be mentioned that the laboratory cyclic strength is almost coincident with the value of cyclic strength computed from the observed acceleration at the time of strong shaking during the Kobe earthquake.

1.2 Undrained Behavior of Soil and State Concept

Introduction

During the Hyogoken-Nambu earthquake of January 17, 1995 ($M=7.2$), widespread liquefaction occurred in the landfills along the seaside area of Kobe city leading to large ground deformation and severe damage to engineering structures. The liquefaction was particularly extensive and devastating in the reclaimed lands of the man-made Port Island, resulting in an average settlement of 40 to 50 cm in the unimproved areas (Yasuda et al., 1996; Ishihara et al., 1996) and lateral displacements of the quay walls toward the sea from 2 to 5 m (Inagaki et al., 1996). The seaward displacement was associated with lateral spreading and cracking of the ground which progressed inland to a distance of about 150 m from the revetment line (Ishihara et al., 1996).

Port Island was constructed in the period between 1966 and 1980 by transporting soils from Suma, Rokko Mountains, in bottom dump-barges and dumping the soil under water at the reclamation site. The thickness of the relcaimed deposit is from 15 m to 20 m. The fill material used for the reclamation of Port Island is a decomposed granite known as Masado. It is a well-graded soil containing a fairly large portion of gravel. Defying expectations, a large part of the reclaimed area liquefied during the Hyogoken-Nambu earthquake.

In essence, all of the originally reclaimed soils or untreated deposits liquefied.

To investigate the characteristics of the undrained behavior of the reclaimed soil, multiple series of triaxial tests on reconstituted and undisturbed samples of Masado soil recovered from Kobe Port Island were carried out (Tsukamoto et al., 1998). This paper is an attempt to provide a proper interpretation for these test results and characterize the undrained behavior of the reclaimed Masado soil within a more general framework through the use of the state concept proposed previously (Ishihara, 1993). The employed approach enables a rational interpretation to be made for the behavior of both reconstituted and undisturbed samples of the Masado soil.

Laboratory Tests

Materials Tested: After the earthquake, a batch of reclaimed Masado soil was recovered from the Minatojima Tunnel site on Port Island. By sieve analyses, two soils with distinct gradation curves, denoted as *A* and *B* in Fig. 1.7, were provided from the recovered original soil. Soils *A* and *B* were obtained by removing particles greater than 16.0 mm and 6.73 mm in diameter, respectively. Most of the laboratory tests discussed in the following were carried out on reconstituted samples from these two soils.

Besides the tests on soils *A* and *B*, results of tests on reconstituted samples from soil *C* (Goto et al., 1997) are also considered. Soil *C*, with a gradation curve shown in Fig. 1.7, was obtained by eliminating particles greater than 2 mm in diameter or the gravel fraction of a reclaimed Masado soil from Port Island. Thus, three soils with different gradation characteristics but all originating from the same reclaimed Masado soil at Port Island were used for tests on reconstituted samples. The grain-size characteristics of the tested soils are summarized in Table 1-1. The maximum and minimum void ratios were

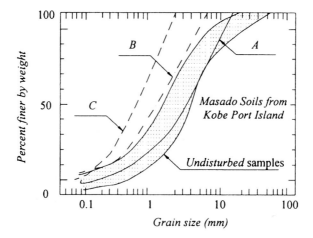

Fig. 1.7. Gradation curves of undisturbed samples and soils used for reconstituted samples

Table 1-1. Grain-size characteristics of Masado soils used for reconsituted samples

Soil	D_{max}	D_{50}	U_C	P_G	ρ_S	e_{max}	e_{min}
(mm)	(mm)			(%)	(g/cm³)	(D < 2 mm)	
A	16.00	2.51	35.3	55.0	2.638	1.043	0.485
B	6.73	0.83	21.7	30.8	2.638	1.043	0.485
C	2.00	0.57	8.7	0.0	2.654	1.046	0.576

determined according to the method of JSSMFE for sand fractions of soils with particles smaller than 2 mm in diameter. The parameter P_G in Table 1-1 indicates the percentage of the gravel fraction ($D \geq 2$ mm) by weight. As illustrated in Fig. 1.7 and Table 1-1, there is a significant difference in the grain-size characteristics regarding the coarse fractions or gravel contents whereas the sand fractions and the fines contents of the soils are similar.

In addition to the tests on reconstituted samples, triaxial tests on undisturbed samples recovered from three sampling sites at Port Island were also conducted. The samples from the Minatojima Tunnel sampling site (UD-samples) were recovered by block sampling (Tsukamoto et al., 1998). Gradation curves of the undisturbed samples lie within the shaded area indicated in Fig. 1.7. The gravel fraction P_G of these samples ranges between 37.1% and 63.2% while the fines content ($D < 0.074$ mm) is from 4.5% to 12.6% by weight.

Reconstituted Samples: Reconstituted samples from soils A and B were formed either by water sedimentation or wet tamping. This choice of method for preparation of reconstituted samples was based on the reasoning that water sedimentation might resemble the actual deposition process exercised on Port Island at the time of reclamation when the soils were dumped from barges. On the other hand, it is envisioned that the materials might have settled in large blocks with water soaking into pores afterwards. This process might be duplicated in the laboratory by the method of wet tamping. Having this in mind, wet tamping was also adopted as an alternative method for preparing samples. In the case of water sedimentation, the soil was first mixed with de-aired water at a water content of 4%, and then poured from just above the water surface to sediment through a height of 4 cm under water. In this manner the loosest possible samples were prepared by water sedimentation. Denser samples were prepared by imparting compacting energy to samples; this was achieved by hitting the side of the pedestal with a hammer. For samples prepared by the wet tamping method, the soil was first mixed with de-aired water to yield a water content of 5% and then strewn by hand to a predetermined level. The sample was prepared in 6 layers and tamping with a small flat-bottom tamper was lightly applied for each layer successively. By changing the tamping energy or the number of tampings during the compaction, different initial densities of the samples were achieved.

Using the above procedures, cylindrical samples with a diameter of about 8 cm and a height of approximately 16 cm were prepared. The samples were put in a freezing chamber for a couple of days and prior to testing the frozen surface to each sample was smoothed by filling in the peripheral voids with a moist fine sand. This treatment of the sample surface was done in order to reduce the effects of membrane penetration. The samples were set in the triaxial chamber and allowed to thaw under a confining stress of 30 kPa. Afterwards they were saturated by percolating carbon dioxide gas and de-aired water until the *B* parameter exceeded a value of 0.95. The samples were isotropically consolidated to a confining stress of 98 kPa, 196 kPa or 392 kPa, and eventually they were monotonically sheared under undrained condition either in triaxial compression or triaxial extension, in a strain-controlled manner.

Undisturbed Samples: The recovered undisturbed samples were trimmed from the frozen solid body to form samples for testing 7.5 cm in diameter and 15 cm in height approximately. The treatment of the surfaces of the samples as well as the testing procedures applied to the undisturbed samples were identical to those of the reconstituted samples. Following saturation, the undisturbed samples were isotropically consolidated to a confining stress of 98 kPa or 118 kPa and then subjected to monotonic loading in triaxial compression or triaxial extension, under undrained condition.

Undrained Behavior of Reconstituted Samples

Steady States: Results of the tests on reconstituted samples are discussed in the following through the use of a normalization procedure within the framework of the state concept. The steady states of all reconstituted samples are shown in terms of the void ratio *e* plotted versus the mean normal stress, $p' = (\sigma'_v + 2\sigma'_h)/3$, in Fig. 1.8. Here each point represents the steady state of a given sample or the ultimate *e-p'* state of the sample attained in a triaxial compression or triaxial extension test. At the steady state, the samples exhibited shear deformation under a very small and practically negligible change in the stresses. Typically, the axial strain at the steady states shown in Fig 1.8 was about 10% and 20%, for the triaxial extension and compression tests respectively. The data plotted in Fig. 1.8 include results from triaxial compression tests (soils *A*, *B* and *C*) and triaxial extension tests (soil *A*) on samples prepared by water sedimentation (soils *A*, *B* and *C*) and wet tamping (soils *A* and *B*). Conditions of the tests regarding the method of sample preparation and mode of deformation are listed in Table 1-2. It can be seen in Fig. 1.8 that in spite of the different methods of sample preparation and modes of deformation a steady state line can be defined for each tested soil with a reasonable degree of consistency. For this reason, in the following, no distinction is made with respect to the sample preparation method and mode of deformation when the steady state of the Masado soil is considered. The steady state data shown in Fig. 1.8 define a distinct steady state line for each of the tested soils with different grain compositions. Apparently, these steady state lines are parallel to each other. It is important to note that the

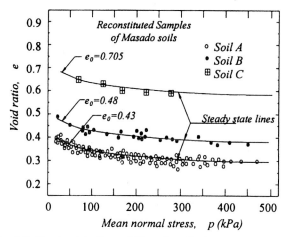

Fig. 1.8. Steady state lines of Masado soils (reconstituted samples)

location of the steady state line in the *e-p′* plot is closely related to the grain size characteristics of the soil, i.e., as the material becomes coarser the steady state line is located lower in the *e-p′* plot. A threshold void ratio e_0, which is defined by Ishihara (1993) as a void ratio above which samples exhibit zero strength upon reaching the steady state in undrained shearing, was determined for each tested soil as the point of intersection between the extended steady state line and the ordinate. As indicated in Fig. 1.8, the threshold void ratio e_0 takes values of 0.43, 0.48 and 0.705 for soils *A, B* and *C*, respectively.

Table 1-2. Tests on reconstituted samples

| Soil | Number of tests | | | |
| | Triaxial compression | | Triaxial extension | |
	WS[1]	WT[2]	WS[1]	WT[2]
A	30	27	15	23
B	11	18	–	–
C	5	–	–	–

WS[1]—Water sedimentation; WT[2]—Wet tamping

The fact that the steady state lines are parallel amongst soils with different grain composition implies that these lines can be brought to a common steady state line if translated along the vertical axis. A convenient way of achieving this is shown in Fig.1.9 where the void ratio axis *e* is replaced by the void ratio difference *e- e_0* (Cubrinovski and Ishihara, 1998). In other words, each steady state point in Fig. 1.8 is shifted downwards along the vertical axis by a distance which is equal to the value of the threshold void ratio e_0. This manipulation results in a single steady state line for all soils encompassing *A, B* to *C*. The

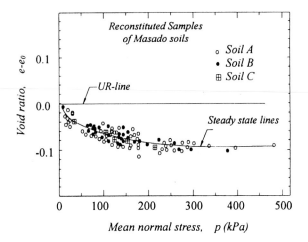

Fig. 1.9. Reference lines of Masado soils in a modified *e-p'* diagram

analytical approximation of the steady states, shown with the solid line in Fig. 1.9, may be given by

$$e_s - e_0 = 0.034 - 0.051 \log p',$$

$$\text{for } p' > 10 \text{ kPa} \tag{1-4}$$

where e_s is the void ratio of the steady state line. This expression provides an excessively steep rise in the analytical steady state line at small normal stresses, and therefore for $p' \leq 10$ kPa, the expression of Eq. (1-4) is replaced with the linear approximation

$$e_s - e_0 = -0.017 \, p',$$

$$\text{for } p' \leq 10 \text{ kPa} \tag{1-5}$$

which is indicated with dashed lines in Figs. 1.8 and 1.9. In addition, this allows greater flexibility in the evaluation of the threshold void ratio e_0 which may be accompanied by difficulties and uncertainties in obtaining experimental steady state data at very small normal stresses, say less than 10 kPa.

Initial States : In the state concept, the state of the soil relative to some characteristic states in the *e-p'* plane is used for specifying deformation behavior during shearing. In this context, Ishihara (1993) proposed the state index I_s as a measure for quantifying the initial state of the soil regarding whether the soil exhibits similar behavior or not in subsequent undrained loading. In defining the state index, two reference lines in the *e-p'* plane were used, i.e., the quasi steady state line (QSS-line) and an upper reference line (UR-line). The UR-line is taken as being either the threshold void ratio e_0 or the isotropic consolidation line for the loosest state. The QSS-line was difficult to define properly in the majority of the tests conducted in the present study, and therefore, the steady state reached at large strain was adopted as a line of reference to determine I_s. Thus, in what follows, the steady state line and upper reference line shown in Fig. 1.9 are used as reference lines for evaluating the state index I_s. In this manner a unique pair of reference lines is defined for the Masado soil irrespective of the grain-size characteristics, method of sample preparation and mode of deformation.

The relative initial states of the reconstituted samples are shown in the modified e-p' diagrams in Fig. 1.10. Here, each symbol represents the state of the soil after consolidation or prior to shearing. During the subsequent undrained shearing, each point moves along a horizontal path until a corresponding steady state point is reached. In general, the samples with initial states above the steady state line exhibit contractive behavior and move leftward whereas those below the steady state line are dilative and move towards the right in Fig. 1.10. To quantify the relative initial states, the state index I_s is calculated for each sample according to (Ishihara, 1993)

$$I_s = \frac{(e_0 - e_0) - (e_c - e_0)}{(e_0 - e_0) - (e_s - e_0)} = \frac{e_0 - e_c}{e_0 - e_s} \tag{1-6}$$

where e_c is the void ratio of the sample after consolidation to an initial mean normal stress p'_c, e_0 is the corresponding threshold void ratio and e_s is the void ratio of the steady state line at p'_c . The void ratio of the steady state line was calculated using Eqs. (1-4) or (1-5). The definition of I_s is schematically illustrated in Fig. 1.10 (c).

The range of the calculated I_s-values for each soil and method of sample preparation is listed in Table 1-3. It is noteworthy that the definition of I_s yields the conditions $I_s = 0$ and $I_s = 1$, for the initial states at the UR-line ($e_c = e_0$) and at the steady state line ($e_c = e_s$), respectively.

Quantification of Undrained Behavior via State Index : According to the definition of I_s, any two samples with different initial states will exhibit similar

Table 1-3. State index values of reconstituted samples

Soil	I_s-values	
	Water sedimentation	Wet tamping
A	0.67-1.60	0.16-1.41
B	0.89-1.21	0.03-0.78
C	0.92-1.68	–

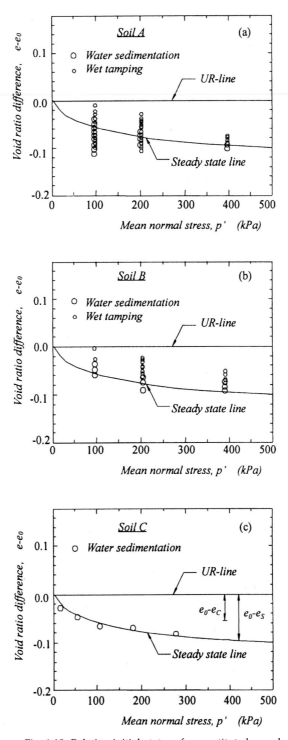

Fig. 1.10. Relative initial states of reconstituted samples

behavior during subsequent undrained shearing provided that the initial states of the samples possess an identical value I_s. In other words, if two samples have similar I_s values, they also have similar effective stress paths and stress-strain curves. This link between the state index and the characteristics of the undrained behavior is examined in Fig. 1.11(a) to Fig. 1.11(d) where effective stress paths from the tests on soil A are comparatively shown. Effective stress paths measured in three triaxial compression tests on samples prepared by water sedimentation are shown in Fig. 1.11(a). These samples have similar relative initial states with I_s-values ranging between 0.80 and 0.85. It may be seen in Fig. 1.11(a) that all of the samples exhibited very similar undrained behavior. Data from compression tests on samples prepared by wet tamping are shown in Fig. 1.11(b). Although each of the samples has a distinct initial *e-p′* state, they can be classified in two groups with respect to the relative initial states, i.e., with $I_s = 0.64$-0.67 and $I_s = 0.84$-0.86. Apparently, the measured effective stress paths within each set of tests are very similar. The samples with smaller state index ($I_s = 0.64$-0.67) are more contractive and approach the steady state with descending stress paths in the *q-p′* diagram, which is a typical manifestation of the flow nature of deformation. Results from triaxial extension tests confirming the link between I_s and the characteristics of undrained behavior are as shown below in Fig. 1.11(c) and 1.11(d).

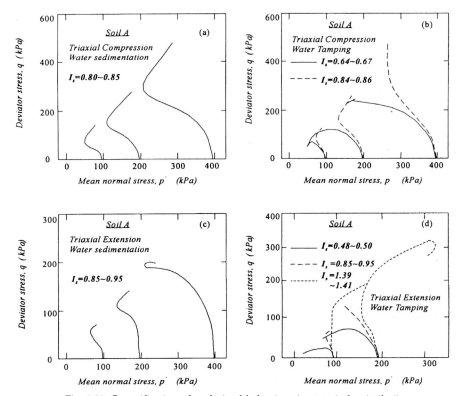

Fig. 1.11. Quantification of undrained behavior via state index (soil A)

The data shown in Fig. 1.11 demonstrate an aptness to quantify the undrained behavior of soil A via the state index. In essence, the link between 4 and the undrained behavior has the following features: 1) samples with similar I_s values exhibit similar undrained behavior; 2) samples with small I_s values are contractive; 3) as the state index increases the contractive behavior gradually transforms to dilative, with a further increase in I_s leading to even stronger diagram.

Effects of Grain Size Characteristics: In spite of the difference in the gradation curves of the Masado soils (Fig. 1.7) and the corresponding difference in the void ratio of initial states (Fig. 1.8), it is now possible to compare the undrained behavior of these different soils on a common criterion based on the state index I_s. Results of compression tests on samples prepared by water sedimentation are displayed for comparison sake in Fig. 1.12 (a) and Fig. 1.12 (b), for soils A, B and C. It is evident that, though the grain composition is different, the effective stress paths are similar if the I_s-values are identical or close to each other. Similar results from tests on samples prepared by wet tamping, are shown in Fig. 1.12 (c). Although there are some differences in this comparison, they are within the range of scatter observed for a single material. Thus, it might be concluded that, regardless of the gradation characteristics of the soil, the undrained behavior of the Masado soil is uniquely related to the initial state as quantified by the state index I_s if the method of sample preparation is the same. It is of interest to note that the state index fails to be an index property if the mode of deposition is different, but it is useful to quantify the behavior of soils with a different grain composition when the soils have been derived from the same origin.

Effects of Sample Preparation Method: The fact that there is no distinction for the steady states with respect to the mode of deposition of the Masado soil implies that, for samples prepared by water sedimentation and wet tamping, the behavior at large strains is either identical or similar. This is illustrated in Figs. 1.13 and 1.14 where the undrained behavior of samples prepared at similar I_s-values by water sedimentation and wet tamping is presented. It is to be noticed that, unlike the behavior at large strains, there is a distinct difference in the behavior at the intermediate stage of deformation, prior to reaching the steady state of deformation. The samples prepared by water sedimentation are initially more contractive and give rise to a higher pore pressure, resulting in smaller deviator stresses at phase transformation.

Accordingly, the stress-strain curves of the water-sedimented samples diverge largely from the initial sharp ascent observed in the samples prepared by wet tamping. This continues to be so until an axial strain of approximately 10% develops.

Returning to the initial states of the samples shown in Fig. 1.10, it should be noted that the uppermost state of the water-sedimented samples indicates the loosest possible state which is achieved by this method, and it is located below the corresponding state of the samples prepared by wet tamping. In other words, looser samples or states with a higher void ratio can be produced by means of

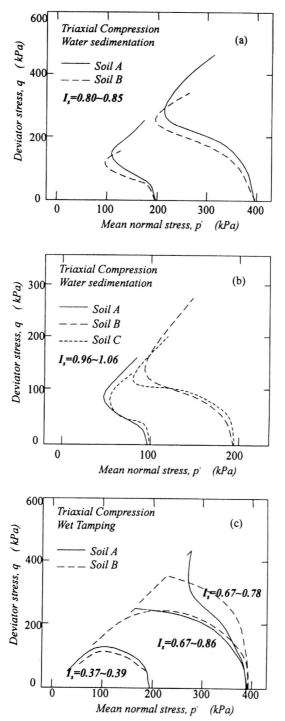

Fig. 1.12. Undrained behavior of samples with similar I_s-values
(comparative plots for soils *A, B* and *C*)

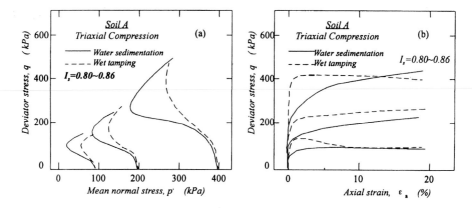

Fig. 1.13. Effects of sample preparation method on undrained behavior
in triaxial compression:
(a)Effective stress paths, (b) Stress-strain curves

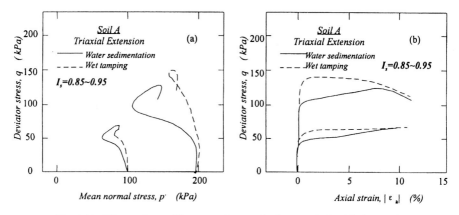

Fig. 1.14. Effects of sample preparation method on undrained behavior in
triaxial extension:
(a) Effective stress paths, (b) Stress-strain curves

wet tamping. The behavior of loose samples prepared by wet tamping is fully
contractive, as seen in the sample behavior for the smallest I_s-values in Figs.
1.11(d) and 1.12(c). On the other hand, even the loosest samples prepared by
water sedimentation, with a state index of 0.67 (Table 1-1), do not exhibit any
drop in shear stress upon undrained shearing. However, as illustrated in Figs.
1.8 and 1.9, when the relative initial states of samples prepared by water
sedimentation and wet tamping are identical, the samples prepared by water
sedimentation are more contractive.

Effects of Deformation Mode: A comparison between undrained behavior in
the triaxial compression and extension tests on samples prepared by water
sedimentation is shown in Fig. 1.15. The axial strains in the triaxial extension

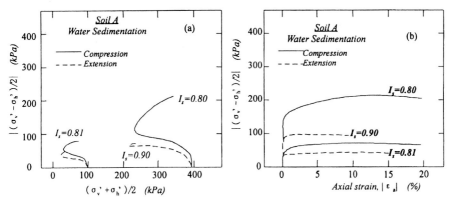

Fig. 1.15. Effects of deformation mode on undrained behavior:
(a) Effective stress paths, (b) Stress-strain curves

are smaller than those in the triaxial compression since the extension tests were terminated once necking of the samples was observed. Typically, the behavior is seen as being more contractive in the extension tests. While the compression tests on even the loosest water-sedimented samples $(I_s = 0.67)$ show a monotonic increase in the shear stress upon undrained shearing, in the extension tests, samples with an initial state with $I_s < 0.85$ show marginal behavior between drop and no-drop in the shear stress. The tendency of sands to be more contractive in extension than in compression has also been reported by several other researchers (e.g., Vaid et al., 1990; Been et al., 1991; Yoshimine, 1996). Similar results regarding the stress path dependency of the undrained behavior were also observed in the tests on samples prepared by wet tamping. In this case, the threshold values of the state index below which the samples show drop in shear stress and exhibit fully contractive behavior are $I_s = 0.65$ and $I_s = 0.85$, for triaxial compression and extension, respectively.

Undrained Behavior of Undisturbed Samples

The undisturbed samples recovered from the three sampling sites at Port Island, as described previously, vary over a wide range of void ratio and gravel content, with $e_c = 0.268$ to 0.461 and $P_G = 37.1\%$ to 61.0%, respectively. The void ratio after consolidation and the gravel content of each sample are listed in Table 1-4. It is likely that the variation in void ratio is partially due to differences in the gradation curves and gravel contents of the samples (see Fig. 1.7 and Table 1-4). In addition, differences in void ratio arose since the sampling site for the FS-samples is located in an improved area, whereas the UD-site and Fud-site are in intact as-deposited areas without any treatment. Characteristic effective stress paths measured in the tests on undisturbed samples are displayed with solid line in Fig. 1.16. The behavior of the UD-samples is shown to be fully contractive with an initially increasing shear stress towards a peak and subsequent strain development with a descending shear stress (see Fig. 1.16 (a)). Unlike the UD-

Table 1-4. Void ratio after consolidation and gravel content of undisturbed samples

Sample	e_c	P_G
FS 2-6	0.268	52.7
FS 1-5	0.270	55.5
FS 3-004	0.354	53.2
FS 3-13	0.410	46.3
FS 1-14	0.415	39.9
FS 1-12	0.433	50.1
FS 4-12	0.461	37.1
Fud 2	0.349	44.7
Fud 3	0.415	61.0
UD 6	0.386	50.2
UD 1	0.401	45.9
UD 5	0.404	44.3
UD 4	0.406	45.1
UD 7	0.417	47.4
UD 3	0.442	42.0

samples, the behavior of the FS-samples shown in Fig. 1.16(b) and 1.16(c) is strongly dilative. Like the tendency observed in the tests on reconstituted samples, the behavior in the triaxial extension is less than that in the compression.

Back Calculation of Threshold Void Ratio e_0: As discussed above, normalization along the void ratio axis and quantification of the undrained behavior of reconstituted samples via I_s makes it possible to concurrently consider deformation behavior of soils with different gradation characteristics. Therefore, it would be of interest if the behavior of undisturbed samples could be examined within the same framework, since each of these samples has apparently different grain size distribution.

An assumption is now made that the steady state line and the UR-line established from the tests on reconstituted samples are applicable to intact in-situ soils as well. This would be considered reasonable because of the similarity of the stresses at the largely deformed steady state of the soil, irrespective of the mode of deformation and fabrics reflecting the method of soil deposition. However, the location of the reference lines in the e-p' plot is not fixed yet, since the threshold void ratios e_0 of the recovered undisturbed samples are not known. It may be assumed that the fabric formed by the in-situ deposition exhibits behavior somewhere between the water-sedimented and wet-tamped samples. With these assumptions, it is possible to back-calculate the threshold void ratios of the undisturbed samples as described below.

The behavior observed in the tests on undisturbed samples is compared to that of the reconstituted samples in Fig. 1.16. The two kinds of test data are regarded as exhibiting similar behavior in the stress space and therefore considered to have an identical value for the state index. In this way, based on the known I_s-values of the reconstituted samples and similarity in the effective

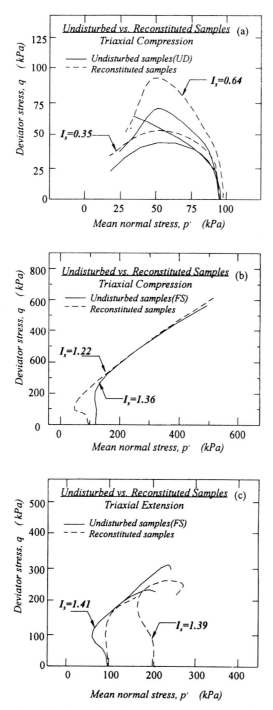

Fig. 1.16. Estimation of I_s-values of undisturbed samples via comparison with undrained behavior of reconstituted samples: (a) UD-samples, (b) FS-samples (triaxial compression), (c) FS-samples (triaxial extension)

stress paths, the state index is estimated for each of the undisturbed samples. In general, the behavior of the reconstituted samples prepared by water sedimentation is used as a basis for evaluation of I_s. In cases where an undisturbed sample shows contractive behavior, behavior of the wet-tamped samples is employed. Once the I_s value of the undistrubed samples has been estimated, the threshold void ratio for each sample is calculated as

$$e_0 = I_s(e_0 - e_s) + e_c \qquad (1\text{-}7)$$

where e_c is the void ratio after consolidation of the undisturbed sample and $(e_0 -e_s)$ is the distance between the two reference lines which is calculated for the corresponding consolidation stress p'_c according to Eq. (1-4).

For reconstituted samples, it was shown that the threshold void ratio e_0 is related to the grain size characteristics of the soil, with the e_0-value being smaller for coarser material. A correlation between the threshold void ratio e_0 and the gravel content P_G including data from both reconstituted and undisturbed samples is presented in Fig. 1.17, where the e_0 of the undisturbed samples are those back-calculated using the above procedure. In spite of some scatter of the data, a clear tendency for an increase in e_0 with decreasing gravel content is apparent. The solid line in Fig. 1.17 shows the best-fit linear approximation of the data given by

$$e_0 = 0.69 - 0.5 \frac{P_G}{100} \qquad (1\text{-}8)$$

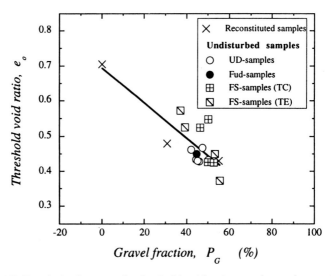

Fig. 1.17. Correlation between the threshold void ratio e_0 and gravel content P_G

Intial States: Having thus found the threshold void ratio e_0, it is now possible to identify the initial state of a given undisturbed sample relative to the reference lines. Figure 1.18 shows data for each of the tested undisturbed samples based on the calculated value of e_0 and known *e-p'* state after consolidation. In this figure, each large symbol indicates the initial state of an undisturbed sample, whereas a small symbol denotes the corresponding steady state in the modified *e-p'* diagram. The fact that the steady states of the undisturbed samples agree very well with the steady state line of the reconstituted samples confirms that the estimated I_s-values and computed threshold void ratios e_0 of the undisturbed samples are reasonable. It is to be noted that most of the data of the dilative undisturbed samples or samples with initial and ultimate states below the steady state line are those obtained at strains of approximately 10% and do not represent actual steady states. It is anticipated that if further shearing had been executed, these data would have moved towards the right due to additional dilation and would have approached even closer to the steady state line shown in Fig. 1.18.

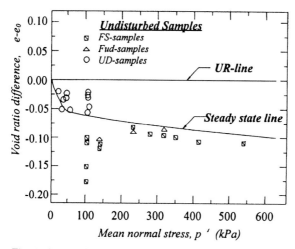

Fig. 1.18. Initial and steady states of undisturbed samples

A clear difference between the initial states of the UD-samples and those of the FS-samples can be seen in Fig. 1.18. These initial states are in accordance with the fact that the UD-samples are from originally deposited Masado soil while the FS-samples are from a densified deposit. Hence, the vertical distance between the UD-samples and the FS-samples in Fig. 1.18 indicates the decrease in the void ratio caused by the densification of the soil. The location of the Fud-samples relative to the reference lines is lower than expected since these samples are from an undensified deposit of Masado soil. This point remains to be studied.

In-situ Characteristics of Reclaimed Masado Soil

Characteristic States : Since the reconstituted and undisturbed samples are shown to have identical UR-lines and steady state lines in the modified *e-p'* diagram, it is resonable to adopt these lines as reference lines for the in-situ Masado soil as well. As discussed earlier, these reference lines are independent of the gradation characteristics of the soil. Shown in Fig. 1.19 with thick solid lines are isotropic consolidation curves for the loosest samples prepared by water sedimentation. Soils *A* and *B* are shown to have a common consolidation line which is expressed as

$$e_c - e_0 = 0.03 - 0.04 \log p', \qquad p' > 10 \text{ kPa} \qquad (1\text{-}9)$$

This consolidation line is located above that of soil *C.* Under the assumption that the water-sedimented samples possess a fabric replicating that of the in-situ deposit, the isotropic consolidation lines shown in Fig. 1.19 can be interpreted as consolidation lines of undensified or untreated deposits of Masado soil. Thus, it may be considered that the shaded zone in Fig. 1.19 denotes the range of possible in-situ initial states of a reclaimed Masado soil in an untreated deposit. As can be seen in Fig. 1.19, most of the initial states in the shaded area are on the right of the steady state line, and hence, upon monotonic undrained shearing they will be contractive and produce positive excess pore pressures at the steady state.

The dashed line shown in Fig. 1.19 indicates the initial state line $I_s = 0.85$. As discussed before, the initial states with $I_s \leq 0.85$ or above the dashed line displayed in Fig. 1.19 show marginal behavior upon undrained shearing in the triaxial extension. Unlike the behavior in the extension, all of the initial states within the shaded area show a monotonically increasing shear stress in the triaxial compression loading.

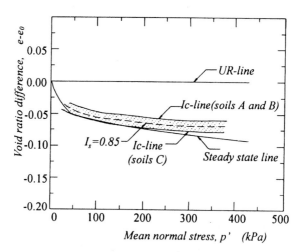

Fig. 1.19. Characteristic states of in-situ soils in a modified *e-p'* diagram

It is important to recognize that along the consolidation line of soils A and B the state index is nearly constant at $I_s = 0.72$-0.74 for confining stresses of 50 to 200 kPa. On the other hand, the I_s-values along the consolidation line of soil C decrease with increasing consolidation stress p'. In other words, as the confining stress increases, soil C gradually becomes more contractive. Except for the range of initial states at small normal stresses, say $p' \leq 50$ kPa, this tendency is not strongly exhibited in the undrained behavior. Another interesting observation from Fig. 1.19 is that the coarser Masado soils A and B with gravel contents of $P_G = 30.8\%$ and 55% have an initial state with smaller I_s-value, and hence, they are more contractive during shearing than soil C, which does not contain gravel.

Behavior Upon Undrained Shearing: The findings elaborated in the preceding sections may be summarized and put in a generalized perspective for characterization of the reclaimed Masado soil at Kobe Port Island. The procedure is schematically illustrated by a flow-chart in Fig. 1.20 and is executed through the following steps:

Step 1: From a given gradation curve of a Masado soil, the percentage of the gravel fraction P_G is read off (Fig. 1.20(a)).

Step 2: The threshold void ratio e_0 is determined from the correlation between e_0 and P_G (Figs. 1.17 and 1.20(b); Eq. (1-8)).

Step 3: The locations of the upper reference line, the steady state line (Eqs. (1-4) and (1-5)) and the consolidation line (Eq. (1-9)) are established in the e-p' plane (Fig. 1.20 (c)).

Step 4: For a known value of the mean normal stress and assumed consolidation line for the reclaimed deposit, the in-situ void ratio e_c at a given depth is estimated (Fig. 1.20 (c)).

Step 5: The state index I_s is calculated according to Eq. (1-6) and it is used to assess the undrained behavior of the reclaimed Masado soil at the considered depth of the deposit (Fig. 1.20 (c)).

Based on the definition of the characteristic states given in the preceding section and the previously established link between the undrained behavior of the Masado soils and I_s, the monotonic undrained behavior of untreated in-situ Masado soils may be characterized as follows. Assuming that the in-situ states of an undensified reclaimed deposit lie along the consolidation line of soils A and B(Fig. 1.19 and Eq. 1-9)), the initial states along the depth of the deposit are set at $I_s = 0.73$. Next, the results of the triaxial tests on water-sedimented samples are used to define representative effective stress paths and stress-strain curves for $I_s = 0.73$ by plotting average lines through the experimental data having I_s-values in the range between 0.67 and 0.81. Figure 1.21 shows the representative monotonic undrained behavior for $I_s = 0.73$ evaluated in this manner, where the lower and the upper boundaries of the shaded area denote behavior in the triaxial extension and triaxial compression, respectively. It can be seen in the figure that the undensified Masado soils exhibit moderately dilative behavior in the triaxial compression whereas they show marginal behavior between contractive and dilative in the triaxial extension.

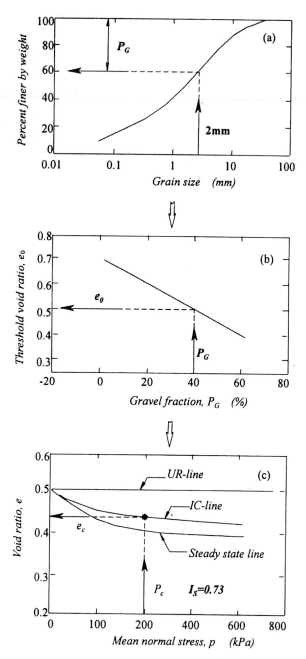

Fig. 1.20. Flow-chart of a generalized procedure for assessment of void ratio and state
index I_s throughout the depth of a reclaimed deposit of Masado soil

Results of multiple series of triaxial tests on Masado soils with a wide range of
gradation characteristics were examined through the use of the state concept.

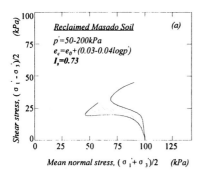
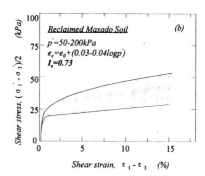

Fig. 1.21. Undrained behavior of in-situ Masado soils in untreated areas: (a) Effective stress paths, (b) Stress-strain curves

The key findings can be summarized as follows:

(a) A unique steady state line for Masado soil exists in the modified e-p' diagram. This steady state line is independent of the gradation characteristics, fabric of the soil and mode of the deformation.

(b) The undrained behavior of different Masado soils is uniquely related to the state index when the above steady state line and the corresponding upper reference line are used as reference states for evaluation of the state index I_s. It is found that the state index can quantify the measured undrained behavior to a reasonable level of accuracy.

(c) The threshold void ratio e_0 or the location of the steady state line in the e-p' plane mainly depends on the grain size characteristics of the soil. A linear relationship between e_0 and the gravel contents P_G provides a rational link that embodies this correlation. The threshold void ratio e_0 decreases with increasing P_G.

(d) There is no strong evidence to decide which of the employed sample preparation methods can best duplicate the in-situ process of deposition. In general, however, water sedimentation appears to be closer to the actual depositional process and fabric in the field.

(e) An increase in the gravel content of Masado soils was found neither to increase the strength nor improve the performance of the soil upon undrained shearing. In fact, some of the data suggests that for fill materials with gravel contents of up to 60% the initial state is associated with smaller I_s-values with increasing P_G. In other words, the soil which contains more gravel is more contractive upon shearing.

(f) It is estimated that the in-situ initial state of an undensified deposit of Masado soil is roughly at I_s=0.73. The reclaimed soil at this state exhibits marginal behavior between contractive and dilative upon monotonic loading in the triaxial extension, but the behavior is shown to be moderately dilative in the triaxial compression.

(g) A generalized procedure is established which enables a quick assessment to be made of in-situ void ratios and undrained behavior throughout the depth of an originally reclaimed Masado deposit if the gradation curve of the fill materials is known.

1.3 Modeling of Soil Based on State Concept

Introduction

When a sand is subjected to a monotonic or cyclic shear load it exhibits very complex behavior which is governed by the coupling between the shear and the volumetric strains. The link between the distortional and volumetric strains is certainly the most distinctive feature of the deformation characteristics of granular materials. In accordance with this feature the behavior of sand is profoundly affected by the initial density and fabric of packing, as well as by the initial normal and shear stresses. In other words, a given sand behaves either in contractive or dilative manner and shows different cyclic strength or resistance to liquefaction depending on the initial packing and stress conditions.

Roscoe and Poorooshasb (1963) were the first to emphasize the importance of the combined influence of density and normal stress on soil behavior. On the basis of both theoretical and experimental considerations, Roscoe and Poorooshasb introduced the critical state of the soil as a reference for quantification of the effects of the initial state on soil behavior. Here, the initial state refers to the void ratio and the mean normal stress of the soil prior to shearing. In a similar vein, Been and Jefferies (1985), Bolton (1986), Ishihara (1993) and Verdugo (1992) have reported extensive studies on the combined influence of the initial density and normal stress on sand behavior. A distinctive feature of these studies is that particular indices and parameters serving as measures for the effects of the initial state on sand behavior have been proposed. The proposed relative dilatancy index I_r (Bolton, 1986), the state parameter ψ (Been and Jefferies, 1985) and the state index I_s (Ishihara, 1993; Verdugo, 1992) integrate the effects of the density and normal stress into a single parameter and as such, they are different from the commonly used parameter, the relative density D_r.

Recently, several attempts have been made to use the state parameter ψ and the state index I_s for sand modeling (Jefferies, 1993; Cubrinovski, 1993; Wood et al., 1994), Jefferies (1993) incorporated the state parameter ψ as a hardening parameter in the Cam clay model to describe monotonic behavior of sand and Wood et al. (1994) used the same parameter for modeling drained monotonic stress-strain behavior associated with strain softening. Described in this paper is a stress-strain-dilatancy model for sand based on the state index I_s (Cubrinovski, 1993). This model is developed and designed for modeling the monotonic and cyclic undrained behavior of sand, and targets analysis of liquefaction problems. In general, however, the concept applies to drained behavior as well (e.g., Wood et al., 1994). The principal feature of the present model is that the influence of the initial *e-p* state (void ratio-mean normal stress state) on the plastic behavior is taken into consideration through the use of the state index I_s. Unlike conventional sand models in which each density is considered as an index property for a separate material, the adopted modeling concept enables us to offer an integral representation of sand behavior over the relevant range of density and stress states. Besides the physical relevance of the concept, its practical benefit is that a single set of material parameters can be used to model and behavior for any, practically relevant, initial void ratio and confining stress.

Stress-Strain Model Based on State Index

Background : Based on results from a multiple series of undrained triaxial compression tests on sandy soils, Ishihara (1993) and Verdugo (1992) concluded that any two samples of a given soil will have similar stress-strain relations as well as pore water pressure response provided that the relative initial state is the same for each sample. To identify the relative initial state, Ishihara and Verdugo used some characteristic states of sand in the *e-p* plane, and defined a property index, termed the state index I_s , as a direct measure for quantifying the relative initial state. Two reference line in the *e-p* plane are employed in the definition of I_s, i.e., the quasi steady state line (QSS-line) and the upper reference line (UR-line). The quasi steady state (Alarcon-Guzman et al., 1988) represents a particular case of the state of phase transformation (Ishihara et al., 1975), and occurs following a temporary drop in the shear stress upon undrained shearing of loose sands. As shown in Fig. 1.22, when the initial mean normal stress is $p \le p_0$, the upper reference line is a horizontal line corresponding to void ratio e_0. On the other hand, for initial mean normal stresses in excess of p_0 $(p > p_0)$, the isotropic consolidation line for the loosest state (IC-line) is used as the upper reference line. The void ratio e_0 is defined as a threshold value at or above which the initial states are associated with a zero steady state strength. In other words, upon monotonic undrained shearing of samples with void ratios $e \ge e_0$, both shear and normal stressed drop to zero and the soil exhibits 'static liquefaction'. The void ratio e_0 can be evaluated as the point of

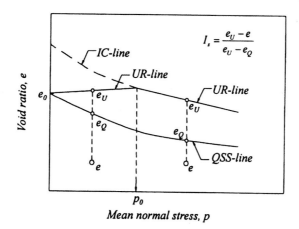

Fig. 1.22. Definition of state index I_s (Ishihara, 1993)

intersection between the quasi steady state line and the y-axis. Alternatively, the steady state line may be used instead of the quasi steady state line, since for any initial state with $e \geq e_0$ the steady state and the quasi steady state are coincident. The mean normal stress p_0 denotes the normal stress at which the e_0-horizontal line intersects the isotropic consolidation line for the loosest state. As illustrated schematically in Fig. 1.22, for a given initial void ratio e and a mean normal stress p, the state index is defined as

$$I_s = \frac{e_U - e}{e_U - e_Q} \qquad (1\text{-}10)$$

where e_Q is the void ratio of the QSS-line and e_U is the void ratio of the UR-line, both at the initial mean normal stress p. Equation (1-10) yields the conditions $I_s=0$ and $I_s=1$, for the initial states at the UR-line ($e= e_U$) and at the QSS-line ($e=e_Q$), respectively. The choice of the reference states in the definition of I_s is elaborated by Ishihara (1993). It is important to note that the above frames of reference are inherent in the fabric of the sand formed during its deposition, and therefore, they are fabric-dependent.

Experimental Evidence : In order to further examine the link between the state index I_s and the characteristics of the stress-strain curves, results from a series of drained torsional shear tests on hollow cylindrical samples of Toyoura sand (Cubrinovski, 1993) were used. The Japanese standard Toyoura sand is classified as a uniform fine sand and consists of sub-angular particles with mostly quartz composition. It was found to have a specific gravity of $G_s = 2.65$, maximum void ratio of $e_{max} = 0.988$ and minimum void ratio of $e_{min} = 0.616$. Sieve analyses indicated a mean diameter of $D_{50} = 0.19$ mm and a uniformity coefficient of $U_c = 1.7$.

Uniform samples of clean Toyoura sand were prepared by dry pluviation with a constant height of fall. By using different heights of pluviation, various initial preconsolidation void ratios of the samples were achieved ranging between $e_i = 0.731$ ($D_r=69.1\%$) and $e_i = 0.916$ ($D_r=19.4\%$). Following saturation, the samples were isotropically consolidated up to normal stress of $p = 30, 50, 100, 200, 300$ or 400 kPa. The e-p states of the samples attained at the end of the consolidation, i.e., prior to shearing, are shown in Fig. 1.23. Applying torque, the samples were sheared in a monotonic stress-controlled manner under drained p-constant condition.

For each initial state, the state index I_s, was calculated according to Eq. (1-10) by using the QSS-line and the UR-line shown in Fig. 1.23. These reference lines were obtained by Yoshimine (1996) from a series of undrained torsional shear tests on samples of Toyoura sand prepared by dry deposition, which corresponds to dry pluviation under zero height to fall. The initial e-p states shown in Fig. 1.23 include various states with respect to the reference line (relative initial states) and have I_s-value between $I_s= -0.4$ and $I_s=9.8$. In the following, the link between the state index and the stress-strain curves measured in the drained torsional tests is examined. For the purpose of modeling, normalized stress-strain curves associated with the plastic response are considered. In general, however, the discussion applies to the stress-strain response expressed in terms of total strains as well.

A set of tests with considerably different initial states, including samples from loose sand under relatively high normal stress (D_r=32%, p= 300 kPa) to dense sand under low normal stress (D_r = 70%, p = 30 kPa), is shown in Fig. 1.24. Each of these tests has a distinct relative initial state, and hence, different I_s-value. The solid lines in Fig. 1.24(a) are used to illustrate the quantitative difference in the relative initial states. Along each line the state index is constant or the relative initial states are identical; the lines I_s=0 and I_s=1 denote the UR-

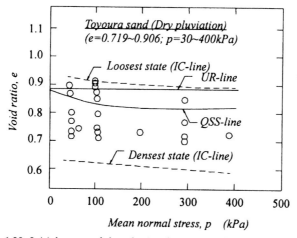

Fig. 1.23. Initial states of dry-pluviated samples of Toyoura sand in drained torsional shear tests

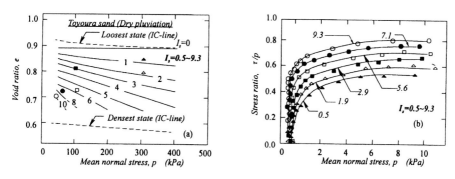

Fig. 1.24. Samples of dry-pluviated Toyoura sand with distinct relative initial states (drained *p*-constant torsional shear tests): a) Initial states; b) Measured stress-strain curves

line and the QSS-line. It can be seen in Fig. 1.24(a) that the selected tests have uniformly distributed relative initial states in the range between $I_s=0.5$ and $I_s=9.3$. The stress-strain curves measured in these tests are displayed in Fig. 1.24(b), where the state index value for each test is also indicated. The link between the stress-strain curve and the state index is apparent; as the state index increases, the stress-strain curve gradually becomes higher, with a larger initially modulus and peak stress ratio.

Other data from the drained torsional tests on Toyoura sand are presented in Fig. 1.25(a) and Fig. 1.25(b), where the initial states and the measures stress-strain curves are shown respectively. Although the initial states of the samples are considerably different, these samples can be classified in two sets with respect to the relative initial states. The samples within each set have similar relative initial states, with I_s being 1.9-2.9 and 4.0-5.6, for each set respectively. It is evident in Fig. 1.25(b) that a nearly unique stress-strain curve was measured for each set of tests, indicating that any two samples will have similar normalized stress-strain curves provided that the samples have similar I_s-values.

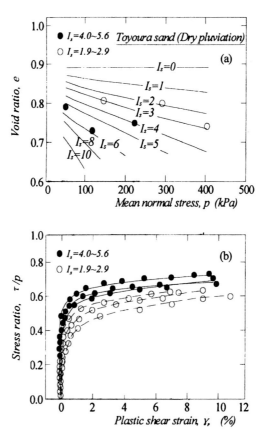

Fig. 1.25. Samples of dry-pluviated Toyoura sand with similar relative initial states (drained *p*-constant torsional shear tests): a) Initial states; b) Measured stress-strain curves

The threshold void ratio e_0 which is used for the definition of the UR-line (see Fig. 1.22) is associated with extreme undrained response. Namely, the samples with initial states that lie at or above the UR-line ($p \leq p_0$) exhibit zero residual strength (Ishihara, 1993). To examine the response associated with such relative initial states in the case of drained shearing, a series of drained torsional p-constant tests on samples of Toyoura sand with 10-15% fines by weight was performed (Cubrinovski, 1993). In this series of tests, the soil was mixed with de-aired water at a water content of 3% and uniform samples with various densities were prepared by wet tamping. Except for the sample preparation method, the testing procedures were identical to those previously described.

A specific feature of the tests discussed in the following is that the initial states of the samples lie either close to or above the UR-line, with I_s-values ranging between –4.3 and 0.2. This is illustrated in Fig. 1.26(a) where the initial states of the samples and the reference lines are shown in a modified e-p diagram; in this diagram, the void ratio axis is replaced by the void ratio difference (e-e_0). The advantage of this diagram is that the UR-lines ($p \leq p_0$) of different materials merge into a single line, which enables us to comparatively display relative initial states of different soils. In spite of the considerable differences in the initial void ratios and the mean normal stresses of the samples, Fig. 1.26(b) shows that a nearly unique stress-strain curve was measured in these tests. This result suggests that the normalized stress-strain curve is not significantly affected by initial states that have either zero or negative values of the state index. In other words , the normalized stress-strain curves of samples with initial states that lie at or above the UR-line are similar; this curve is the lowest in the stress-strain plot among the family of curves for all the initial e-p states of practical importance for a sand with a given fabric, and hence, it does represent an extreme drained response.

The link between the state index and the characteristics of the stress-strain curves measured in the drained torsional tests might be summarized as follows: 1) all the initial states in the e-p diagram that have identical or similar state index values tend to have similar normalized stress-strain curves: 2) the gradual change in the state index is associated with gradual change in the characteristics of the stress-strain curve; as the state index decreases, both the initial shear modulus and the peak strength of the sand decrease as well; 3) the degradation of the stress-strain curve ceases once the state index drops to $I_s = 0$ or the initial approaches the UR-line. For this reason, the normalized stress-strain curve for the initial states at the UR-line ($I_s = 0$) can be used as an approximation for the lowest curve in the stress-strain plot of a sand with a given fabric.

The combined influence of the density and the normal stress on the stress-strain behavior is a key feature of the observed response in the drained torsional tests. The ability of the state index to associate its value with the characteristics of the stress-strain curve clearly illustrates its aptness to neatly combine, in a single parameter, the effects of density and normal stress on the stress-strain relation of sands. According to its definition, the state index assesses sand

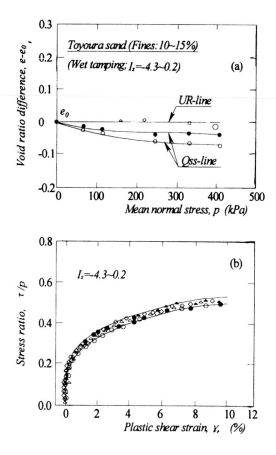

Fig. 1.26. Drained torsional shear tests on wet-tamped samples of Toyoura sand with
10-15% fines: a) Relative initial states; b) Measured stress-strain curves

behavior based on the relative position of the initial *e-p* state with respect to the
characteristic states. Therefore, what counts is the relative initial state of the
soil. The presented experimental evidence further elucidates the definition of I_s
and reveals its possible role in sand modeling.

Modified Hyperbolic Relation: The mathematical formulation of a stress-strain
curve is based on the hyperbolic relation in which change in the shear stress
ratio τ/p is linked with the plastic shear strain γ_p as

$$\frac{\tau}{p} = \frac{G_N \gamma_p}{\left(\dfrac{\tau}{p}\right)_{max} + G_N \gamma_p}$$

(1-11)

where $G_N = G/p$ and $(\tau/p)_{max}$ are the normalized initial plastic shear modulus and the peak stress ratio. It is well known that if these two parameters are constant, the relation cannot fit well a given stress-strain curve over a wide range of strains. Numerous modifications of the two-constant hyperbolic relation have been proposed to overcome this deficiency (e.g., Tatsuoka and Shibuya, 1992). The modification adopted in the present model is based on the assumption that, for a given initial e-p state, the peak stress ratio $(\tau/p)_{max}$ is constant, while the normalized initial shear modulus G_N varies as a function of the plastic shear strain γ_p as

$$G_N = (G_{N,max} - G_{N,min}) \exp\left(-f\frac{\gamma_p}{\gamma_0}\right) + G_{N,min} \qquad (1\text{-}12)$$

where $G_{N,max}$ and $G_{N,min}$ are the best-fit normalized initial moduli at small-strains and at large strains (say greater than 1%) respectively, γ_0 is the plastic shear strain at which G_N approaches $G_{N,min}$ and f is a constant. The value of the constant f is generally greater than 3, and it increases with any increase in the relative size of the difference $(G_{N,max} - G_{n,min})$ with respect to the value of $G_{N,min}$. Equation (1-12) was also used by Tsujino (1992). The accuracy of the modified hyperbolic relation over a wide range of strains is demonstrated in Fig. 1.27(a) and 1.27(b), where three simulations are compared with the measured stress-strain curve in the drained torsional shear test on Toyoura sand with $e = 0.906$ and $p = 100$ kPa. The initial state of this test is also indicated in Fig. 1.22. The simulations shown in Fig. 1.27 were conducted by using Eq. (1-11) with the initial shear modulus G_N being equal either to $G_{N,max}$, $G_{N,min}$ or the expression given in Eq. (1-12). It should be noted that the same level of accuracy as that shown in Fig. 1.27 is not always attainable. This is especially the case when there is a large difference between $G_{N,max}$ and $G_{N,min}$, or the soil stress-strain curve deviates remarkably from the two-constant hyperbolic relation.

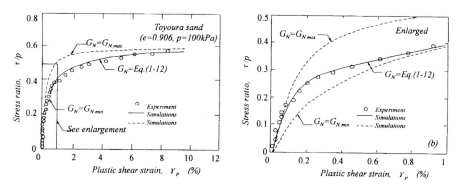

Fig. 1.27. Comparison between modified hyperbolic relation and two-constant hyperbolic relations): a) Large strain level; b) Small strain level

To clarify the need for degradation of the normalized initial shear modulus with the increase of the plastic shear strain, Eq. (1-11) is solved for G_N and rewritten as

$$G_N = \frac{1}{\gamma_p} \frac{\frac{\tau}{p}\left(\frac{\tau}{p}\right)_{max}}{\left(\frac{\tau}{p}\right)_{max} - \frac{\tau}{p}}, \qquad \frac{\tau}{p} < \left(\frac{\tau}{p}\right)_{max} \qquad (1\text{-}13)$$

When data from a measured stress-strain curve is introduced in Eq. (1-13), an experimental relationship between G_N and γ_p can be established. Figure 1.28 shows such a relation for the measured stress-strain curve displayed in Fig. 1.27. Each symbol in Fig. 1.28 indicates the value of the normalized initial shear modulus G_N in Eq. (1-11) that most accurately simulates the measured stress-strain curve at the corresponding strain level γ_p. For example, $G_{N,max}$ is the best-fit initial shear modulus at small strains, whereas $G_{N,min}$ provides accurate simulation of the measured curve at shear strains larger than 1%. It is apparent that, if a given stress-strain curve has to be accurately modeled with Eq. (1-11), the normalized initial shear modulus G_N should degrade with the plastic shear strain. Equation (1-12) is an analytical approximation of such degradation of G_N.

It is important to recognize that $G_{N,max}$ and $G_{N,min}$ as maximum and minimum values of the initial shear modulus G_N, correspond to the initial shear moduli of the 'original' and Kondner's hyperbolic relations (Tatsuoka and Shibuya, 1992; Kondner and Zelasko, 1963). Since Kondner's relation accurately simulates soil stress-strain curves for strains larger than 1%, the shear strain γ_0 is set to be constant and equal to 0.01. It is assumed that, for a given fabric, the parameter

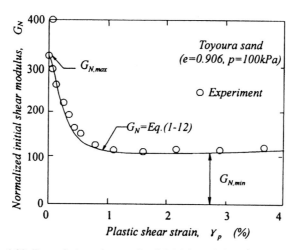

Fig. 1.28. Degradation of normalized initial modulus with shear strain

in Eq. (1-12) is also constant, thus reducing the parameters of the modified hyperbolic relation to three, i.e., $(\tau/p)_{max}$, $G_{N,max}$ and $G_{N,min}$. The modified hyperbolic relation significantly improves the accuracy of the two-constant hyperbolic relation in modeling a given stress-strain curve of soil; yet, it remains simple and has parameters with a clear physical meaning.

Stress-Strain Parameters as Function of State Index: It was demonstrated in the previous sections that there is an obvious link between the characteristics of the normalized stress-strain curve and the state index I_s, and that the modified hyperbolic relation provides accurate modeling of a given stress-strain curve. This implies possible correlation between the stress-strain parameters $(\tau/p)_{max}$, $G_{N,max}$ and $G_{N,min}$ and the state index I_s. The drained torsional p-constant tests on dry-pluviated samples of Toyoura sand, discussed in the previous sections, are used in the following to scrutinize such correlation.

Based on the reference lines and the initial states of the samples shown in Fig. 1.23, I_s was calculated for each test according to Eq. (1-10). For each of the measured stress-strain curves in the drained tests, the stress ratio at approximately 10% shear strain was defined as the peak stress ratio $(\tau/p)_{max}$ and the maximum and minimum normalized initial shear moduli, $G_{N,max}$ and $G_{N,min}$ were determined from plots like that shown in Fig. 1.28. In order to achieve better overall simulation of the measured stress-strain curves, $G_{N,max}$ was evaluated as the best-fit initial shear modulus at a shear strain of approximately 10^{-4} instead of the smallest measured strain level. The experimental correlation between the stress-strain parameters and the state index is shown with the symbols in Fig. 1.29; the solid lines in this figure are the best-fit linear regression lines. It is apparent that a remarkably good linear correlation exists between the state index I_s and each of the stress-strain parameters $(\tau/p)_{max}$, $G_{N,max}$ and $G_{N,min}$ as exemplified by the coefficients of linear correlation of 0.96, 0.84 and 0.92 respectively. Therefore, the parameters of the modified hyperbolic relation can be expressed as linear functions of the state index

$$\left(\frac{\tau}{p}\right)_{max} = a_1 + b_1 I_s, \qquad I_s \geq 0 \qquad\qquad (1\text{-}14)$$

$$G_{N,max} = a_2 + b_2 I_s, \qquad I_s \geq 0 \qquad\qquad (1\text{-}15)$$

$$G_{N,min} = a_3 + b_3 I_s, \qquad I_s \geq 0 \qquad\qquad (1\text{-}16)$$

where a_1, b_1, a_2, b_2, a_3 and b_3 are linear coefficients that represents the stress-strain parameters of the model. Numerical values of the coefficients for dry-pluviated Toyoura sand are listed in Table 1-5.

In accordance with the earlier suggestion that the stress-strain curve is not affected by the initial state $I_s \leq 0$, it is postulated that Eqs. (1-14)-(1-16) apply to

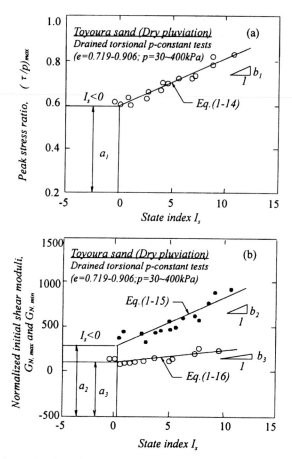

Fig. 1.29. Ralationship between stress-strain parameters and state index: a) Peak stress ratio $(\tau/p)_{max}$; b) Normalized initial shear moduli $G_{N,max}$ and $G_{N,\,min}$

initial states with non-negative values of I_s. For all the initial states with negative values of I_s a unique stress-strain curve which is equal to the curve for $I_s=0$ is assumed. This assumption is illustrated with the dashed horizontal lines in Fig. 1.29, and can be mathematically expressed by omitting the I_s–terms of Eqs. (1-14)-(1-16). Using Eqs. (1-9)-(1-16) the stress-strain curve can be evaluated for any initial e-p state of a sand with a given fabric. Figure 1.30 demonstrates this feature of the model where simulations of measured stress-strain curves in the drained p-constant tests, for samples with various initial states (D_r=31-70% and p = 30-400 kPa), are shown.

Table 1.5. Stress-strain coefficients for dry-pluviated Toyoura sand

$(\tau/p)_{max}$	$G_{N,max}$	$G_{N,min}$
a_1=0.592	a_2=291	a_3=98
b_1=0.021	b_2=55	b_3=13

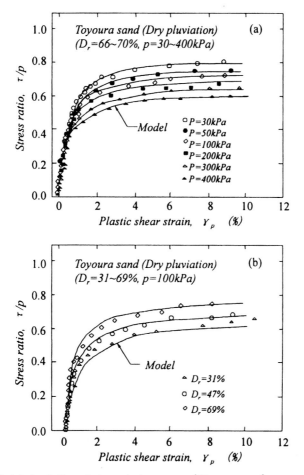

Fig. 1.30. Model simulation of stress-strain curves of Toyoura sand measured in drained *p*-constant tests: a) Effects of mean normal stress; b) Effects of density

State Index as a Current Variable : In the process of evaluation of the linear coefficients a_1, a_2, a_3 and b_1, b_2, b_3 (Eqs. (1-14)-(1-16)) the state index is used as an initial index parameter, while during undrained shearing it is employed as a current variable. The state index as a current variable is calculated from an expression equivalent to Eq. (1-10) in which the current state is considered rather than the initial state of the sand; i.e., e_Q and e_U in Eq. (1-10) are replaced with the void ratios of the QSS-line and UR-line at the current normal stress *p*. The application of the state index as a current variable is described in the following.

Let us consider typical responses of dense and loose samples of sand upon monotonic undrained loading. Figure 1.31 schematically shows the behavior of the two samples in characteristic diagrams. As illustrated by the *A-B-C* paths in the *e-p* and *τ-p* diagrams (Fig. 1.3(a) and Fig. 1.3(c)), the dense sand is contractive

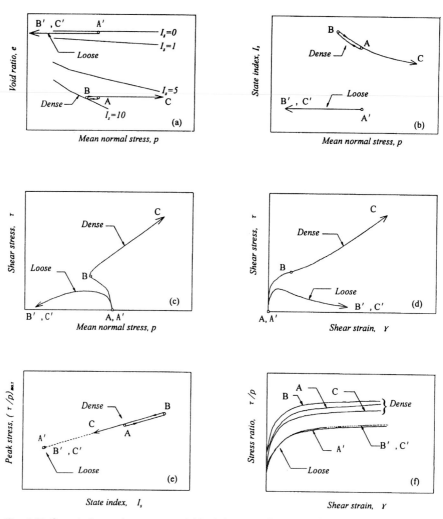

Fig. 1.31. State index and a current vriable (schematic illustration for monotonic undrained loading): a) e-p paths; b) Variation of current state index; c) Effective stress paths; d) Stress-strain curves; e) Variation of peak stress ratio $(\tau/p)_{max}$; f) Stress-strain curves in drained p-constant shearing

from the initial state A to the state of phase transformation B and subsequently it becomes dilative toward the steady state or critical state. The corresponding stress-strain curve along the A-B-C path is shown in Fig. 1.31(d). During the continued shearing, the current state index changes with the current e-p state; initially, it increases along the A-B path, and subsequently decreases along the B-C path or the dilative phase of the response (Fig. 1.31(b)). If I_s in Eqs. (1-14)-(1-16) is replaced with the current state index, the change in the current state index along the undrained A-B-C path will be reflected on the stress-strain parameters; for instance, the peak stress ratio will change as shown in Fig. 1.31(e).

Consequently, in the course of undrained shearing the stress-strain characteristics will gradually change according to the characteristics of the drained *p*-constant stress-strain curves for the current *e-p* state (Fig. 1.31(f)).

Unlike the response of the dense sand, the monotonic undrained response of the loose sand is fully contractive (Fig. 1.31(a) and 1.31(c)) and is characterized by a very small, practically negligible, change in the current state index (Fig. 1.31(b)). As a result, the stress-strain parameters along the *A'-B'-C'* path remain nearly constant (Fig. 1.31(e) and Fig. 1.31(f)). For these reasons, along an undrained path, the change in the characteristics of the stress-strain curve will be most pronounced for dense sand, whereas for very loose sand the parameters of the stress-strain curve will remain practically unchanged.

It is customary in soil models to specify the dependence of the initial modulus upon the normal stress by an expression of the form

$$G = G_r \left(\frac{p}{p_r} \right)^n \tag{1-17}$$

where p_r is the reference normal stress, G_r is the initial modulus at p_r, and n is an exponent determining the rate of variation of G with p. To illustrate the difference between the present model and the specification in Eq. (1-17) or the conventional modeling, let us consider a dense sample and a loose sample of dry-pluviated Toyoura sand, with initial void ratios of e=0.70 (D_r=77.4%) and e =0.86 (D_r=34.4%) respectively. Figure 1.32 comparatively shows the dependence of the initial plastic modulus of the dense sand on the mean normal stress, at small strains (G_{max}) and at large strains (G_{min}) as specified by the present model and by Eq. (1-17) with different values of the exponent n. Identical plots for the loose sand are shown in Fig. 1.33. It is apparent in Figs. 1.32 and

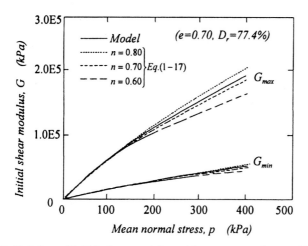

Fig. 1.32. Variation of initial shear modulus with mean normal stress (dense sand)

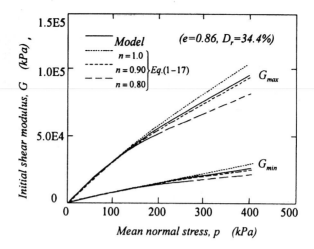

Fig. 1.33. Variation of initial shear modulus with mean normal stress (loose sand)

1.33 that the behavior of the model corresponds to that of the expression given in Eq. (1-17) with a variable exponent n. There are two important tendencies in the behavior with respect to the exponent n. First, the model specification is equivalent to a gradual increase in n with the plastic shear strain, i.e., n corresponding to G_{min} is larger than that corresponding to G_{max}. Second, the behavior of the model is equivalent to using an exponent n that is dependent on the sand density; the corresponding exponent n for the loose sand ($n = 0.91$-0.93) is greater than that for the dense sand ($n = 0.74$-0.78). Identical tendencies of the exponent n with respect to both shear strain and density are reported by Silver and Seed (1971), who compiled experimental data for several sands in terms of total strains. Unlike the present model and contrary to the existing experimental evidence, the conventional soil models assume that the exponent n in Eq. (1-17) is independent of both density and strain level. Most commonly it is assumed that $n = 0.5$, which is a typical value for the elastic response or behavior at very small strains.

The peak stress ratio, i.e, the second principal parameter that defines the stress-strain curve, is also affected by the adopted modeling concept. This is illustrated in Fig. 1.34 for the two densities considered above. In this figure, the thin straight lines denote constant peak stress ratios of $\tau/p = 0.62$ and 0.85 whereas the bold lines show the change of the peak stress ratios with p as modeled for the loose sand ($e = 0.86$) and the dense sand ($e=0.70$), respectively. As shown in Fig. 1.34, the peak stress ratio of the loose sand is nearly constant at small normal stresses ($\tau/p \approx 0.62$), and only slightly decreases with an increase in p. This characteristic of the model for the loose sand is due to the small change in the state index and consequently in the peak stress ratio with p, as illustrated in Fig. 1.31(b) and 1.31(e). On the other hand, Fig. 1.34 shows that

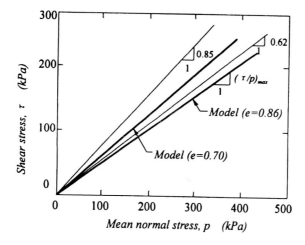

Fig. 1.34. Variation of peak stress ratio with mean normal stress

the modeled peak stress ratio of the dense sand notably decreases with increasing p, from a value of about 0.85 at small normal stress, and it gradually approaches the 0.62 peak stress ratio line as the mean normal stress becomes larger. In general, the model provides a varying degree of curvature of the peak stress ratio line, from a zero curvature or a straight peak stress ratio line for very loose sand, towards increasing curvature with increase in the density of the sand. This trend in the curvature of the peak strength line is in agreement with the experimental evidence of sand behavior presented by many researchers (e.g. Lade, 1978).

Stress-Dilatancy Relation

The well known energy based approach (Roscoe et al., 1963) is adopted to define the link between the plastic volumetric strain increment $d\varepsilon_v^p$ and the shear strain increment $d\varepsilon_q^p$. This link is usually expressed through a stress-dilatancy relation in which the ratio of the volumetric plastic strain increment due to shear distortion and the plastic shear strain increment is related to the shear stress ratio. Here, the following stress-dilatancy relation is used

$$\frac{d\varepsilon_v^p}{d\varepsilon_q^p} = \mu - \frac{q}{p}c \qquad (1\text{-}18)$$

where $q = (\sigma_1 - \sigma_3)/2$ is the shear stress, c is a non-coaxiality term (Gutierrez et al., 1993), μ is a dilatancy parameter dependent on the shear strain ε_q^p and $d\varepsilon_q^p = d\varepsilon_1^p - d\varepsilon_3^p$. For an assumed coaxial behavior ($c=1$), and entirely frictional dissipation of energy, which yields the condition $\mu = M =$ constant, where $M = (q/p)_{cs}$ is the critical state stress ratio, the stress-dilatancy relation given in Eq. (1-18) reduces to the original Cam-clay flow rule. The non-coaxiality term c in

Eq. (1-18) is defined as $c=\cos2\beta$, where β is the angle of non-coaxiality or the angle between the principal stress and the principal plastic strain increment directions. This term was introduced by Gutierrez et al. (1993) based on experimental evidence that for loading involving rotation of principal stress directions the plastic flow is affected by the stress increment direction. Another important role of the non-coaxiality term in Eq. (1-18) is to control the sign of the stress ratio term upon reversal in cyclic loading which is important in cases when Eq. (1-18) is used in conjunction with a very small yield surface. Except for the two cases discussed, for all practical purposes $c = 1$ may be assumed.

The dilatancy parameter μ specifies the rate at which the energy is dissipated during the shear distortion of a unit volume of soil under mean stress p

$$\mu = \frac{\left(\dfrac{dW}{p}\right)}{d\varepsilon_q^p} = \frac{d\Omega}{d\varepsilon_q^p} \tag{1-19}$$

where W denotes the dissipated energy and Ω is the normalized shear work defined as $d\Omega=dW/p$. The increment of normalized shear work was used as a parameter for shear deformation by Moroto (1976), where it was also suggested that $d\Omega$ is a function of the plastic shear strain, as indicated in Eq. (1-19). In the present model the following expression for the parameter p proposed by Kabilamany and Ishihara (1990) is adopted

$$\mu = \mu_0 + \frac{2}{\pi}(M - \mu_0)\tan^{-1}\left(\frac{\varepsilon_q^p}{S_c}\right) \tag{1-20}$$

where μ_0 and M are slopes of the normalized shear work versus plastic shear strain relationship $(\Omega - d\varepsilon_q^p)$ at small strains and large strains respectively, and S_c specifies the shear strain at which $\mu =(\mu_0 +M/ 2)$ is attained. Kabilamany and Ishihara derived the expression given in Eq. (1-20) from experimental evidence that the ratio $d\Omega/ d\varepsilon_q^p$ tends to increase with the shear strain. Figure 1.35 schematically illustrates this feature of the $(\Omega - \varepsilon_q^p)$ relationship through a typical result from cyclic undrainded torsional shear test on a sample of dry-pluviated Toyoura sand. The data are shown in terms of cumulative shear strains in order to elongate the small strain response region observed prior to cyclic softening and development of large shear strains. It is worth noticing that p increases with the shear strain within each half-cycle, as indicated with the change of the slope between the arrows which denote reversal points.

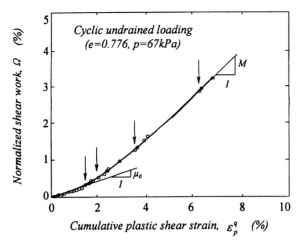

Fig. 1.35. Normalized shear work versus cumulative shear strain relation measured in cyclic undrained torsional shear test on dry-pluviated sample of Toyoura sand

Momen and Ghaboussi (1982) compiled data from a number of triaxial tests on six different sands and found that the $(\Omega - \varepsilon_q^p)$ relationship is independent of the relative density, mean pressure, overconsolidation ratio and inherent anisotropy. A similar result is shown in Fig. 1.36 where the $(\Omega - \varepsilon_q^p)$ relationship measured in the drained torsional tests on Toyoura sand introduced in the previous section is given. In spite of the various initial states including relative densities of D_r=22-70% and mean normal stresses of p= 30-400 kPa (Fig. 1.23), a nearly unique $(\Omega - \varepsilon_q^p)$ relation was obtained in these tests.

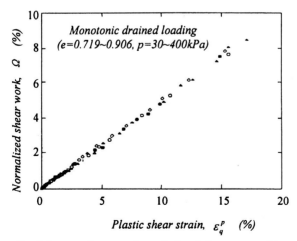

Fig. 1.36. Normalized shear work versus shear strain relation measured in monotonic drained torsional shear tests on dry-pluviated samples of Toyoura sand with various initial *e-p* states

For this reason, it is assumed that the parameters of the stress-dilatancy relation, i.e., μ_0, M and S_c are independent of both initial density and normal stress of the sand. It should be recognized, however, that the effects of the initial e-p state on the stress-dilatancy relation are implicitly incorporated though the dependence of the dilatancy parameter p on the shear strain ε_q^p (Eq. 1-20), in conjunction with the link between the shear strain development and the e-p state of the sand established through the stress-strain model.

Material Parameters

The stress-strain-dilatancy model has three groups of material parameters which are the states index, stress-strain and dilatancy parameters. Table 1-6 lists the material parameters and indicates the basic relations as well as the expressions where they are used and defined. Except for the mean normal stresses of the reference lines, the parameters are dimensionless. It is to be recalled that the reference lines are affected by the sand fabric, and therefore the material parameters are fabric-dependent.

A series of 12 to 15 laboratory tests is needed to determine the material parameters. The tests include monotonic undrained tests (UR-line; QSS-line; M= critical state stress ratio; dilatancy parameter μ_0), monotonic drained p-constant tests (stress-strain parameters) and cyclic undrained or liquefaction resistance tests (dilatancy parameter S_c). In general, torsional shear tests are preferred since in this type of test the required test conditions are easy to replicate and, more importantly, the deformation mode is relatively close to the undrained simple shear deformation which is considered a prevalent mode of deformation for level ground subject to an earthquake excitation. When using test results from undisturbed samples, however, it is more practical to evaluate the model parameters using conventional triaxial tests. In the latter case, an interpretation of the effects of the intermediate principal stress as represented by the b-value $\{b=(\sigma_2-\sigma_3)/(\sigma_1-\sigma_3)\}$ and the anisotropic characteristics of the in-situ soil is needed.

A unique feature of the model is that it has a single set of material parameters for a sand with a given fabric. In other words, once the parameters listed in Table 1-6 are determined, sand behavior can be modeled for any given initial e-p state of sand. This feature of the model is qualitatively illustrated in Figs. 1.37 and 1.38, where two sets of simulated effective stress paths and stress-strain curves for monotonic undrained loading are shown. It is apparent that the simulated behavior gradually changes with the initial e-p state and exhibits typical characteristics of observed sand behavior in laboratory tests under identical conditions. These simulations can be obtained by incorporating the stress-strain-dilatancy model in an elementary elasto-plastic model that features linear isotropic elastic behavior and isotropic 'hardening' or evolution of the loading ('yield') surface.

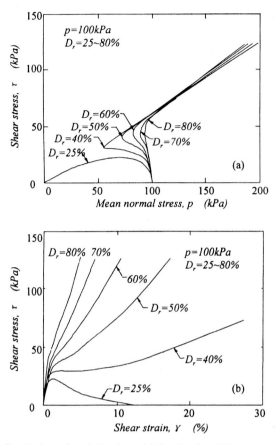

Fig. 1.37. Simulation of undrained sand behavior for different relative densities:
a) Effective stress paths; b) Stress-strain curves

Table 1-6. Material parameters

Relation	Material parameter		Expression
	Reference lines		
	UR-line (Void ratios and normal		
State index I_s	stresses)	(e_U, p_U)	Eq. (1-10)
(UR-line; QSS-line)	QSS-line (Void ratios and normal	(e_Q, p_Q)	Eq. (1-10)
	stresses)		
	Stress-strain parameters		
Stress-strain curve	Peak stress ratio coefficients	a_1, b_1	Eq. (1-14)
$\{(\tau/p)_{max}, G_{N,max}, G_{N,min}\}$	Max. shear modulus coefficients	a_2, b_2	Eq. (1-15)
	Min. shear modulus coefficients	a_3, b_3	Eq. (1-16)
	Degradation constant	f	Eq. (1-12)
	Dilatancy parameters		
Stress-dilatancy	Dilatancy coefficients (small strains)	μ_0	Eq. (1-20)
(μ_0, M, S_c)	Critical state stress ratio	M	Eq. (1-20)
	Dilatancy strain	S_c	Eq. (1-20)

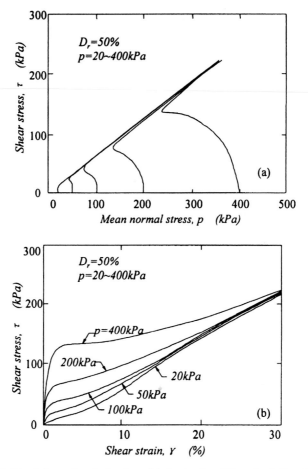

Fig. 1.38. Simulation of undrained sand behavior for different relative densities:
a) Effective stress paths; b) Stress-strain curves

Summary

A stress-strain-dilatancy model for sand which is based on the idea of integral modeling over the relevant density and normal stress states has been presented. The model is established within the framework of the state concept and uses the relative initial *e-p* state to assess the combined influence of density and normal stress on sand bahavior.

Results from a series of torsional tests on Toyoura sand demonstrated the ability of the state index I_s to quantify the link between the relative initial state and the normalized stress-strain curve of sand. It is shown that: 1) initial states that have similar I_s-values have similar stress-strain curves; 2) as the state index decreases, both the initial shear modulus and the peak stress ratio decrease as well; 3) the degradation of the stress-strain curve ceases once I_s approaches the UR-line ($I_s=0$). It is also shown that a rational relationship between the stress-strain parameters and the state index can be established when physically meaningful parameters are used. The strong correlation between the stress-strain parameters and I_s is enhanced by the fact that the employed characteristic states are not mere reference lines but rather principal states of sand bahavior.

Employing the state index as a current variable enables us to gradually change the characteristics of the stress-strain curve according to the current *e-p* state and achieve a density-dependent influence of the mean normal stress. The advantage of adopting this modeling concept is illustrated through a comparison with conventional modeling and simulations of drained and undrained behavior of sand. The model is shown to capture the most salient features of observed sand behavior over a wide range of densities and confining stresses. It is worth noting that principal parameters of both stress-strain relation and stress-dilatancy relation are expressed as functions of the shear strain, which enables modeling of the strain-dependent properties of sand behavior.

2. PILE DAMAGE

Ground Settlement Following Liquefaction

Following the Kobe earthquake, a large amount of silt-sand-laden water spurted through pavement joints and in the shrubbery zone along the roads and spread over on the roads and pavemant areas. The settlement of the ground surface sufficiently far from the waterfront was estimated by surveying the difference in elevation between supposedly subsided flat ground areas and objects such as pile-supported buildings which were apparently free from any settlement. The results of such surveys made at many locations in the flat area of Port Island and Rokko Island are presented in Fig. 2.1. It may be seen that the settlements observed varied in a wide range with a maximum of 90 cm on Port Island. The average value of settlements was 50 cm on Port Island and 40 cm on Rokko Island.

The procedures for estimating ground settlements resulting from liquefaction have been developed by various researchers. Figure 2.2 shows a

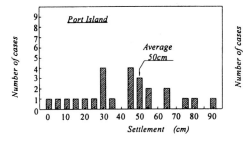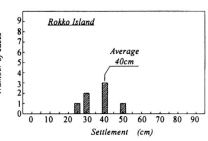

Fig. 2.1. Settlements observed on the ground surface at Port and Rokko Islands

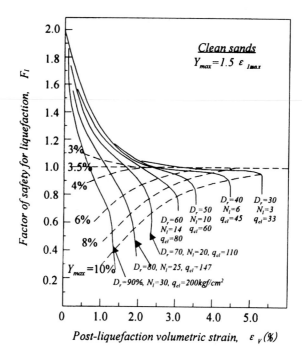

Fig. 2.2. Post-liquefaction volumetric strain as functions of factor of
safety against liquefaction

chart prepared by Ishihara and Yoshimine (1991) for estimating the post-liquefaction settlements of sand deposits. In this procedure, the volumetric strain is estimated as a function of the factor of safety F_l if the density of a deposit is made known in terms of SPT N-value or CPT q_c-value. If the near-surface soil deposit is assumed to have developed liquefaction with the development of single-amplitude shear strain of about τ_{max} =3.5%, the factor of safety is estimated from the chart as having been near unity at the onset of liquefaction. Entering the chart with these values, the post-liquefaction volumetric strain is estimated to be approximately e =2~4%. Since the liquefaction appears to have penetrated to a depth of 15m, the post liquefaction settlement is calculated as (0.02~0.04)×15 m=30~60 cm. This value is approximately in the range of observed settlements shown in Fig. 2.1. It should be noted, however, that the chart in Fig. 2.2 has been established based on the laboratory test data on clean sands and a more exact chart like the one in Fig. 2.2 should be established for the gravel-containing silty sand such as that encountered in the area affected by the earthquake in Kobe.

Family Court House Building

In the area of extensive liquefaction at the time of the Niigata Earthquake in 1964, lateral flow of liquefied sands took place exerting deleterious effects on foundations of structures. Among many of those injured by the lateral spreading, the reinforced concrete piles (RC-piles) supporting a three-story reinforced concrete building were those investigated in details by excavating the surrounding soil deposits to a depth of about 10 m (Yoshida and Hamada, 1991). The location of the building is shown in Fig. 2.3. The movement of the ground as identified by virtue of the air-photo interpretation is also reported in the paper as shown in Fig. 2-4, where it may be seen roughly that the movement

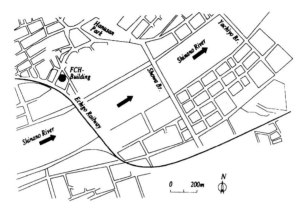

Fig. 2.3. Location of the building in Niigata investigated by excavation

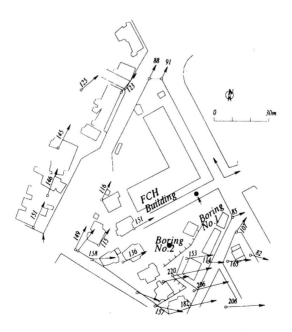

Fig. 2.4. Permanent ground displacement (cm) near the FCH Building (Yoshida and Hamada, 1991)

of the ground consequent upon liquefaction was about 1.0 m in the north-east direction at the place where the Family Court House was located. The layout of the piled foundations are displayed in Fig. 2.5 where the exact locations of two investigated footings (No. 1 and No. 2) are indicated. These footings 1 m×1 m in size were supported by single reinforced concrete piles 35 cm in diameter as indicated in the inset of Fig. 2.5. The investigations were conducted in 1989, 25 years after the earthquake. Such an opportunity present itself when the building was completely demolished. Sheet piles were driven to enclose the foundation pile and while being dewatered, the ground was excavated to expose the complete body of the pile as it had been buried.

Building in Fukaehama

The damage to foundation piles of a 3-story building at the time of the 1995 Kobe Earthquake is reported by Tokimatsu et al. (1997). The building was situated 6 m inland from the quay wall on a reclaimed fill in Higashi-Nada, Kobe. It was constructed early in 1980's and supported by prestressed concrete piles (PC-piles) 40 cm in diameter and about 20 m long.

The building sustained a tilt of 3 degrees due to the lateral spreading of the soils. The piles beneath the foundation were exposed and detailed investigations were carried out in-situ to see features of injury on precast reinforced concrete piles. The investigations included visual observation of cracks around inside walls of the hollow cylindrical piles by means of a television camera and measurements of tilt through the depth by using a slope indicator lowered into the inside hole of the piles.

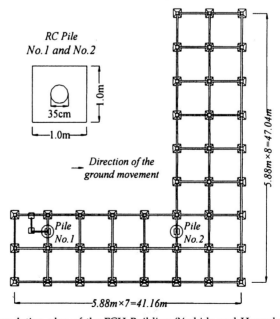

Fig. 2.5. Foundation plan of the FCH Building (Yoshida and Hamada, 1991)

3. DESIGN OF PILES IN LIQUEFIED GROUND

Introduction

Design of piles for the effects of seismic motions is generally performed by using a soil-pile interaction model in which a vertically-placed beam is supported by a series of spring elements. The beam represents the performance of the pile, and soil properties are represented by the spring constants. The effects of horizontal seismic motions on piles are allowed for by incorporating a horizontal force at the top of the pile, which is equivalent to the inertia force from the superstructure. For this type of analysis the spring constants are determined for conditions of no softening of the soils due to liquefaction. When liquefaction is of concern, the stiffness of the liquefied soils is dramatically reduced, and these effects need to be considered. The Japanese Code of Highway Bridge Design stipulate, for example, that for the majority of cases the spring constants be reduced by a factor of $1/6$ to $2/3$ depending upon the degree of safety against liquefaction. However, when it comes to the effects of lateral spreading of once liquefied soils, there has been no requirement stipulated in this code for the design of pile foundations. Thus, concerns have been kindled on these effects since the Kobe earthquake in 1995 because of the extensive occurrence of damage to foundation piles apparently due to the lateral spreading. When piles are subjected to the lateral flow of once liquefied soils, lateral forces would be applied directly to the pile body throughout the depth of liquefaction. In assessing this force in the design, there would be two approaches. The first method consists of assessing directly the lateral force on the pile body either based on empiricism or by means of the concept of viscous flow (Chaudhuri et al., 1995; Hamada and Wakamatsu, 1998). This may be called the "Force-based approach". In either way, it would be difficult to introduce a parameter which is indicative of the degree of destructiveness of the ground failure. Thus, the specification of the lateral force would have to be made irrespective of whether the ground displacement is destructively large or small. In the second method, the lateral displacement of the ground is specified through the depth of the deposit where the lateral spreading is induced. This prescribed displacement is applied to the spring system inducing lateral forces acting on the pile body. This procedure may be called the "Displacement-based approach". One of the advantages of this method is that it allows to specify the magnitude of ground displacement which is indicative of the degree of destructiveness or severity of the lateral spreading. In this method however, the choice of the spring constants has a profound influence on the magnitude of the lateral force induced, and as such difficulty is encountered in evaluating correctly this value for design purpose. It is expected that the spring constant in laterally spreading soils is much smaller than that in the case of the back and forth movement of soils as stipulated in the Japanese Code of Highway Bridge Design as mentioned above. While the code basically stipulates $1/6$ to $2/3$ reduction in the coefficient of subgrade reduction, the reduction is anticipated to be much more drastic if the effects of lateral spreading are allowed for. Thus, it becomes necessary to know the order of magnitude by which the conventionally used coefficient of subgrade reaction should be degraded to account for the interaction phenomenon taking place in the course of lateral spreading of liquefied soils.

Calculations in the above context were reported and discussed in previous papers (Ishihara, 1997; Ishihara and Cubrinovski, 1998). As a result, it was found that the stiffness of soils in laterally deforming deposits is reduced by a factor or 2×10^{-4} to 1×10^{-2}, and the degree of this reduction depends upon the relative displacement between the pile and the surrounding ground. However, in these studies, back-analyses were made only for the case of failure of relatively low-stiffness precast reinforced concrete piles having a diameter of 30-40 cm. These piles are used generally for foundation of medium-weight structures such as buildings and warehouses. In contrast to the above, the cast-in-place reinforced concrete bored piles are commonly used for supporting a large body of footings of piers for highway bridges. These piles have a diameter of 1.0-2.0 m, and are constructed by what is known as the benoto method. Thus, it is felt necessary to perform similar kind of back-analysis for such large-diameter piles to examine the stiffness degradation characteristics of the surrounding liquefied soils. The outcome of the studies in this context is described in the following pages of this paper.

3.1 Characteristics of Piled Foundation

Majority of the piles damaged by the lateral flow of liquefied deposits at the time of the Kobe earthquake in 1995 may be divided into two groups, that is, the precast reinforced concrete piles and the cast-in-place reinforced concrete bored piles. The precast concrete pile is hollow-cylindrical and has a diameter of 30-40 cm. The length is generally in the range of 10-20 m. The piles are arranged generally in a group of 4-6 piles which are embedded at their tops into a common footing slab about 0.5-1.0 m thick. The top of the piles are connected to the footing slab in different ways, and therefore it is difficult to identify whether the top was rigidly connected or not. Especially because these piled foundations were constructed more than 20 years ago, and construction details are not known. The footing slab having a thickness of 0.5-1.0 m is generally embedded into the surface soil layer above the ground water table which is therefore free from liquefaction. In as much as the thickness or depth of embedment is relatively small, the lateral pressure acting on the sidewall of the slab in the non-liquefied surface layer is considered to be relatively small and therefore the presence of the footing slab may not be pronounced.

In contrast to the above, cast-in-place reinforced concrete bored piles are commonly used for supporting a large body of footings of piers for highway bridges. These piles have a diameter of 1.0-2.0 m, and are constructed by the benoto method. In this type of structures, the footing is constructed of massive reinforced concrete and has a thickness of 3 to 4 m. The whole body is embedded in the ground to a depth of 3-4 m where the ground water table is encountered. Thus, the effects of the lateral force in the non-liquefied surface layer acting on the sidewall of this embedded footing may not be ignored when making the back analyses for the behavior of the underlying piles subjected to lateral flow.

3.2 Basic Assumptions and Rules of Load Partitioning

When making the back-analyses for the large-diameter group piles with a massive footing, it is necessary to assume, in one way or another, the magnitude and direction of the lateral force applied to the footing slab from the non-liquefied surface layer, and to set up hypotheses as to how the total load on the footing is apportioned among the individual piles to which analysis is to be conducted.

If the ground surrounding the footing moves by an amount larger than the footing slab itself, it is apparent that the lateral force applied to the footing is oriented in the direction of the ground flow as illustrated in Fig. 3.1. It may be assumed that the lateral force would be equal to or smaller than the passive earth pressure. The total lateral force, P, due to this earth pressure which is deemed as the maximum possible value is given by

$$P = 1/2 \gamma K_p H_1^2 B \tag{3-1}$$

where $K_p = \tan^2(45° + \phi/2)$, ϕ is the angle of internal friction, H_1 is the thickness of the unliquefied surface layer, and B is the width of the footing. In apportioning this total force to the individual piles, there would be two concepts, as follows, which are regarded as two extremes within which the actual conditions lie.

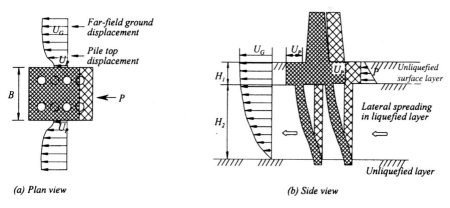

(a) Plan view   (b) Side view

Fig. 3.1. Pattern of lateral displacements of the ground and pile during lateral spreading of liquefied soil

(1) Single pile hypothesis

Suppose there are nine piles arranged at an equal spacing as shown in Fig. 3.2. The simplest concept would be to assume that the total lateral force is carried equally by each pile. Thus, the back-analysis may be made for a pile with a

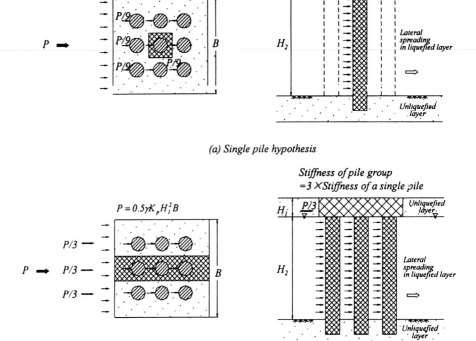

(a) Single pile hypothesis

(b) Pile row hypothesis

Fig. 3.2. Two hypotheses for partitioning a load from the footing to piles

given stiffness as illustrated in the side view of Fig. 3.2(a).

(2) Pile row hypothesis

The three piles immediately adjacent to the upstream wall are considered to carry large portion of the lateral force as compared to the other two piles located downstream, as illustrated in Fig. 3.2(b). However, it would be acceptable to postulate that one-third of the total load is transmitted equally to each row of the pile alignment in the direction of the ground deformation, as illustrated in Fig. 3.2(b). In this modeling, the flexural stiffness of the pile row as a whole is assumed to be three times the flexural stiffness of a single pile, and the modeled three-pile unit is assumed to be subjected to the lateral force of $P/3$ near the head of the pile.

The above is an illustration for the case of 9 piles equally spaced. For the footing having more piles with complicated arrangements in plan, other hypotheses with similar context will be made in the following back-analyses. It is to be noted that, no matter which is the rule of load partitioning, the piles in the liquefied deposits are assumed to behave independently as a single pile. Thus, group effects of interaction amongst the piles are not taken into consideration in the present analysis.

3.3 Scheme of Back Analysis

The behavior of piles is assumed to be represented by the model in which the lateral force, F, acting on the piles is proportional to the relative displacement between the pile and the soil in far-field condition. This may be written as

$$F = \beta kd(U_G - U_P) \tag{3-2}$$

where U_G and U_p are lateral displacements of the ground and pile respectively, d is the effective area, and k is the coefficient of subgrade reaction. The model is schematically illustrated in Fig. 3.3. If the soil is brought to a state of liquefaction and consequent lateral flow, the stiffness of the soil would be reduced drastically leading to a reduction in the k-value. The degree of this stiffness reduction is expressed by β in Eq. (3-2), which will be referred to as the "Stiffness degradation parameter". The main aim of the present back-analyses is to pursue the range of this degradation parameter. The steps of the analysis to be followed are described below.

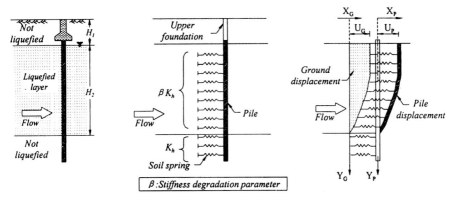

Fig. 3.3. Soil-pile model and numerical scheme

(1) The stiffness of the spring is assumed to decrease by a factor β upon liquefaction and lateral spreading of the soil through the depth of liquefaction, H_2. The spring constant in the underlying non-liquefied zone is assumed not to be degraded. It is further assumed that the stiffness degradation occurs uniformly throughout the depth H_2 where lateral spreading is taking place.

(2) Generally, there is a non-liquefied layer to a certain depth H_1 near the surface. This depth may be roughly defined to be equal to the depth to the ground water table. The movement of this non-liquefied soil mass may be modeled in different ways. One is to assume the surface layer to move in unison with the underlying liquefied stratum. In this case, it may be assumed that the passive earth pressure is applied to the wall of the footing on the upstream side in the same direction as the flow of the underlying liquefied soil layer, as illustrated in Fig. 3.1. In this type of surface soil movement, the displacement of the pile head is considered always smaller than the overall movement of the surrounding soil.

(3) The ground displacement due to lateral spreading is specified and given to the springs in the liquefied portion of the soil deposit; displacements,

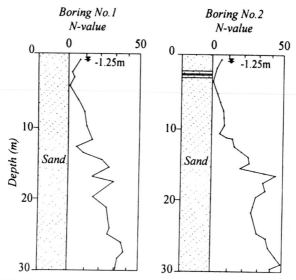

Fig. 3.4. Soil boring data in the vicinity of FCH Building (Yoshida and Hamada, 1991)

bending moments and lateral forces acting on the pile body are calculated, whereby the pile is assumed to deform in an elasto-plastic manner where the moment-curvature relation is represented by a trilinear relationship.

3.4 Case Studies

Analyses for Family Court House Building

Soil profiles were investigated in 1987 by means of borings at two sites near the building as indicated in Fig. 2.4. Results of the soil investigation are shown in Fig. 3.4 where it can be seen that deposits of clean sand prevail to a depth of 30 m with SPT *N*-value less than 10 to a depth of about 12 m.

The outcome of field measurements on the lateral deformation of the two piles made in the excavated pits is demonstrated in Fig. 3.5. According to the design chart, both piles were designed to have a length of 14 m, but perhaps because of a stiff sand layer encountered, the pile driving for the Pile No. 1 must have been stopped at a depth of 10-12 m. This situation is conjectured from the absence of any injury to the Pile No. 1 around the bottom of the excavated pit (Yoshida and Hamada, 1991). The Pile No. 2 appeared to have been driven to a deeper stratum with the design length of 14 m as evidenced by many cracks that had developed near the bottom of the pit. This is indicative of a large bending moment having exerted during the lateral flow of the liquefied sand. The Pile No. 2 was completely chopped off about 2 m below the pile head, as indicated by the discontinuity in the measured pile deformation shown in Fig. 3.5.

When excavation was conducted in 1989, the pit was dug first to a depth of 2.0 m while breaking and removing the footing slab. According to the design

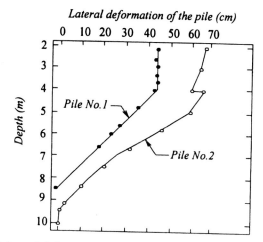

Fig. 3.5. Lateral deformation of Piles No. 1 and No. 2 in FCH Building
(Yoshida and Hamada, 1991)

chart, the pile top was located at an elevation 1.65 m below the ground surface. The elevation of the ground water table could be somewhat variable, but based on the design chart and soil investigations it may be assumed to have been about 1.0-1.5 m deep as accordingly indicated in Fig. 3.4. The feature in which the pile head was connected with the footing slab is not known unfortunately. When the piles suffered injury due to the lateral flow at the time of the 1964 Niigata Earthquake, the soil mass above the ground water table may be assumed to have moved together with the liquefied soil deposit underlying the ground water table. This appears to have been the case, however, in the free field conditions where there was no constraint near the piles. In the foundation system of FCH Building, a number of footing slabs were connected to each other by underground horizontal beams about 1.0 m in height and 0.8 m in width. Existence of such beams is considered to have exercised some constraint on the movement of the surface soil deposit in the vicinity of the single piles being studied. Thus, it may well be assumed that the deformation pattern has occurred at the time of the lateral spreading in 1964 Niigata Earthquake. If this assumption is valid the lateral force must be acting in the direction opposite to that of the flow and its magnitude may be assumed to be equal to the passive earth pressure $K_p \gamma H_1^2 / 2$. It is also envisaged that the near surface deposit might have been ravaged by violent motions of the ground due to liquefaction involving cracking and spurting of mud water. Thus, the earth pressure acting on the pile would have been much smaller than that mobilized in the passive earth pressure conditions. As an extreme case, there would have been no lateral force acting on the pile-top. This could be the case where the pile and surrounding surface-soil move together without any relative displacement. With the considerations as above, it will be of interest to see the effects of the lateral force by assuming two cases, that is, zero earth pressure and passive earth pressure $K_p \gamma H_1^2 / 2$.

Though the connection between the footing and pile top is not known, it may be acceptable to assume for simplicity that the pile-footing system is modeled by a single continuous beam throughout the depth. The single pile is postulated to be supported by a series of springs and subjected to the zero or passive earth pressure to a depth of $H_l = 1.25$ m as illustrated in Fig. 3.6. Regarding the lateral force below the depth H_l, it is assumed to have been applied through the springs which are deformed in consistence with an amount of displacement specified at each depth of the liquefied layer.

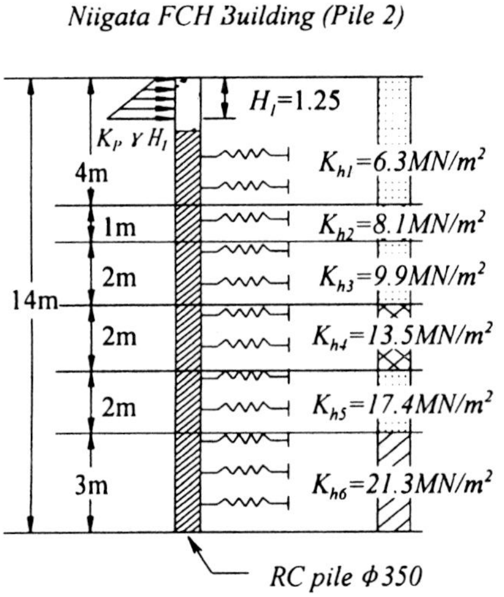

Niigata FCH Building (Pile 2)

Fig. 3.6. Spring constants for the RC-Pile No. 2 in FCH Building, Niigata

In the analysis of the Piles No. 1 and No. 2 in the building of the Family Court House, the spring constants were evaluated by using an empirical formulas for the coefficient of subgrade reaction k stipulated in the Japanese Code of Bridge Design.

The spring constant K_h was determined by multiplying the coefficient of subgrade reaction k by an effective width of the pile

$$K_h = kd \qquad (3-3)$$

The value of K_h evaluated in this way with reference to the SPT N-value of Fig. 3.4 is shown in Fig. 3.6 for the representative soil profile at the site of Family Court House. In making the analysis, it is necessary to specify the relation between the bending moment and curvature of the beam representing the flexural characteristics of the reinforced concrete pile. This relation was established based on the cross section and deformation characteristics of the piles as shown in Fig. 3.7. The analysis of the piles undergoing the lateral displacement were performed by a series of computations as described below.

(1) First, the lateral displacement is prescribed throughout the depth. In the case of the Family Court House site, the sand deposits are postulated to have developed liquefaction to a depth of 9.0 m, and the entire soil mass above it to have moved

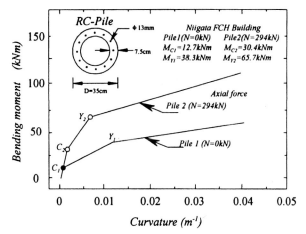

Fig. 3.7. Bending moment versus curvature relation for the RC-piles in the FCH Building

with a cosine distribution of lateral displacement with its maximum value of 1.0 m at a depth of 1.25 m. The displacement of the ground as postulated above was applied to the equivalent linear spring model and analysis was made for the spring-supported beam to obtain the lateral displacement of pile. In conducting the analysis, the pile head was assumed to be free to move horizontally whereas the lower end of the pile was assumed to be free only to rotate for the Pile No. 1 at the depth of 9.5 m and fixed for the Pile No. 2 at the depth of 14.0 m. In the analysis, the spring constant for the supposedly liquefied layer was reduced in a wide range taking the stiffness degradation parameter of $\beta = 5 \times 10^{-4} \sim 1 \times 10^{-2}$ as defined by Eq. (3-2).

(2) As briefly mentioned above, the passive earth pressure was applied in the upper portion of the pile to a depth of $H_1 = 1.25$ m, where the angle of internal friction of $\phi = 300$ and $\gamma_t = 17.6$ kN/m³ were assumed to be the case. In the remaining portion of the pile to the depth of 9.0 m, the displacement with a cosine distribution was applied to the springs. Analysis was made as well for the case of no earth pressure in the upper portion of the piles with the same conditions otherwise.

(3) From the layout of the piled foundations shown in Fig. 2.5, the Pile No. 1 and No. 2 may be considered as single piles and as such the analysis was conducted for a pile with a width of $d = 35$ cm.

The outcome of the analysis for the Pile No. 1 is presented in Fig. 3.8 through Fig. 3.13 and the results for the Pile No. 2 are shown in Figs. 3.14 through 3.19.

As shown in Figs. 3.8 and 3.11, the lateral displacement for the Pile No. 1 is assumed to be zero at the depth of 9.5 m. Assuming this to be reasonable, Fig. 3.8 shows that the case of analysis with passive earth pressure using the degradation parameter $\beta = 8 \times 10^{-3}$ can provide the best fit to the displacements observed in the in-situ investigation. It is to be noticed that the displacement of the pile at the top of the liquefied layer or at the bottom of the unliquefied surface layer was shown to be 0.6 m which is smaller than the value of 1.0 m of the ground displacement prescribed in the analysis at this elevation. The degradation parameter $\beta = 8 \times 10^{-3}$ is plotted in Fig. 3.28 versus the relative displacement $U_G - U_p$ normalized to the thickness of the liquefied layer, H_2.

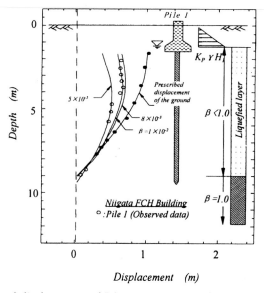

Fig. 3.8. Lateral displacements of Pile No. 1 computed for different β-values: Analysis with passive earth pressure (Niigata FCH Building)

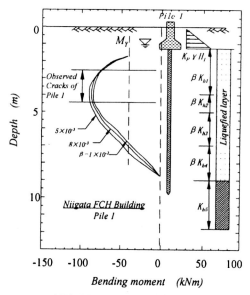

Fig. 3.9. Bending moments of Pile No. 1 computed for different β-values: Analysis with passive earth pressure (Niigata FCH Building)

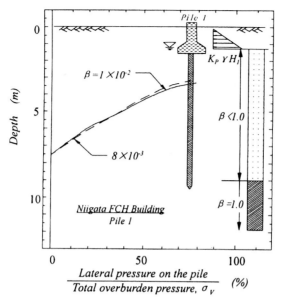

Fig. 3.10. Lateral pressure on Pile No. 1 computed for different β-values: Analysis with passive earth pressure (Niigata FCH Building)

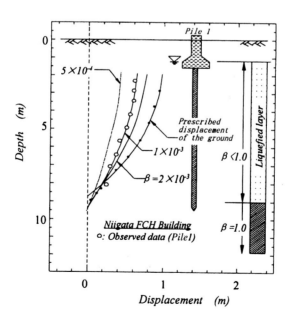

Fig. 3.11. Lateral displacements of Pile No. 1 computed for different β-values: Analysis with passive earth pressure (Niigata FCH Building)

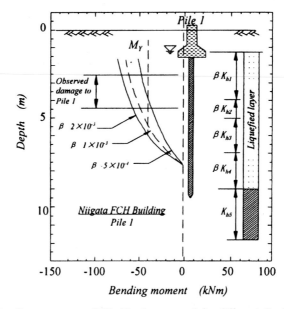

Fig. 3.12. Bending moments of Pile No. 1 computed for different β-values: Analysis without earth pressure (Niigata FCH Building)

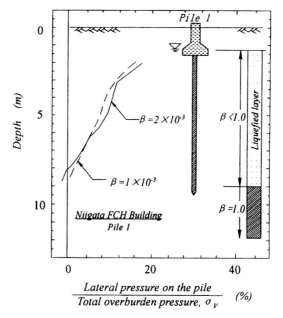

Fig. 3.13. Lateral pressure on Pile No. 1 computed for different β-values: Analysis without earth pressure (Niigata FCH Building)

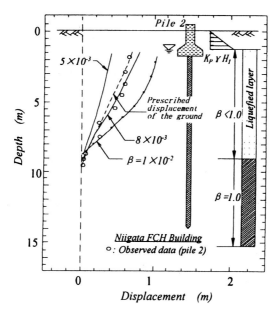

Fig. 3.14. Lateral displacements of Pile No. 2 computed for different β-values: Analysis with passive earth pressure (Niigata FCH Building)

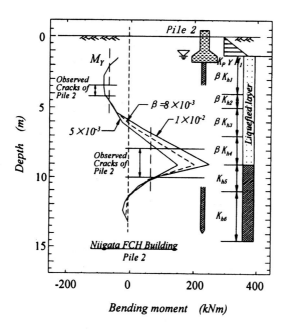

Fig. 3.15. Bending moments of Pile No. 2 computed for different β-values: Analysis with passive earth pressure (Niigata FCH Building)

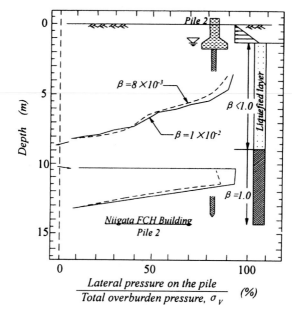

Fig. 3.16. Lateral pressure on Pile No. 2 computed for different β-values: Analysis without passive earth pressure (Niigata FCH Building)

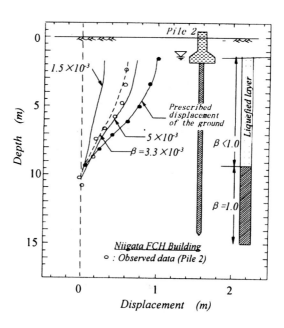

Fig. 3.17. Lateral displacements on Pile No. 2 computed for different β-values: Analysis without passive earth pressure (Niigata FCH Building)

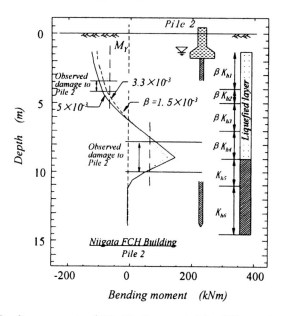

Fig. 3.18. Bending moments of Pile No. 2 computed for different β-values: Analysis without earth pressure (Niigata FCH Building)

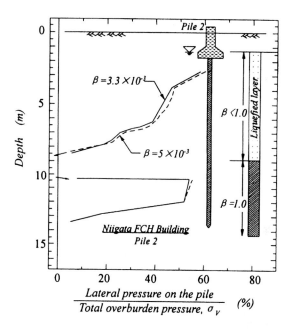

Fig. 3.19. Lateral pressure on Pile No. 2 computed for different β-values: Analysis without earth pressure (Niigata FCH Building)

In the case of the analysis with no earth pressure in the surface layer, the results of calculations are presented in Fig. 3.11. It may be seen that the best fit between the observed and calculated behavior is achieved with $\beta = 1\times10^{-3}$, where the lateral pile-top displacement is obtained also as 0.6 m in the analysis. The value of the normalized relative displacement $(U_G - U_p)/H_2$ is plotted in Fig. 3.28 versus the value of β.

Thus, if the liquefied layer is assumed to have induced lateral spreading overall in the area of FCH Building without any constraint from surrounding objects, a relative displacement of $(1.0-0.6) = 0.4$ m is considered to have occurred at the top of the liquefied layer. Then, the stiffness degradation parameter is reckoned to have taken a value between $\beta = 1\times10^{-3} \sim 8\times10^{-3}$ which depends generally upon the magnitude of the lateral pressure acting on the upper portion of the pile in the unliquefied surface layer.

The bending moment induced in the pile body is shown in Fig. 3.9 for the case of passive earth pressure application. It may be seen that at depths of 1.5 m to 7.5 m the induced bending moment far exceeds the yield value M_y at which larger cracks are considered to develop in the concrete. Observation of the in-situ investigation revealed that predominant cracks existed at the depths of 2.5 m to 4.5 m as accordingly indicated in Fig. 3.9. In the case of no earth pressure application, the range of depth at which the calculated bending moment exceeds the yield value coincides reasonably well with the depth of observed cracks as indicated in Fig. 3.12, but the value of the calculated moment for this case is smaller as compared to the moment computed by considering the passive earth pressure. It is to be noticed that the bending moments are the least affected by the degradation parameter β, but they are significantly influenced by the condition whether or not there is a lateral pressure from the surface layer. It would be of interest to see how much lateral force is applied to the pile body in the lower layer from the surrounding soil deposit when it is undergoing lateral spread. For this reason, the lateral pressure per unit area of the pile body was computed and normalized to the total overburden pressure σ_v at each depth. The results of such analysis are demonstrated in Figs. 3.10 and 3.13 where it may be seen that, while the influence of the stiffness degradation parameter β is very small, effects of the lateral pressure in the unliquefied surface layer is significant on the lateral force. In the case of no earth pressure application, the lateral pressure on the pile is found to be somewhere between 5 and 20% of the total overburden pressure whereas that computed by considering the passive earth pressure is predominantly in the range between 10% and 70% of σ_v.

The outcome of the analysis for the Pile No. 2 at FCH Building foundation is demonstrated in Figs. 3.14 through 3.19. With respect to the pile deformation, the best degree of coincidence is obtained, as shown in Figs. 3.14 and 3.17, when a value of $\beta=8\times10^{-3}$ and 3.3×10^{-3} is assumed for the degradation parameter. Regarding the bending moment and lateral stress, the value of β does not generate much difference, but the magnitude of the earth pressure in the unliquefied surface layer is a dominant factor. In both cases being considered, the computed bending moment is in excess of the yielding moment and far above the crack-inducing moment within the depth range in which cracks were observed at the time of the in-situ investigation. Therefore, the bending moment would not be regarded as a quantity by which validity of the assumption is checked regarding the influence of the earth pressure in the surface layer.

For the Pile No. 2, Figs. 3.16 and 3.19 indicate that the lateral pressure on the pile in the liquefied deposit is in the range of 10 to 80% and 5 to 50% of the total overburden pressure for the cases with and without earth pressure application, respectively.

Analyses for Building in Fukaehama

The plan view of the building foundation is shown in Fig. 3.20 with precise locations of the piles investigated. The soil profiles of the building foundation obtained at locations of boreholes No. 1 and No. 2 are presented in Fig. 3.21 where it may be seen that the reclaimed fills of the Masado soil exist to a depth of about 8.5 m. Liquefaction and lateral spread seem to have occurred in this reclaimed deposit. The results of the ground survey reported by Tokimatsu et al. (1997) are shown in Fig. 3.22 in terms of the lateral displacement of the ground surface plotted versus the distance from the waterfront. It may be seen that the ground displacement at the place of Pile S-7, 6 m inland, was about 1.2 m towards the sea. From an independent survey involving measurements of the inclination of the building structure and displacement of the roof, the displacement of head of the Pile S-7 was estimated to be about 80 cm.

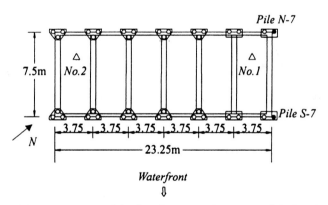

Fig. 3.20. Arrangements of the foundations in plane view of the building in Higashi-Nada, Kobe (Tokimatsu et al., 1997)

The scheme of the back analysis for the performance of Pile S-7 is indicated in Fig. 3.22. The pile 20 m long is assumed to be rigidly-connected to the overlying footing slab and embedded to stiff strata extending from 9 m to 20 m which did not develop liquefaction. The soil above the ground water table may be assumed to have moved together with the underlying liquefied soil, but because of the constraint produced by the horizontal underground beams between the front and back rows of the footing slabs, the soil in the unliquefied surface layer appears not to have been free to move with the underlying liquefied soil stratum.

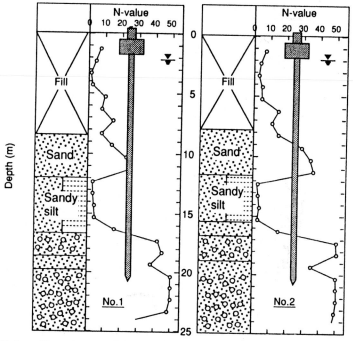

Fig. 3.21. Soil profile at the site of the building in Higashi-Nada, Kobe (Tokimatsu et al., 1997).

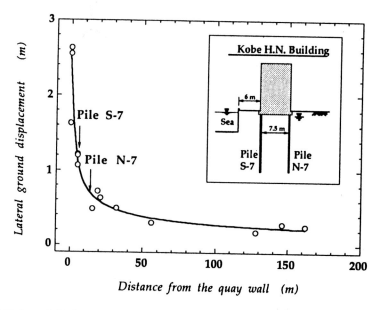

Fig. 3.22. Lateral displacement of the ground surface at the site of the building in Higashi-Nada, (Tokimatsu et al., 1997).

Kobe H.N. Building

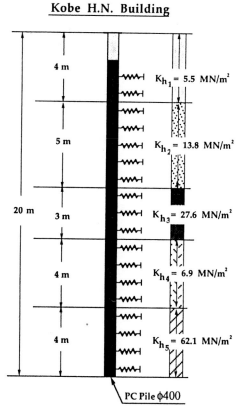

PC Pile φ400

5 Fig. 3.23. Spring constants for the P- pile of the building in Higashi Nada, Kobe.

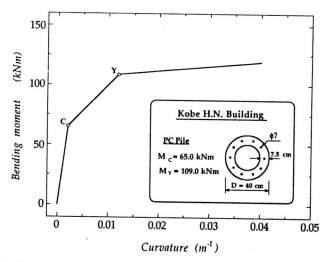

Fig. 3.24. Bending moment versus curvature relation for the P- pile of the building in Higashi Nada, Kobe.

Thus, it appears highly likely that the pile in the unliquefied layer has moved nearly in unison with the surrounding soil stratum. With this assumption, the earth pressure in the surface layer may be neglected. If the pile is assumed to have moved ahead of the surrounding soil, then the passive earth pressure acting against the movement of the pile will be induced. The computation for this case is not presented in this paper.

The model of the beam-spring system used in the analysis is illustrated in Fig. 3.23, where K_h-values estimated by the empirical formula in the design code are indicated. The nonlinear stiffness characteristics of the PC-pile are shown in Fig.3-24 in terms of the bending moment versus curvature relation.

Displacement distributions computed in analyses of Pile S-7 for the case of zero earth pressure in the upper layer are shown in Fig. 3.25. Superposed in this diagram is the displacement of the pile measured by means of the inclinometer. Comparison of the observed displacement with the computed one appears to indicate that the stiffness degradation parameter should take a value of $\beta = 5\times10^{-3}$ in order for the displacement of the pile head to become equal to the observed value of 0.8 m. The value of $\beta = 5\times10^{-3}$ obtained above is shown in Fig. 3.28 for comparison sake plotted versus the normalized relative displacement $(U_G - U_p)/H_2$. Distribution of the computed bending moment is shown in Fig. 3.26 versus depth where it may be seen that the moment in excess of the yield value M_y occurs at two depth ranges which are roughly coincident with the depths of crack development observed by virtue of the video camera Fig. 3.27 shows the lateral pressure acting on the pile which is normalized to the total overburden pressure σ_v. It may be seen that the lateral pressure on the pile is less than 50% of the total overburden pressure.

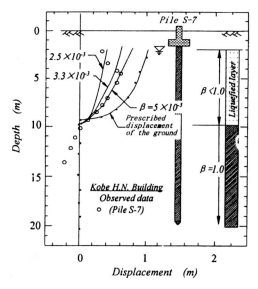

Fig. 3.25. Lateral displacements of the Pile S-7 computed for different β-values: Analysis without earth pressure (building in Higashi-Nada, Kobe)

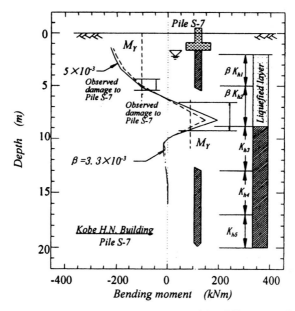

Fig. 3.26. Bending moments of the Pile S-7 computed for different β-values: Analysis without earth pressure (building in Higashi-Nada, Kobe)

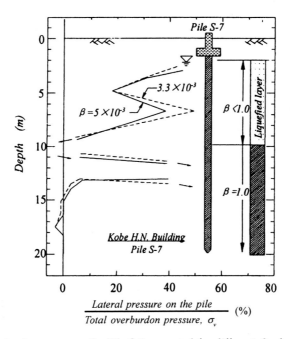

Fig. 3.27. Lateral pressure on the Pile S-7 computed for different β-values: Analysis without earth pressure (building in Higashi-Nada, Kobe)

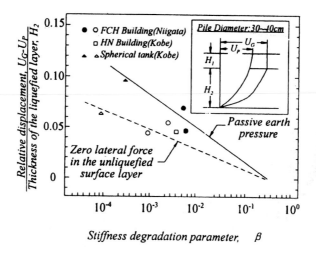

Fig. 3.28. Normalized relative displacement between the ground and pile versus the stiffness degradation parameter β: Summary plot for the three case histories

Analyses for Foundation of Spherical Tan

At the time of the Kobe Earthquake in 1995, spherical tanks in a man-made island were damaged by extensive liquefaction and consequent lateral flow of the ground. Details of the damage feature and accounts of pile behavior were introduced elsewhere (Ishihara, 1997) with reference to the back analysis made for the damaged piles. The back analysis was conducted in the same fashion as described above in order to assess the range of the stiffness reduction factor for piles subjected to lateral flow of the liquefied deposit. In that analysis scheme, the earth pressure on piles in the upper unliquefied layer was assumed to be a passive earth pressure. It was possible to perform the back analysis by assuming zero earth pressure on the piles in the surface layer. The results of the analysis for the two extreme cases as above are presented in Fig. 3.28 in terms of the relation between the normalized relative displacement and the stiffness degradation parameter.

DISCUSSIONS

In the routine practice, the design of piles is often made through the use of a Winkler-type model in which soil-pile interaction is represented by the coefficient of subgrade reaction k. When soil deposits are identified to be softened as a result of liquefaction, this coefficient is reduced drastically. For the level ground conditions the Japanese Code of Highway Bridge Foundation recommends that the effect of liquefaction should be taken into account by reducing the value of k by a factor ranging between $1/6$ and $2/3$ for design of embedded foundations. If this norm is to be extended to include sloping grounds undergoing lateral spreading, it would be logical for all practical purposes to seek how much reduction in the k-value would farther be necessary to allow for the effect of lateral spreading on the design of piles embedded in such ground. While many laboratory model tests have been performed using 1g shaking table to evaluate effects of level-ground liquefaction on the k-value, little has been known on the effects of the lateral spreading on the reduction of the k-value.

As it is generally difficult to conduct model tests, it would be worthwhile trying to investigate these effects based on back-analysis of known behavior of piles in the field during earthquakes. Thus, the above data of back-analyses were arranged so as to find out governing factors influencing the stiffness degradation parameter. In an effort in this vein, it was discovered that the stiffness degradation parameter, β, is closely related with the amount of pile deflection relative to the displacement of the ground. It was further shown that the deflection of the pile does depend on the thickness of the liquefied stratum H_2. Thus, the normalized relative displacement $(U_G - U_p)/H_2 -$ was chosen as a parameter influencing the value of β. In this context, plots were made of the back-analyzed data as shown in Fig. 3.28. It may be seen in this diagram that the value of β tends to decrease with increasing normalized relative displacement. Note that the above relation is dependent on the flexural characteristics of the pile being considered. Thus, the relation shown in Fig. 3.28 does hold valid only for the case of reinforced concrete piles with a diameter of 30-40 cm. If the pile of greater stiffness is considered such as cast-in-place concrete pile with a diameter of 1.0-1.5 m, the relation would be different.

As seen in Fig. 3.28, the correlation between the normalized relative displacement and the stiffness degradation parameter tends to change depending upon the lateral force being applied in the upper portion of the pile in the non-liquefied surface soil deposit. It is apparent that, if the passive earth pressure is applied in the direction opposite to the flow, the value of β giving an equal normalized displacement would become larger as compared to the case of no lateral force.

In all the cases studied above, the piles being considered were located within the space enclosed by neighboring piles or close to the horizontal connecting beams. Thus, strong constraints appear to have been present for the upper portion of the piles under consideration. This indicates that the surface unliquefied soil layer must have moved together with the piles in question thereby producing no earth pressure against the upper portion of the piles. With this fact in mind, the relation for the case of no earth pressure in Fig. 3.27 is regarded as reflecting actual situations developed in the field during the earthquakes. It is to be noted that the piles in the liquefied layer must have been displaced behind the soil deposit, producing a fairly large relative displacement. Then attention should be drawn solely to the soil-pile interaction in the liquefied deposit. It is apparent that the greater the reduction in the stiffness of the liquefied soil, the larger the relative displacement would be between the soil and piles.

REFERENCES

Alarcon-Guzman, A., Leonards, G. A. and Chameau, J. L. (1988): "Undrained monotonic and cyclic strength of sands", ASCE, *J. of Geotech. Engrg. Division*, Vol. 114, No. 10, pp. 1089-1109.

Been, K., Jefferies, M. G. and Hachey, J. (1991): "The critical state of sands", *Geotechnique*, Vol. 41, No. 3, pp. 365-381.

Been, K. and Jefferies, M. G. (1985): "A state parameter for sands", *Geotechnique*, Vol. 35, No. 2, pp. 99-112.

Bolton, M. D. (1986): "The strength and dilatancy of sands", *Geotechnique*, Vol. 36, No. 1, pp. 65-78.

Chaudhuri, D., Toprak, S. and O'Rourke, T. D. (1995): "Pile response to lateral spread: A benchmark case", Lifeline Earthquake Engineering, Proceedings 4th U. S. Conference, San Francisco, California, pp. 1-8.

Cubrinovski, M. (1993): "A constitutive model for sandy soils based on a stress-dependent density parameter", Dr. Eng. Thesis, Univ. of Tokyo.

Goto, S., Morii, H., Cubrinovski, M. and Sueoka, S. (1997): "Mechanical characteristics of Masado soil and simulations by an effective stress model", Proc. of 24th Earthquake Engrg. Symp. of JSCE, Vol. 1, pp. 473-476 (in Japanese).

Gutierrez, M., Ishihara, K., and Towhata, I. (1993): "Model for the deformation of sand during rotation of principal stress directions", *Soils and Foundations*, Vol. 33, No. 3, pp. 105-117.

Hamada, M. and Wakamatsu, K. (1998): "A study on ground displacement caused by soil liquefaction", *JSCE Journal of Geotechnical Engineering*, No. 596/III-43, pp. 189-208.

Hanshin Highway Authority (1996): "Investigation on the seismic damages of bridge foundations in the reclaimed land" (in Japanese).

Inagaki, H., Iai, S., Sugano, T., Yamazaki, H. and Inatomi, T. (1996): "Performance of caisson type quay walls at Kobe Port", *Soils and Foundations*, Special Issue on Geotechnical Aspects of the January 17, 1995 Hyogoken-Nambu Earthquake, pp. 119-136.

Ishihara K. and Cubrinovski M. (1998): "Soil-pile interaction in liquefied deposits undergoing lateral spreading", Proc. 11th Danube-European Conference on Soil Mechanics and Geotechnical Engineering, Porec, Croatia, pp. 51-64.

Ishihara, K. and Takatsu, H. (1979): "Effects of over-consolidation and K_0 conditions on the liquefaction characteristics of sand", *Soils and Foundations*, Vol. 19, No. 4, pp. 59-68.

Ishihara, K. and Yoshimine, M. (1991): "Evaluation of settlements in sand deposits following liquefaction during earthquakes", *Soils and Foundations*, Vol. 32, No. 1, pp. 173-188.

Ishihara, K. (1997): "Geotechnical aspects of the 1995 Kobe Earthquake", Proceedings 14th International Conference on Soil Mechanics and Foundation Engineering, Terzaghi Oration, Hamburg.

Ishihara, K. (1993): "Liquefaction and flow failure during earthquakes", 33rd Rankine lecture, *Geotechnique*, Vol. 43, No. 3, pp. 351-415.

Ishihara, K., Tatsuoka, F. and Yasuda, S. (1975): "Undrained deformation and liquefaction of sand under cyclic stresses", *Soils and Foundations*, Vol. 15, No. 1, pp. 29-44.

Ishihara, K., Yasuda, S. and Nagase, H. (1996): "Soil characteristics and ground damage", *Soils and Foundations*, Special Issue on Geotechnical Aspects of the January 17, 1995 Hyogoken-Nambu Earthquake, pp. 109-118.

Jefferies, M. G. (1993): "Nor-Sand: a simple critical state model for sand", *Geotechnique*, Vol. 43, No. 1, pp. 351-415.

Kabilamany, K. and Ishihara, K. (1990): "Stress dilatancy and hardening laws for rigid granular model of sand", *Soils Dynamics and Earthquake Engrg.*, Vol. 9, No. 2, pp. 66-77.

Kondner, R. L. and Zelasko, J. S. (1963): "A Hyperbolic stress-strain formulation for sands", Proc. 2nd Pan-American Conf. on SMFE, Brazil, Vol. 1, pp. 289-324.

Labe, P. V. (1978): "Prediction of undrained behavior of sand", ASCE, *J. of Geotech. Engrg. Division*, Vol. 104, GT6, pp. 721-735.

Momen, H. and Ghaboussi, J. (1982): "Stress dilatancy and normalized work for sands", Proc. of IUTAM Conf. on Deformation and Failure of Granular Materials, Delft, pp. 265-274.

Moroto, N. (1976): "A new parameter to measure degree of shear deformation of granular material in triaxial compression tests", *Soils and Foundations*, Vol. 16, No. 4, pp. 1-9.

Nagase, H., Shinji, R., Kimura, K. and Tsujino, S. (1995): "Liquefaction strength of undisturbed sand subjected to overconsolidation", Proc. 1995 Annual Convention of Japanese Society of Soil Mechanics and Foundation Engineering.

Roscoe, K. H. and Poorooshasb, H. B. (1963): "A fundamental principle of similarity in model tests for earth pressure problems", Proc. 2nd Asian Regional Conf. on SMFE, Tokyo, Vol. 1, pp. 134-140.

Roscoe, K. H., Schofield, A. N. and Thurairajah, A. (1963): "Yielding of clays in states wetter than critical", *Geotechnique*, Vol. 13, No. 3, pp. 211-240.

Silver, M. L. and Seed, H. B. (1971): "Deformation characteristics of sands under cyclic loading", *ASCE, J. of SMFE Division*, Vol. 97, SM8, pp. 1081-1098.

Tatsuoka, F. and Shibuya, S. (1992): "Deformation characteristics of soils and rocks from field and laboratory tests", Keynote Lecture for Session No. 1, Proc. 9th Asian Regional Conf. on SMFE, Bangkok, 1991, Vol. 12, pp. 101-170.

Toki, K. (1995): Committee of Earthquake Observation and Research in the Kansai Area.

Tokimatsu, K., Oh-oka, H., Shamoto, Y., Nakazawa, A. and Asaka, Y. (1997): "Failure and deformation modes of piles caused by liquefaction-induced lateral spreading in 1995 Hyogoken-Nambu Earthquake", Geotechnical Engineering in Recovery from Urban Earthquake Disaster, KIG FORUM '97 Kobe, Japan, Kansai Branch of the Japanese Geotechnical Society, pp. 239-248.

Tsujino, S. (1992): "Constitutive model for sandy soils and its application to seismic response analysis", Dr. Eng. Thesis, Univ. of Tokyo (in Japanese).

Tsukamoto, Y., Ishihara, K. and Nonaka, T. (1998): "Undrained deformation and strength characteristics of soils from reclaimed deposits in Kobe", *Soils and Foundations*, Special Issue on Geotechnical Aspects of the January 17, 1995 Hyogoken-Nambu Earthquake, No. 2, pp. 47-55.

Vaid, Y. P., Chung, E. K. F. and Kuerbis, R. H. (1990): "Stress path and steady state", *Can. Geotech. J.*, Vol. 27, pp. 1-7.

Verdugo, R. (1992): "Characterization of sandy soil behavior under large deformation", Dr. Eng. Thesis, Univ. of Tokyo.

Wood, D. M., Belkheir, K. and Liu, D. F. (1994): "Strain softening and state parameter for sand modelling", *Geotechnique*, Vol. 44, No. 2, pp. 335-339.

Yasuda, S. (1990): from private pile.

Yasuda, S., Ishihara, K., Harada, K. and Shinkawa, N. (I 996): "Effect of soil improvement on ground subsidence due to liquefaction", *Soils and Foundations*, Special Issue on Geotechnical Aspects of the January 17, 1995 Hyogoken-Nambu Earthquake, pp. 99-107.

Yoshida, N. and Hamada, M. (1991): "Damage to foundation pile and deformation pattern of ground due to liquefaction-induced permanent ground deformations", Proceedings 3rd Japan-U.S. Workshop on Earthquake Resistant Design of Lifeline Facilities and Countermeasures for Soil Liquefaction, National Center for Earthquake Engineering Research, NCEER-91-0001, pp. 147-156.

Yoshimine, M. (1996): "Undrained flow deformation of saturated sand under monotonic loading conditions", Dr. Eng. Thesis, Univ. of Tokyo (in Japanese).

3

The Leaning Tower of Pisa— Present Situation

M.B. Jamiolkowski
Technical University of Turin, Italy

1. INTRODUCTION

1. Aim of this paper is to present the current condition of the leaning Tower of Pisa, updated till the end of year 1998.

    A brief summary history of the Monument will introduce the information concerning the subsoil condition and its structural features, followed by the presentation of the monitored data documenting the progressive increase of the Tower inclination.

    On the basis of the above information, a phenomenological outline motivating the reasons for the continuous increase of the Tower's inclination over time, since the completion of its construction, is subsequently presented. At this point it will be possible to attempt to formulate some considerations about the margin of safety relative to the risk of the Tower falling over.

2. Finally, a brief update on the state of knowledge concerning the Monument, an equally concise description of the stabilization works on the Tower foundation, as well as the project to reinforce its structure undertaken by a 14-member International Multi-Disciplinary Commission appointed by the Italian Government in the middle of 1990, will be presented.

2. HISTORICAL BACKGROUND

3. The Monuments of Piazza dei Miracoli, see Fig. 1, including the Tower, are: the Cathedral, the Baptistery and the Monumental Cemetery, which were all erected during the middle ages.

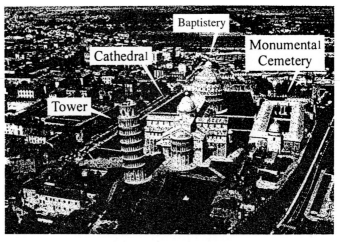

Fig. 1. Piazza dei Miracoli—airial view

In fact, construction of the Cathedral, the first monument to be erected, began in late 1000.

The design of the Tower is ascribed to the Architect and Sculptor Bonanno Pisano. The Tower consists, see Fig. 2, of a hollow masonry cylinder, surrounded by six loggias with columns and vaults merging from the base cylinder.

Inside the annular masonry body a helicoidal stair case leads to the bell chamber located at the top of the Monument.

Its construction started in August 1173 but after five years the works were interrupted at middle of the fourth order as shown in Fig. 3. The construction was resumed in 1272 under the lead of the Architect Giovanni Di Simone who brought the Tower almost to completion, up to seventh cornice (Fig. 3) in six years. The construction of the Tower was finally completed when Architect Tommaso di Andrea Pisano added the bell chamber between 1360 and 1370.

4. It was during the second construction phase that the curvature in the axis of the Tower began to appear, see Fig. 4, reflecting the attempt of the masons, charged with the construction works, to compensate against the ongòing manifestation of tilting.

This compensation was attempted by the progressive change in thickness of properly hand cut stone blocks of each"ricorso" (tiers of stones of which the Monument facing is made) while moving from North, southwards.

By measuring the thickness of blocks within each "ricorso", the evolution of inclination during the construction period can be inferred.

The position is which the bell chamber was added, by Tommaso di Andrea Pisano, testifies a further attempt to correct the geometry of the structure and to compensate for the occurring inclination.

$$V \cong 142 \ MN, \quad M \cong 327 \ MNm, \quad e \cong 2.3 \ m$$

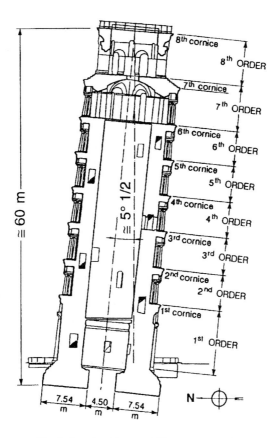

Fig. 2. Leaning Tower of Pisa—cross-section

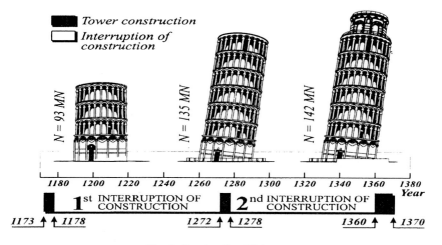

Fig. 3. Construction History

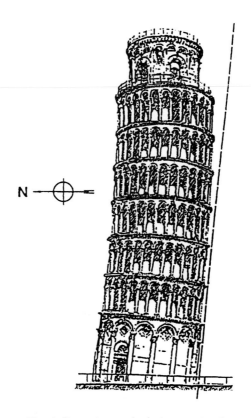

Fig. 4. Correction made during construction

5. Our timeline of the Tower's history is based on the variation of the thickness of "ricorsi" and on other historical evidence such as:
    -   the fresco by Antonio Veneziano of 1384 showing the funeral of Saint Ranieri;
    -   the work life of Arnolfo by Vasari;
    -   the measurements of the tilt performed in 1818 with the plump line by two English Architects E. Cresy and G. L. Taylor;
    -   Measurement similar to those mentioned above carried out by the French Rouhault De Fleury in 1859. There is no record of an inclination measurement but only mention of an appreciably larger inclination than that recorded by the two English Architects.

    The increased rate of inclination after the Cresy and Taylor measurements is usually attributed to the works by architect Della Gherardesca who, in 1838, excavated an annular ditch around the Tower called "catino" as shown in Fig. 5. The aim of the catino was to uncover the basis of the columns, originally from the upper portion of the foundation plinth, which sank into the ground as a consequence of settlement. Given that the bottom of the catino is below the ground water table, it has been

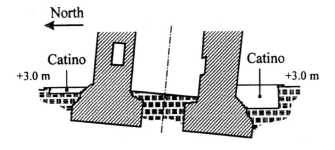

Fig. 5. Catino cross-section

necessary, since 1838 to continuously dewater it triggering an increase in the Tower tilt rate.

The reconstruction of the history of the Monument tilt shown in Fig. 6 has also been possible because of the geodetic measurements of the inclination, started in 1911. It must be pointed out however, that all information dated prior to the start of systematic modern monitoring concerning the inclination, should be considered as approximate, highly qualitative and, to some extent, subjective.

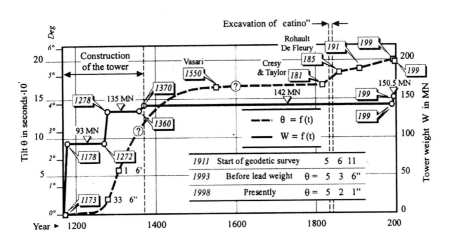

Fig. 6. Evolution of rigid tilt with time

3. SUBSOIL CONDITIONS

6. Many geotechnical investigations have been performed at different times around the Tower. The most relevant and comprehensive among them are described in detail in:

a. Three volumes published by the Ministry of Public Works Commission, MPW (1971) whose data are summarized and updated in the work by Croce et al. (1981).

b. Works by Jamiolkowski (1988), Berardi et al. (1991), Lancellotta and Pepe (1990, 1990a), which report the results of soil investigation carried out in the mid-eighties by the Design Group appointed by the Ministry of Public Works chaired by Finzi and Sanpaolesi.

c. The investigation carried out in the years 1991 through 1993 by the International Committee presently charged for the project on safeguarding the Monument. The results of this investigation have been only partially published and the relevant results can be found in works by: Calabresi et al. (1993), Lancellotta et al. (1994), Costanzo (1994) and Costanzo et al. (1994).

7. Based on all the above mentioned geotechnical investigations, it is possible to determine the following soil profile[1], see also Fig. 7, starting from the ground surface at an elevation of approximately +3.0 a.m.s.l.,

-    Horizon A: ≅10 m thick, consists of interbedded silt, clay and sand layers as well as lenses covered by ≅3 m thick layer of man-made ground. This Horizon can be subdivided in the following layers:

■ Layer A_0: from elev. +3.0 to ±0.0, man-made ground containing numerous archeological remainings dated from the 3rd Century B.C. to

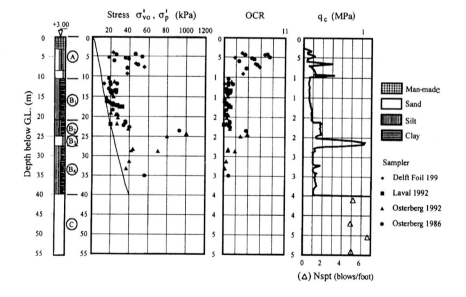

Fig. 7. Subsoil conditions

---

[1] according to the designations of main Horizons adopted in MPW (1971).

the 6th Century A.C.

- Layer A_1: from elev. ±0.0 to −3.0, yellow silty sand and sandy silt.
- Layer A_2: from elev. −3.0 to −5.0, yellow clayey silt.
- Layer A_3: from elev. −5.0 to −7.0, uniform medium grey sand.

Recent borings and piezocone (CPTU) tests performed in the vicinity of the Tower suggest that moving from the South perimeter of the Tower catino northwards, the layer A_2 becomes increasingly sandy. Overall, a comparison of the cone resistance (q_c) yielded by CPTUs reveals that resistance of Horizon A is markedly lower at South when compared to the North side, see Fig. 8a. The CPTUs also showed that the q_c profiles on the East site yielded an average lower cone resistance than on the west site, see Fig. 8b. The above mentioned trends are confirmed by the exam of penetration pore pressure [Pepe (1995)], resulting from CPTUs.

Furthermore, it is worthwhile reporting the results of five seismic CPTUs

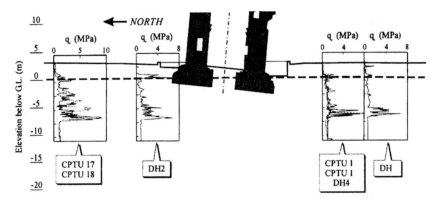

Fig. 8a. Cone resistance in Horizon A, North-South cross-section.

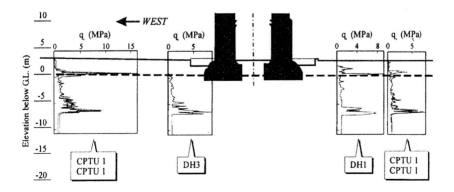

Fig. 8b. Cone resistance in Horizon A, West-East cross-section.

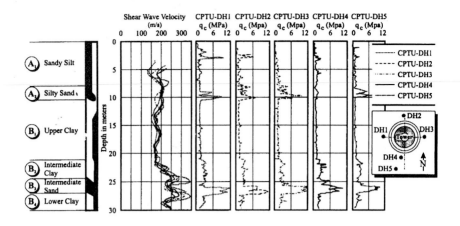

Fig. 9. Results of seismic cone penetration test

performed in the close vicinity of the Monument, see Fig. 9. A part from the profile of shear wave velocity V_s, the figure shows the trend of q_c vs. depth which confirms what emerges from Fig. 8a and 8b.

- Horizon B: $\cong$30 m consists of clay with an interbedded layer of sand. Within this Horizon B the following four layers can be recognized:

  ■ Layer B_1: from elev. –7.0 to –18.0, upper clay, locally named Pancone clay.
  ■ Layer B_2: from elev. –18.0 to –22.5, intermediate clay.
  ■ Layer B_3: from elev. –22.5 to –24.5, intermediate sand.
  ■ Layer B_4: from elev. –24.5 to –37.0, lower clay.

  The highly comprehensive literature review of soil investigation data, produced by Calabresi et al. (1993) has allowed a further subdivision of each layer of Horizon B into a number of sublayers. However, it is beyond the scope of the present paper to elaborate on these findings.

- Horizon C has been recently investigated to the depth of 120 m (elev. –117 b.m.s.l.). Three distinct layers have been found.

  ■ Layer C_1: from elev. –37.0 to –65.0; medium to coarse grey sand in some spots rich of fossils and shells, containing randomly distributed and quite rare lenses of peat.
  ■ Layer C_2: from elev. –65.0 to –75.0; greenish clayey and silty sand.
  ■ Layer C_3: from elev. –75.0 down to the maximum explored depth, grey changing to a shade of green in lower sand.

8. Figure 10 is representative of the groundwater conditions. Three different piezometric levels exist in the Horizon A layer B_3 and Horizon C. The latter has presently a mean phreatic water level within Horizon A has an average seasonal variation of elev. between +1.5 and +2.0 a.m.s.l.

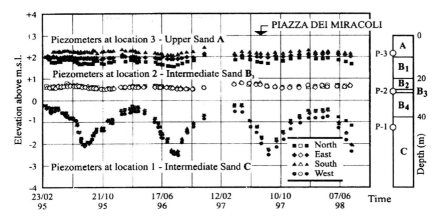

Fig. 10. The typical groundwater conditions

The piezometric level in the intermediate sand layer B_3 is approximately located at the elevation of +0.70 a.m.s.l. and is subject to a minor seasonal fluctuation, $\cong 0.10$ to 0.20 m, which at reduced scale and with some time lag, mimics the one observed in Horizon C.

The above outline of the groundwater scheme indicates that the pumping from Horizon C, began around the fifties, triggered the consolidation of the clay layers belonging to Horizon B, causing the subsidence of the whole Pisa plane.

This phenomenon, now greatly attenuated, had become quite severe in early seventies when the mean piezometric level in the Horizon C decreased to elev. −6.0 b.m.s.l. causing an acceleration of the Tower tilt due to the differential subsidence over the Piazza dei Miracoli. For greater details see Croce et al. (1981). This resulted in the closure of a number of wells in the vicinity of the square and led to a substantial attenuation of the phenomenon in the early eighties. Further information regarding this aspect of the problem can be found in the work by Schiffmann (1995).

Table 1. Grading of main soil layers

Horizon	Layer	Sand fraction %		Silt fraction %		Clay fraction	
A	A_3	31.7	4.7	61.1	12.3	13.0	4.9
	A_4	74.6	16.5	17.9	14.9	4.2	3.3
B	B_1	< 5		42.4	13.3	58.0	13.0
	B_2	6.0	4.2	51.1	15.7	38.9	13.7
	B_3	77.0	8.1	19.8	14.6	8.4	3.1
	B_4	< 5		52.9	17.0	43.1	17.2
C	C	82.5	14.7	7.0	6.2	5.5	4.2

9. Although detailing the geotechnical characterization of the soil underlying the Tower is beyond the scope of this paper, a concise summary of the index and stress-strain-strength properties will follow. However, to obtain a more extensive insight into this aspect of the problem, it should be consulted the MPW (1971), Lancellotta and Pepe (1990, 1990a), Calabresi et al. (1993), Lancellotta et al. (1994) and Costanzo et al. (1994). The mean values and the standard deviations of the index properties can be inferred from Tables 1 and 2, where:

γ = bulk density; G_s = specific gravity; W_n = natural water content; LL = Liquid Limit; PI = Plasticity Index.

Table 2. Index properties of main soil layers

Horizon	Layer	γ (kN/m³)	G, (-)	W_n(%)	LL(%)	PI(%)
A	A_3	19.42 ± 2.03	2.71 ± 0.03	31.6 ± 4.2	35.2 ± 4.7	13.2±3.6
	A_4	18.35 ± 0.61	2.68 ± 0.03	33.6 ± 3.8	-	-
B	B_1	16.64 ± 1.05	2.78 ± 0.03	52.6 ± 7.9	70.8 ± 13.6	42.1 + 12.5
	B_2	19.91 ± 0.50	2.73 ± 0.03	25.8 ± 3.3	51.6 ± 11.7	28.1 ± 11.2
	B_3	18.95 ± 0.45	2.69 ± 0.01	30.2 ± 3.3	-	-
	B_4	19.00 ± 1.00	2.74 ± 0.04	36.1 ± 9.2	55.9 ± 4.8	32.3 ± 13.2
C	C	20.80 ± 0.06	2.66 ± 0.01	18.7 ± 2.4	-	-

Based on the information concerning the piezometric levels and with reference to the values of γ determined in laboratory, the variation of the effective overburden stress (σ'_{vo}) with depth shown in Fig. 7 has been established.

The value of σ'_{vo} in combination with preconsolidation pressure σ'_p as determined by oedometer tests using the Casagrande (1936) procedure, led to the overconsolidation ratio values (OCR) showed in the same figure. The overconsolidation mechanism involved in the case of Pisa subsoil is generally ascribed to aging, due to secondary compression, groundwater fluctuations as well as possibly to a minor removal of the overburden not exceeding 50 to 60 kPa. In addition, in the case of Horizon A and Layer B_2, temporary emersion and related desiccation could have affected the OCR values, see also Calabresi et al. (1993).

The coefficient of earth pressure at rest (K_0), for Pancone Clay, in a normally consolidated (NC) state, ranges between 0.58 and 0.63.

The best estimate of the K_0 in the field, considering the above outlined overconsolidation mechanisms and taking into consideration works by Mesri and Castro (1987), Mesri (1989), Hayat (1992) and Mesri et al. (1997) should be around 0.73 to 0.75. The writer does not have the information necessary to estimate the field K_0 in other clay layers belonging to Horizon B.

10. The mechanical properties stated in the following information provide the reader with a general picture of the subsoil conditions:

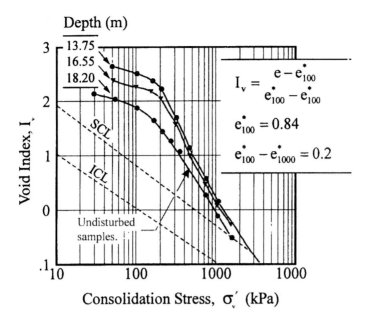

SCL = sedimentation compression line
ICL = intrinsic compression line

Fig. 11. Compression curves of upper pisa clay in term of Void Index.

- The compressibility of the clay layer have been investigated mostly throughout oedometer tests. As an example, Fig. 11 shows the results of incremental loading oedometer tests performed on three high quality undisturbed samples retrieved from Pancone clay. The results are plotted in the plane log σ'_v vs. void index (I_v), the latter defined [Burland (1990)] as follows:

$$I_v = \frac{e - e^*_{100}}{e^*_{100} - e^*_{1000}} = \frac{e - e^*_{100}}{C^*_c}$$

e	=	current void ratio of tested specimen
e^*_{100}	=	void ratio at $\sigma'_v = 100$ kPa determined reconstituted specimen starting from

$$LL \le W_n \le 1.5\, LL$$

e^*_{1000}	=	as above but referring to $\sigma'_v = 1000$ kPa
C^*_c	=	compression index of reconstituted clay.

Figure 11 also locates the positions of Sedimentation (SCL) and Intrinsic (ICL) Compression Lines. These represent compressional characteristics of natural NC sedimentary and reconstituted clay respectively. The compression curves of undisturbed sample at $\sigma_v' > \sigma_p'$ are significantly steeper than SCL and ICL, and only at σ'_v one order of magnitude higher than σ'_p they merge into SCL. This fact highlights the importance of the structure of the Pancone clay at its natural state.

Analogous data as obtained for Layers A_3, B_2 and B_4 may be found in the work by Lancellotta et al. (1994). The results collected by these authors led to the following:

C_{c1}/C_{c2} values of ratio of the tested clays: 1.4 for B_1, 1.0 for B_4, B_2 and A_3 Table 3 summarizes the characteristics of different clay layers tested.

The representative drained peak shear strength characteristics ϕ' and c' of different soil layers encountered under Piazza dei Miracoli are reported in Table 4. Those of clayey layers have been inferred from drained triaxian compression (TX-CD) tests performed on high quality undisturbed samples while those of sands have been estimated on the basis of q_c and standard penetration resistance N_{SPT}.

Table 3. Compressibility indexes from oedometer tests

Horizon	Layer	C_{c1}	C_{c2}	$\dfrac{C_{c1}}{C_{c2}}$	C_s	OCR range	$\dfrac{C_{ae}}{C_{c1}}$
A	A_3	0.243	0.243	1	0.023	2.4 ÷ 4.1	0.011
	B_1	0.909	0.640	1.42	0.072	1.3 ÷ 2.0	0.035
B	B_2	0.266	0.266	1	0.030	2	0.030
	B_4	0.280	0.280	1	0.057	1.3	0.023

C_{c1} = Primary compression index immediately beyond σ_p',
C_{c2} = Primary compression index at $\sigma'_v \gg \sigma_p'$
C_s = Swelling index
C_{ae} = Secondary compression index immediately beyond σ_p',

Table 4. Drained shear strength from TX-CID compression tests.
(TX = triaxial test; CID = consolidated drained tests

Horizon	Layer		c kP
A	A_3	31	0 to 2
	A_4	33	0
	B_1	22	6 to 2
B	B_2	28	12 to
	B_3	34	0
	B_4	27	0 to 5

The angle of friction at critical state ϕ'_{cs} has been determined only for clay of layer B_1 performing TX-CD tests on reconstituted material. These yielded values ranging between 24° and 25°.

- The undrained shear strength (s_u) of clay layers has been determined from K_0 - consolidated undrained triaxial compression tests (TX-CK$_0$U). The tests for specimens reconsolidated under stresses representing the best estimate of those existing in situ, these tests yielded on average the values of normalized s_u as reported in Table 5.

Table 5. Normalized undrained shear strength of upper Pisa clay

s_u/σ'_{vo}	TEST
0.23 (OCR)$^{0.84}$	DSS-CK$_0$U
0.29 (OCR)$^{0.84}$	TX-CK$_0$U

OCR	= overconsolidation ratio
DSS	= direct simple shear
TX	= triaxial test
CK$_0$U	= consolidated in K$_0$ -condition undrained.

- The initial soil stiffness G_0 at strain less than linear threshold strain has been inferred from V_s measurements performed during seismic-CPTU and from laboratory tests on high quality undisturbed samples reconsolidated to the best estimate of existing in situ stresses. Two kinds of laboratory apparatuses were employed; fixed-free resonant column apparatus and a special oedometer instrumented with pressure transducers measuring horizontal stress and bender elements allowing to generate and receive seismic body waves. The results of in situ and laboratory tests compared in terms of G_0 are reported in Fig. 12. Additional information concerning these tests may be found in the work by Jamiolkowski et al. (1994).

4. MOVEMENTS OF THE TOWER

11. The systematic monitoring of the Tower started in 1911 adopting the so-called geodetic method which measures the degree of tilt. It consists in measuring, from a fixed station in Piazza dei Miracoli, the horizontal distance between the South edges of the 7th and the 1st cornices, see Fig. 13. Such measurements were usually performed twice a year, and incorporate the rigid tilt of the foundation as well as the variation of the geometry of the Tower axis, influenced by the environmental conditions, i.e. temperature changes and wind effects.

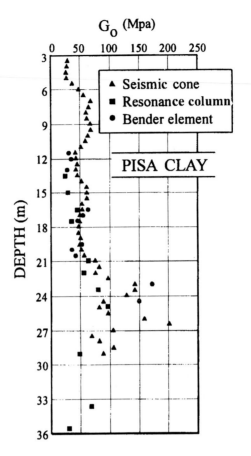

Fig. 12. Maximum shear modulus from in-situ and laboratory tests.

12. In 1934 two additional monitoring devices were installed:
-   Genio Civile (GC) Bubble Level installed in the instrumentation room located at the level of 1st cornice, see Fig. 13.

    It allows to measure, over a span of 4.5 m, the tilt on two orthogonal planes N-S and E-W.

    The measurements till 1992 were taken once a week and they are only moderately affected by wind action and temperature changes.
-   Girometti-Bonecchi Pendulum Inclinometer, 30 m long. It was fixed to the internal wall of the Tower at the elevation of the 6th cornice (Fig. 13). It swings 1.5 m above the instrumentation room floor.

    The continuous measurements reveal simultaneously the displacements of the Tower on the same two orthogonal planes as those relevant to the GC-level. The sensitivity of the instrument is $\cong 0.01$ seconds but the readings are strongly affected by the wind effect and temperature changes.

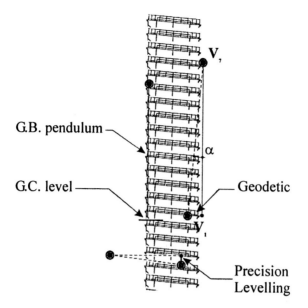

✎ Geodetic measurements, horizontal movements of points V1 and V7, started in 1911.
✎ Precision leveling of 15 points located on foundation (1928, 1929, 1965 through 1986, 1990)
✎ G.C. level, instrumentation room at level of 1st cornice, started in 1934.
✎ G.B. Pendulum Inclinometer 30 m long, fixed to internal wall at 6th cornice, started in 1934.

Fig. 13. Measurements of tower tilt in years 1911 through 1992

13. In 1965 the high precision leveling of fifteen bench marks (Fig.13) located on the foundation plinth was initiated. Due to the lack of deep datum point all settlements measurements are relative because they are based on a bench mark located on the cast iron door of the Baptistery. Given the position of the bench marks in consideration, as well as the insignificant effect of temperature changes, these measurements are more reliable than others and suitable to reflect the evolution of the rigid tilt of the Tower foundation.

14. An overall picture of the Tower tilt on the North-South plane since 1911 is shown in Fig. 14. It is based on geodetic and GC-Level measurements which lead to comparable and reliable results if examined on a long-term basis.

A long-term trend of a steady increase in the Tower inclination emerges from this figure. This trend shows three major perturbations: the first occurred suddenly in 1935, a second one began in the mid-sixties and continued gradually over the following ten years, and the third one occurred in 1985.

The first relevant perturbation occurred in mid-thirties (≅30″) during the works aimed at redoing the catino and the cement grouting into the base of the Tower. During these works, before sealing the water proof joint between

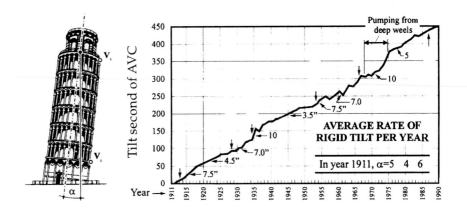

Fig. 14. Rigid tilt of leaning Tower of Pisa

the plinth and the catino, a quite intensive dewatering was put into operation.

The second perturbation was first observed during the site investigation carried out by the Polvani Commission, see Croce et al. (1981), and originated serious concern. It became evident that the increase in the rate of rigid tilt was connected to the exceptionally pronounced drawdown of the piezometric level in the sand aquifer, formation C, which occurred between 1970 and 1974. The lowering of the watertable caused an increase of the tilt of approximately 40 seconds of arc in the North-South direction and of about 20 seconds of arc in the East-West direction. Following these observations, a number of wells in the vicinity of the Tower were closed allowing a partial recovery of the piezometric level reached in 1975 and 1976. Soon afterwards a significant decrease in the rate of tilt was recorded.

The third perturbation occurred after the boring performed in the Northern edge of the foundation in 1985. The increase of tilt was about 7 seconds of arc in the North-South direction.

In order to graph a rate of the Tower inclination, which does not include the consequences of the mentioned events and of the environmental changes, Burland (1990a) attempted to subtract from the GC-level measurements and from the high precision topographical leveling data, the effects of perturbations. The obtained results, reported in Fig.15, show a slow but steady increase in the rate of tilt which implies the future overturning instability of the Tower.

It has only recently [Croce et al. (1981)] been determined that the subsidence of the whole Pisa plain may affect the movements of the Tower as a result of the local phenomena occurring in the Piazza dei Miracoli. Despite the lack of the deep datum point, one can infer that the differential subsidence occurring in the Square might contribute to the present rate of tilting of the Tower.

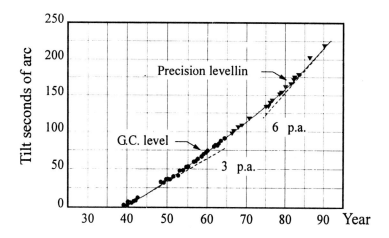

Fig. 15. Net tilt of tower plinth in years 1938 through 1990

15. In early nineties, prior to the stabilization works on the Tower and the consolidation of its masonry, a new monitoring system, having a high degree of redundancy, was implemented to continuously control in real time the movements of the Tower. Details may be found in works by Burland and Viggiani (1994) and Burland (1995).
This system consists in:
   - Eight internal bench marks, 101 trough 109, see Fig. 16, installed at the ground floor level in the entrance to the Tower.
   - These survey points are linked to the previously mentioned fifteen external bench marks, 901 through 915 in Fig. 16, located externally on the Tower plinth.
   - Twenty-four bench marks, 1 through 24, see Fig. 16, used to monitor the movements of Piazza dei Miracoli by means of precision leveling.
   - Deep datum point, DD1 in Fig.16, the most important point of reference for all levelings, reveals the absolute movements of the Tower and the ground surrounding it.
   - Biaxial electrolytic inclinometers, IBIA in Fig.17, are located on the ground floor in the entrance to the Tower. The inclinometers and the automatic hydraulic livelometers, shown in the same figure, allow for the continuous measurement of change in monument tilt over a short term.

The description of additional instrumentation, also installed to monitor the movements of the Tower above the plinth and its masonry, is beyond the scope of this paper.

For convenience of the reader and in relation to the monitoring exposed

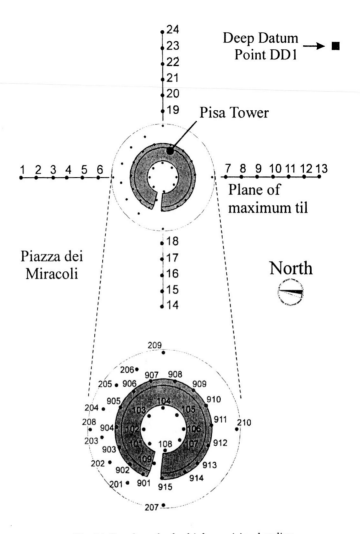

Fig. 16. Benchmarks for high precision leveling

in the section dealing with the stabilizing measures, Fig. 18 shows the reciprocal relationships between the inclination of the monuments (α) and its
overhanging (h) as well as that between the plinth tilt (θ) and the relative
settlement of its South edge (δ).

5. LEANING INSTABILITY

16. The Tower began to lean Southwards during the second construction phase
when the masonry weight exceeded 65% of the monument (Fig. 6). This
phenomenon has continued at a rate of 5 to 6 seconds per annum, a constant

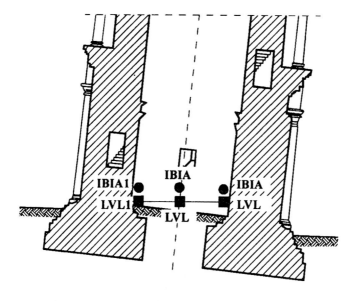

■ LIVELOMETER (LVL)   ● BIAXIAL INCLINOMETER (IBIA)

Fig. 17. Measurements of rigid tilt at tower base

rate for the past few decades without taking into considerations environmental perturbations.

The constant rate of inclination and the relevant increase of the Tower tilt has raised much concern and controversy. Most importantly, it has always been debated the triggering factor for the phenomenon causing the continuing rotation at constant load since the end of the XIV Century as well as the present margin of safety in light of the risk of the Tower falling over.

17. The general consensus over the last decade [Hambly (1985, 1990), Lancellotta (1993, 1993a), Desideri and Viggiani (1994), Veneziano et al. (1995), Pepe (1995), Desideri et al. (I 997)] has been that the behavior of the Tower, since the end of construction can be attributed to phenomenon of the instability of equilibrium. A phenomenon similar to one relevant in the structural mechanics to slender structures, threatens the stability of tall, heavy top, structures seated on compressible soil. This kind of behavior, also called leaning instability, is entirely controlled by the soil-structure interaction phenomena. In the case of the Pisa Tower it was triggered by the initial geometrical imperfection occurred during the second construction stage when the Tower started to lean Southwards. This can be explained in view of the fact that the resisting moment owing to pronounced compressibility and non-linearity of soil support was unable to counteract the overturning

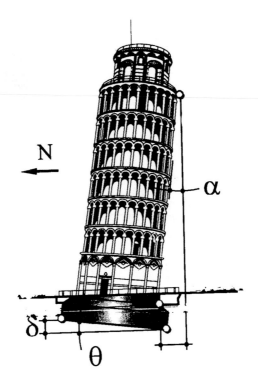

Inclination and overhanging in May 1993 before application of counterweight:

$$q = 5 \ 33 \ 36''$$

$$h = 4.47 \ m$$

θ = inclination
δ = relative settlement of South edge with respect to North edge; $\delta(\theta-1) \cong 0.095$ mm
h = overhanging referred to 7th cornice h $(\alpha=1) \cong 0.22$ mm
θ = $\alpha + 11 \ 5''$
N.B. $1 = 60 = 3600$; $\theta_{1998} = 19971''$

Fig. 18. Inclination of Pisa Tower Terms of Reference

moment generated by the ongoing tilt. The self-driving mechanism was put into operation causing a steady increase of the Tower tilt till nowadays due to a progressive growth of the driving moment generated by the second order effects.

The reasons which have triggered the above depicted phenomenon of the leaning instability [Abghari (1987), Cheney et al. (1991)], are not completely understood. A number of hypotheses have been postulated by very authoritative authors:

-   Differential compressibility and consolidation rate of the soft high plasticity clay layers belonging to Horizon B [Terzaghi (1960)].

- Spatial soil variability combined with differences in compressibility characteristics within Horizon A together with local failure and consequent confined plastic flow developed in upper part of Pancone Clay [Mitchell et al. (1977)].
- Leonards (1979) opted in favor of plastic yield of the soft Pancone clay leading to local shear failure.
- Non-homogeneity of the compressibility and permeability of soils in Horizon C has been postulated by Croce et al. (1981).

In addition, the incipient elastic instability has been suggested by Hambly (1985) as the possible cause for the initial rotation.

In essence, the mechanisms that have caused the initial geometrical imperfection triggering the leaning instability, continue to be uncertain. The writer believes that a combination of more than one of the events envisaged above have contributed to the rise of the initial inclination.

18. The leaning instability problem has been studied by many authors making reference to one and two degrees of freedom mechanical models shown in Fig. 19. For more details see works by: Como (1965), Hambly (1985, 1990), Cheney et al. (1991), Lancellotta (1993, 1993a), Desideri and Viggiani (1994), Veneziano et al (1995), Desideri et al. (1997), Lancellotta and Pepe (1998) and others. Pepe (1995) examined these models from a theoretical point of view and presented the results of physical modeling of the Pisa Tower in the centrifuge which corroborate at phenomenological level the idea that the monument is threatened by the instability of equilibrium.

Even if a detailed discussion of the above studies is beyond the scope of this work, it may beneficial to the readers highlighting the following points:

- As pointed out by Lancellotta (1993, 1993a) and Veneziano et al. (1995) the one degree of freedom scheme, Fig. 19 when coupled with a realistic model of soil restraint, offers a simple but rational approach for evaluating the present margin of safety and its evolution with time.

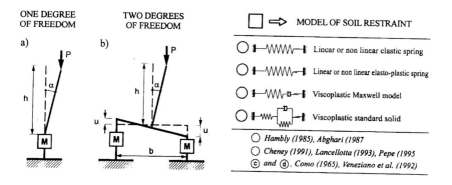

Fig. 19. Leaning instability models.

- The two degrees of freedom model [Pepe (1995), Lancellotta and Pepe (1998)] in addition to what stated above, makes it possible to investigate the effect of some of the stabilization measures that have been considered for a possible implementation on the Tower.
- In order to reproduce, in a realistic manner, the leaning instability phenomenon, the model of soil restraint referred to drained conditions should incorporate at least the following features: non-linearity of moment-rotation relationship, hypothesis about asymptotic value of resisting moment, influence of initial geometrical imperfection and of soil viscosity, variation of overturning moment with time due to secondary order effects.
- All attempts to evaluate the present factor of safety of the Tower against overturning, based on realistic soil models in which the viscous effects have been implicitly [Lancellotta (1993), Pepe (1995)], or explicitly [Veneziano et al. (1995)] considered, led invariably to very low values ranging between 1.1 and 1.2. Veneziano et al. (1995) using two different reological models, positively calibrated against historical rotation measurements, reached the conclusion that "it appears that instability of the foundation is at least several decades away. However, a non-negligible risk that Tower collapse will occur in 40 to 50 years with the risk to be around $2 \cdot 10^{-2}$ and $3 \cdot 10^{-3}$ respectively".

6. STRUCTURAL FEATURES

19. As shown in Fig. 2 the Leaning Tower of Pisa consists of a hollow masonry cylinder, surrounded by six loggias with the bell chamber on the top.

    The Tower is a typical example of the so-called "infill masonry" structure composed of internal and external facings made of San Giuliano marble and of a rubble infill cemented with the San Giuliano mortar, see Fig. 20. A helicoidal staircase allowing the visitors to climb up to the top of the Tower is located inside the annulus of the hollow cylinder.

    The following are the essential characteristics of the Tower:
    - total weight: $N \cong 142$ MN; average foundation pressure: $q \cong 497$ kPa;
    - total height $H = 58.36$ m; height above G.L.: $\cong 55$ m;
    - distance from the center of gravity to the foundation plane $h_g \cong 22.6$ m;
    - annular foundation, inner diameter; $D_i \cong 4.5$ m, outer diameter $D_o \cong 19.6$ m;
    - area of the annular foundation: $A \cong 285$ m²;
    - present inclination: $\alpha = 5^0 \, 32' \, 51''$;
    - present eccentricity of N; $e \cong 2.3$ m.

20. Relevant mechanical properties of the two components of the Tower masonry are summarized in Table 6. Even a preliminary analysis of the Tower structure led to the conclusion that the most dangerous cross-section corresponds to the contact between the first loggia and the base segment where, in addition to the effect of tilt, and the weakening effect of the void represented by the staircases, the diameter of the hollow cylinder

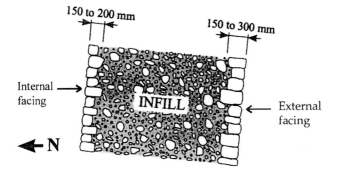

Hollow cylinder, inner and outer surfaces face with highly conpetent San Giuliano marble. Space between these facings is filled with rubble and mortar.

Fig. 20. Cross-section of Tower Masonary

suddenly decreases. At this location on the South side, a compressive stress close to 8.0 MPa has been measured by flat jacks in the external marble facing. An overall picture of the state of stress in the Tower section under discussion attempted by Leonhardt (1991, 1997) is shown in Fig. 21.

Table 6. Mechanical properties of Pisa Tower Masonry

	σ_c (MPa)	σ_t (MPa)	E (MPa)
San Giuliano Marble Facing	110-190	4-8	70.000-90.000
Infill Masonry	4-8	0.3-1.3	5.000-7.500

Thickness of facings: outside ≈200 mm; Inside ≈150 mm
σ_c = Compression strength
σ_t = Tensile strength
E = Elasticity modulus

In these circumstances considering;
- the high compressive stresses in the external facing on the South side;
- the almost no bond strength between rubble infill and facings;
- the presence of voids and inhomogeneities in the rubble infill ascertained by non-destructive geophysical tests, i.e., sonic, infrared and radar tomographies;
- the heavy loaded external facing laying directly on the infill masonry because of the change of the cross-section of the hollow cylinder at the

level of first cornice;
- the deviation of the compressive stress trajectories from the vertical direction in the Tower shaft due to the presence of the staircase and imperfections of the bed joints leading to the appearance of the horizontal force components as evidendiated in Figs. 21 and 22.

The serious concern over the structural safety of the Monument led in 1989 to the decision by the Commission established by the MPW and chaired by Jappelli and Pozzati, to close the Tower to the visitors.

21. The envisaged risk is of a failure due to the local buckling in compression of the external facing of the masonry in the most severely stressed section at the South side of the Tower at the level of the first cornice.

This kind of mechanisms has been responsible for the sudden catastrophic collapses of the Bell Tower in San Marco square in Venice in 1902, and, more recently in 1989, of the Bell Tower of the Cathedral of Pavia, both Towers were made of infill masonry with bricks facings.

Due to the fragility of such structures the local buckling in compression of the facings led to their almost instantaneous collapse with no warnings.

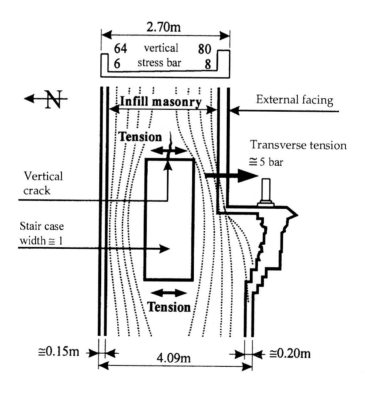

Fig. 21. Cross-section of Tower at first cornice

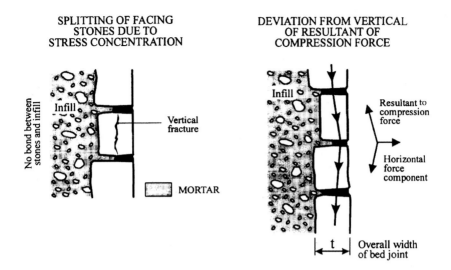

SPLITTING OF FACING
STONES DUE TO
STRESS CONCENTRATION

DEVIATION FROM VERTICAL
OF RESULTANT OF
COMPRESSION FORCE

Fig. 22. Marble stone facing imperfection of bed joints

7. STABILIZATION WORKS

22. In the previous part of the paper it has been evidentiated that the leaning Tower of Pisa is endangered by two phenomena, i.e. instability of equilibrium and risk of fragile structural collapse of the masonry.

    The two phenomena are obviously interdependent. The increasing inclination not only reduces the safety margin of the Monument with respect to the overturning but also causes a further increase of the stresses in the most critical section of the masonry, enhancing the risk of structural collapse.

    In 1989, the MPW Commission chaired by Jappelli and Pozzati pointed out the risk of structural collapse which proved to be realistic when the XIII Century Civic Tower of Pisa [Macchi (I 993)] collapsed without any warning. This event led to the closure of the Pisa Tower to the visitors in January of 1990, and triggered the appointment, by the Italian Prime Minister, of an International Committee for the safeguard and the stabilization of the leaning Tower of Pisa.

    The Committee, the seventeenth in the long history of the Monument [Luchesi (1995)] and the sixteenth in the modern times, has been charged to stabilize the foundation, strengthen the structure and plan the architectural restoration and started its operations in September 1990.

23. The activities of the Committee can be grouped as follows:
    - Numerous experimental investigations and studies dealing with a broad spectrum of problems[2], reflecting the multidisciplinary nature of

---

[2] archeology, history of construction, strength of materials, numerical modeling of structure and foundation soils, in situ and laboratory tests, new monitoring system, methods of structural reinforcement, approach to architectural restoration, etc.

the Committee and aimed at the most comprehensive learning of all the relevant features of the Monument and its environment.

- The design and implementation, in a short time, of the temporary and fully reversible interventions to increase slightly the stability of the Tower foundation and to reduce the risk of structural collapse. This decision was taken in view of the awareness that the selection, the design and the realization of the permanent stabilization and consolidation works would require a long time.

- The studies by means of numerical and physical models as well using field trials, guiding in the selection and design of the final interventions. This task, especially the stabilization of the Tower with regard to the leaning instability, poses serious limitations on the selection of the appropriate solution due to the following circumstances:

  ■ The unanimous decision of the Committee to adopt a solution fully respecting the artistic and cultural value of the Monument.
  It was given preference to the intervention able to stop and reduce the tilt of the Tower plinth acting only on the subsoil without touching the Monument.

  ■ Given the extremely reduced safety margin of the Tower with respect to falling over, any invasive interventions like underpinning, enlargement of the plinth, etc. would represent a serious risk of collapse in the transitory phase during the execution of works.

In these circumstances two possible solutions for stabilizing the foundation have been envisaged, both aimed at inducing differential settlement of the North edge of the plinth with respect to the South.

A brief description of the temporary stabilizing measures as well the studies and the design of the final intervention aimed at stopping-reducing the inclination of the Tower will be given in the next sections.

The temporary, and completely reversible, intervention aimed and improving the structural safety of the most critical cross-section of the masonry at the level of the first loggia has been completed in 1992. It consist of 18 lightly post-tensioned tendons located in the places shown in Fig. 23, their function is to prevent local buckling in compression of the marble stones forming the external facing.

The steady motion of the Tower, increasing its inclination by 5" to 6" per annum, led to the decision to implement a second temporary and fully reversible intervention aimed at reducing the rate or even stopping the progressive increase of inclination. This intervention consisted in placing 6 MN of lead ingots on the North edge of the plinth as shown in Fig. 24. The lead ingots have been placed gradually (Fig. 25) on the prestressed concrete ring shown in Fig. 24 generating a stabilizing moment of 45 MNm. The counterweight placed in the period between May 1993 and January 1994 has determined a very positive response of the Monument, which, for the first time in its history, inverted the direction of the movement reducing slightly the inclination.

The effects of the Tower tilt monitoring during the application of the lead

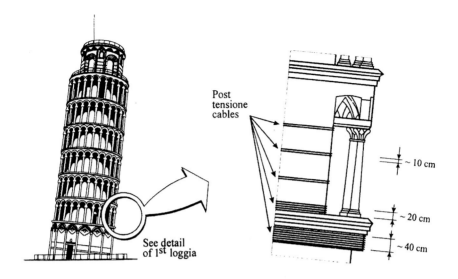

Fig. 23. Temporary structural strengthening

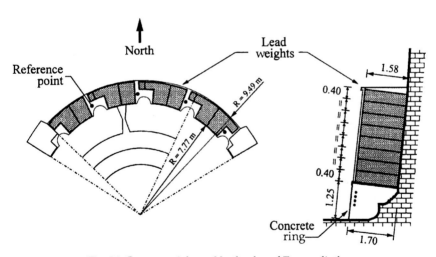

Fig. 24. Counterweight on North edge of Tower plinth

ingots is reported in Fig. 26. It results that during the loading stage the monument reduced its inclination by 34" which grew up to 54" during the following six months.

25. In view of the positive response of the Tower to the counterweight, but considering its visual impact, it was decided to replace the lead ingots by ten deep anchors having each a working load of 1000 kN, see Fig. 27. This intervention was conceived as an intermediate measure between the temporary and the final one and presented the following advantages:

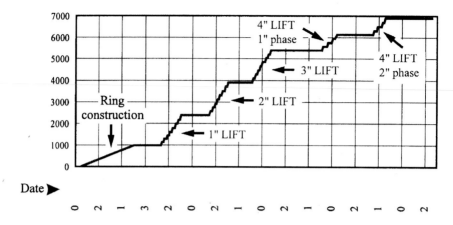

Fig. 25. Counterweight loading sequence

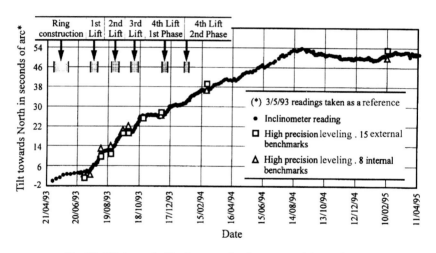

Fig. 26. Tilt towards North as result of counterweight application

- Double the stabilizing moment with an increase of vertical load of only two third of that due to lead ingots.
- Create at the North edge of the plinth, one-directional rotational constraint able to counteract to some extent any tendency of the Tower to tilt southwards.

The implementation of this solution required the construction of a second prestressed concrete ring below that supporting the lead ingots therefore hidden beneath the catino. This, in turn, required an excavation below the perched G.W.L. ranging from 0.3 m at North to 2.0 m, South of the catino.

The design of the ten anchors solution has been developed based on the information gathered by the previous commissions, considering the catino

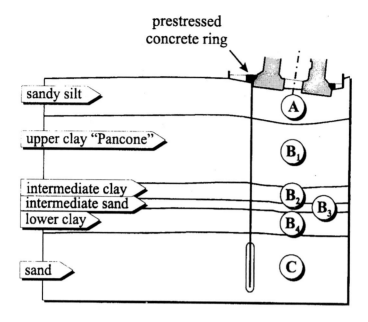

prestressed
concrete ring

sandy silt

upper clay "Pancone"

intermediate clay
intermediate sand
lower clay

sand

A

B₁

B₂ B₃

B₄

C

Fig. 27. Ten anchors solution

statically independent from the Tower plinth. The only known connection was the water-proofing joint located in proximity to the foundation perimeter.

Unfortunately, during the implementation of this solution it was discovered that in the past there had been two attempts to enlarge the Tower foundation:

- The first, probably due to Della Gherardesca, who during the construction of the catino, placed around the Tower plinth, at 0.7 to 0.8 m, a thick layer of mortar conglomerate having the same width of the catino.

- The second one was implemented by the local authority for public works which in mid-thirties had redone the catino. During this intervention involving the cement grouting of the Tower plinth, the under-catino conglomerate was connected to the foundation by means of steel tubes 70 mm and approximately 700 to 800 mm long. Information about this work was never reported in the official documents and was unknown to the professionals dealing with Tower till the summer of 1995.

In view of the above, the hypothesis that the catino is statically independent from the Tower is become no more truthful, see Fig. 28. Moreover, considering that since mid thirties, the South edge of the Tower plinth has settled 20 to 25 mm more than the North one, it is likely that some limited load has been shared since then from the monument to the South part of catino.

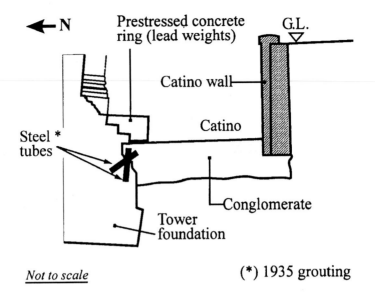

Fig. 28. South section of Catino—actual configuration

In fact, during the first attempt to remove in small segments the South part of the catino to build the prestressed concrete ring for the ten anchors, the Tower started to tilt towards South with a rate of 3" to 4" per day with serious concern for its stability. The phenomenon which occurred in September 1995 was counteracted by applying additional 2700 kN of the lead ingots on the North edge of the plinth. Ever since, the Tower has been motionless as far as its inclination is concerned, see Fig. 29. Subsequently, the design of the ten anchors solution has been modified so that to avoid any modification of the South part of catino. Whether this intervention will be completed or not, has not yet been decided by the Committee. The decision with this respect will depend on the results of the under excavation intervention described in the following.

26. Since 1993, the Committee has undertaken the studies aimed at finding a solution to reduce the inclination of half of degree, acting only on the foundation soils without touching the Tower.

Two possible interventions, able to induce ≅200 mm of settlement of North edge of the plinth with respect to South one, have been taken into consideration. The electro-osmosis aimed at reducing the water content hence inducing a volume change in the most upper part of Pancone clay and the gradual extraction of the soil from the lower part of Horizon A, as postulated many years ago by the Italian civil engineer Terracina (1962), see Fig. 30. The method which has recently been successfully employed to mitigate the impact of very large differential settlements suffered by the Metropolitan Cathedral of Mexico City [Tamez et al. (1992, 1997)].

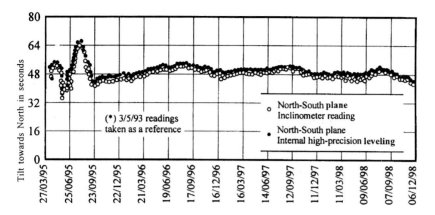

Fig. 29. Tilt of Pisa Tower since May of 1995.

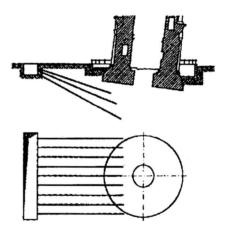

- Reduction of contact pressure on South side.
- Reduction of present inclination (~10%) by 1% would suffice
- Simplest manner, removal of soil under North side by series of borings.
- Regulating number position and diameter of borings, desired reduction of tower inclination can be achieved.

Fig. 30. Underexcavation for correcting inclination of Pisa Tower (Terracina, 1962)

The large scale field trial test performed on the Piazza dei Miracoli evidentiated the non-feasibility of the electro-osmosis, thus all efforts concentrated on investigating the possibility to apply the ground extraction, thereafter named underexcavation. In order to ascertain its feasibility, numerical analyses, physical modeling both in terrestrial gravity field and in centrifuge, as well large scale trial field have been performed. The latter was not only useful as far as the verification of the feasibility of the underexcavation was concerned, but allowed also to test and finalize the technological aspects of the intervention.

27. In order to perform the trial field, a 7-m in diameter circular reinforced concrete footing was built on the Piazza far from the Tower, see Fig. 31, and was loaded eccentrically with the concrete blocks. Both the footing and the underlying soil were heavily instrumented to monitor settlements, rotations, contact pressure and the induced excess pore pressure during the experiment. After a waiting period of a few months, allowing the completion of consolidation settlements, the ground extraction commenced by means of inclined borings having ≅150 mm in diameter as schematically shown in Fig. 31. The under excavation was performed extracting gradually the soil from Horizon A by means of a procedure, shown in Fig. 32, which made it possible to reduce the inclination of the trial plinth by almost 1000" of arc, as documented in Fig. 33.

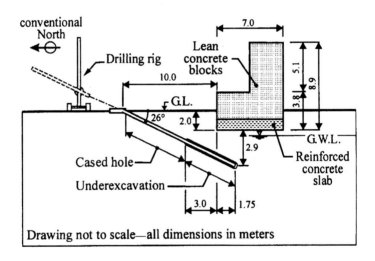

Fig. 31. Underexcavation field trial

During this experiment, the following important lessons were learned:

- A critical penetration exists under the plinth. If the extraction hole exceeds it a rotation of the foundation in the opposite direction is experienced. Such an accident occurred around end of September 1995 and may be detected from Fig. 33.
- Using an appropriate sequence of ground extraction operations it was possible to steer the movements of the plinth both in N-S and W-E plan in the desired way.
- Soon after the completion of the underexcavation, on February 1996, the trial plinth came to rest and up to January 1999 has exhibited negligible movements.

28. Because of the successful validation of the underexcavation by trial field, it was decided to start this intervention under the Tower.

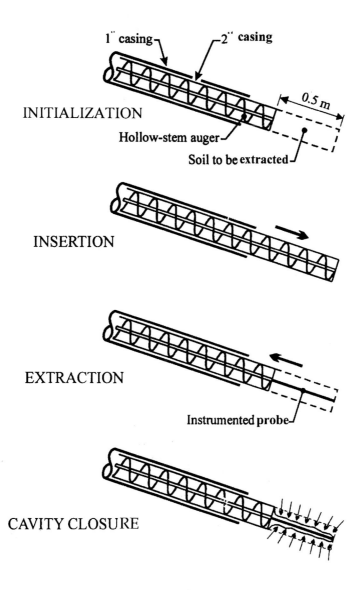

Fig. 32. Soil Extraction Process

A preliminary ground extraction under the monument has been planned, well aware that, by no means, the trial plinth can be considered as a model reflecting completely a possible response of a Tower suffering from the leaning instability. This preliminary intervention will consist in twelve holes whose penetration under the North rim of Tower plinth will not exceed

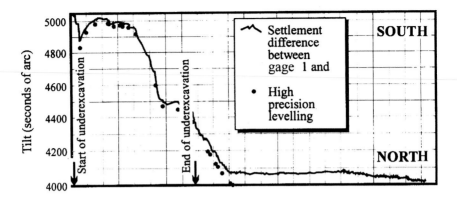

Fig. 33. Underexcavation field trial—Tilt of plinth in North-South plane

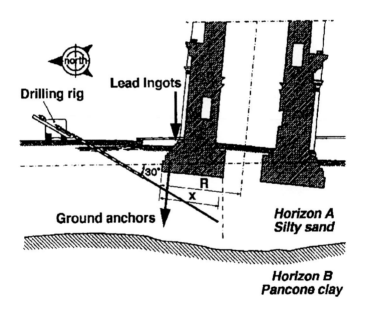

Fig. 34. A hole for ground extraction under the Tower

1 m referring to the scheme shown in Fig. 34. Based on the response of the Monument referring to the scheme shown in Fig. 34, in terms of rotations and settlements to this preliminary intervention, the conclusive decision will be taken on the use of the discussed method as a tool for the final stabilization of the Tower.

To hinder any unexpected adverse movement of the Tower that could occur during this or any other interventions aimed at final stabilization of the Tower, a safeguard structure has been implemented consisting in the cable stay shown in Fig. 35.

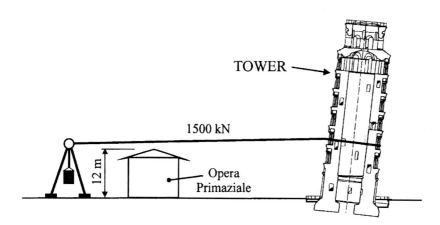

Fig. 35a. Cable stay structure—cross section

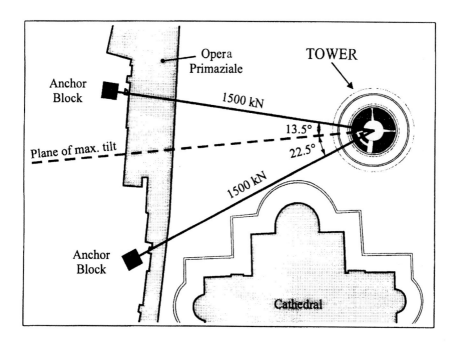

Fig. 35b. Cable stay structure—plan

8. ABBREVIATIONS

ASCE   American Society of Civil Engineers

ECSMFE	European Conference on Soil Mechanics and Foundation Engineering
ICSMFE	International Conference on Soil Mechanics and Foundation Engineering
JGE	Journal Geotechnical Engineering
MPW	Ministry of Public Works of Italy
RIG	Rivista Italiana di Geotecnica

REFERENCES

Abghari, A. (1987). "Leaning Instability of Tall Structures". Ph.D. Thesis, University of California, Davies, Ca.

Berardi, G., Caroti, L., Giunta, G., Jamiolkowski, M. and Lancellotta, R. (1991). "Mechanical Properties of Upper Pisa Clay". Proc. X ECSMFE, Firenze.

Burland, J.B. (1990). "On the Compressibility and Shear Strength of Natural Clays". The 30th Rankine Lecture, Geotechnique, no. 3.

Burland, J.B. (1990a). Personal Communication.

Burland, J.B. (1991). Personal Communication.

Burland, J.B. (1995). "Monitoring the Foundations of the Leaning Tower of Pisa". Proc. Seminar on Instrumentation in Geotechnical Engineering, Hong Kong, Institution of Engineers.

Burland, J.B. and Viggiani, C. (1994). "Osservazioni del compartimento della Torre di Pisa". RIG, no.3.

Calabresi, G., Rampello, S. and Callisto, L. (1993). "The Leaning Tower of Pisa Geotechnical Characterization of the Tower's Subsoil within the Framework of the Critical State Theory". Report of the University of Rome La Sapienza, Report.

Casagrande, A. (1936). "The Determination of the Pre-Consolidation Load and Its Practical Significance Discussion, Proc. I ICSMFE, Harward University. Cambridge, Mass.

Cheney, J.A., Abghari, A. and Kutter, B.L. (1991). "Stability of Leaning Towers". JGE Div. ASCE, no. 2.

Como, M. (1965). "Asymptotic of a Leaning Tower on Slowly Sinking Ground". Giornale del Genio Civile (in Italian).

Costanzo, D. (1994). "Mechanical behaviour of Pisa Clay". Ph.D. Thesis, Technical University of Torino, Italy.

Costanzo, D., Jamiolkowski, M., Lancellotta, R. and Pepe, C. (1994). "Leaning Tower of Pisa— Description of the Behaviour". Banquet Lecture, Settlement '94, College Station, Tx.

Croce, A., Burghignoli, A., Calabresi, G., Evangelista, A. and Viggiani, C. (1981). "The Tower of Pisa and the Surrounding Square: Recent Observations". Proc. X ICSMFEE, Stockholm.

Desideri, A. and Viggiani, C. (1994). "Some Remarks on Stability of Towers". Proc. Symposium on Developments 1936-1994 in Geotechnical Engineering, Bangkok.

Desideri, A., Russo, C. and Viggiani, C. (1997). "Stability of Towers on Compressible Ground". RIG, no. l.

Hayat, T.M. (1992). "The Coefficient of Earth Pressure at Rest". Ph.D. Thesis, University of Illinois at Urbana, Champaign, Ill.

Hambly, E.C. (1985). "Soil Buckling and the Leaning Instability of Tall Structures". The Structural Engineer, No. 3.

Hambly, E.C. (1990). "Overturning Instability". JGE Div., ASCE no. 4.

Jamiolkowski, M. (1988). "Research Applied to Geotechnical Engineering". Proc. Instn. Civ. Engrs., Part 1, Vol. 84, London.

Jamiolkowski, M., Lancellotta, R. and Lo Presti, D.C.F. (1994). "Remarks on the Stiffness at Small Strains of Six Italian Clays". Proc. Int. Symposium on Prefailure Deformation of Geomaterials, Sapporo, Japan.

Lancellotta, R. (1993). "The Stability of a Rigid Column with Non-Linear Restraint, Geotechnique, no. l.

Lancellotta, R. (1993a). "The Leaning Tower of Pisa—Geotechnical Theory and Model Tests". Discussion, Soil and Foundation, no. 3.

Lancellotta, R. and Pepe, C. (1990). "Pisa Tower—Geotechnical Properties of the Horizon A". Research Report 2.1, Technical University of Torino.

Lancellotta, R. and Pepe, C. (1990a). "Mechanical behaviour of Pisa Clay". Research Report 2.2, Technical University of Torino.

Lancellotta, R., Costanzo, D. and Pepe, M.C. (1994). "The Leaning of Tower of Pisa Geotechnical Characterization of the Tower Subsoil". A Summary Report of the Technical University of Torino.

Lancellotta, R. and Pepe, M.C. (1998). "On the Stability of Equilibrium of the Leaning Tower of Pisa". Accademia della Scienza—Atti di Scienze Fisiche no. 132.

Leonards, G.A. (1979). "Foundation Performance of Tower of Pisa". Discussion. JGE Div. ASCE, GT1.

Leonhardt, F. (1991). Personal Communication.

Leonhardt, F. (1997). "The Committee to Save the Tower of Pisa: A Personal Report". Structural Engineering International, no. 3.

Luchesi, G. (1995). "La Torre Pendente—Un Mistero Lungo Ottocento Anni". Edit. Bandecchi e Vivaldi, Pontedero.

Macchi, G. (1993). "Monitoring Medieval Structures in Pavia". Structural Engineering International, no. 1.

Mesri, G. (1989). Personal Communication.

Mesri, G. and Castro, A. (1987). "The C_α/C_c Concept and K_0 during Secondary Compression". JGE Div. ASCE, GT3.

Mesri, G., Hediew, J.E. and Shahien, M. (1997). "Geotechnical Characteristics and Compression of Pisa Clay". Proc. XIV ICSMFE, Hamburg.

Mitchell, J.K., Vivatrat, V. and Lambe, W.T. (1977). "Foundation Performance of Tower of Pisa". Journ. of JGE Div., ASCE, GT3.

MPW Ministero dei Lavori Pubblici (1971) "Ricerche e studi su la Torre pendente di Pisa ed i fenomeni connessi alle condizioni di ambiente". Edit. I.G.M., Firenze.

Pepe, M.C. (1995). "The Leaning Tower of Pisa—Theoretical and Experimental Analysis of the Stability of Equilibrium". Ph.D. Thesis—Technical Univ. of Torino (in Italian).

Schiffmann, R.L. (1995). "Settlements due to Groundwater Fluctuations in the Lower Sand". Report submitted to International Committee for Safeguard of the Leaning Tower of Pisa.

Tamez, E., Santoyo, E. and Cuaves, A. (1992). "The Metropolitan Cathedral of Mexico City and El Sagrario Church". Correcting Foundation Behaviour, Commemorative Volume, Raul, J. Marsal, Edit. The Mexican Society for Soil Mechanics, in Spanish.

Tamez, E., Ovando, E. and Santoyo, E. (1997). "Underexcavation of Mexico City's Metropolitan Cathedral and Sagrario Church". Proc. XIV ICSMFE, Hamburg.

Terracina, F. (1962). "Foundations of the Tower of Pisa, Geotechnique, no. 4.

Terzaghi, K. (1960). "Die Ursachen der Schiefstellung des Turms von Pisa". no.1/2, 1934, pp. 1-4 Reprinted in Bjerrum et al., eds., John Wiley and Sons, Inc., New York, 1960.

Veneziano, D., Van Dyck, J. And Koseki, J. (1995). "Leaning of the Tower of Pisa". Part I: Reological Models, Part II: Future Rotations and Risk of Instability". Draft of papers.

4

Structure-Medium Dynamic Interaction

Liao Zhen-Peng

Institute of Engineering Mechanics, CSB, 9 Xue-fu Road, P. R. China

ABSTRACT

The state-of-the-art of the dynamic interaction studies is first discussed. A practical technique—the decoupling direct method for the studies is then introduced briefly and demonstrated systematically via a series of numerical experiments. The focal point of the technique, that is, the decoupling simulation of infinite media is further clarified in four aspects—its original formulation, its relationship with other schemes of simulating infinite media, its spurious reflection analyses and its stable implementation. Applications of the decoupling direct method to the forward problems of dynamic interaction and a challenge to cope with the related inverse problems in the next century are briefly addressed.

Key Words: Dynamic interaction, modeling of infinite media, local artificial boundary condition, numerical simulation of wave motion.

1. INTRODUCTION

1.1 Purpose

In fields of engineering and sciences we are often interested in interpreting the dynamic response of a bounded structure of finite dimension, which is connected with the surrounding infinite medium. Since existence of the surrounding medium affects the structure response, we need to study the dynamic interaction between the structure and the medium for an appropriate interpretation of the structure response. The dynamic interaction problems are encountered in various fields, such as civil engineering, geophysics, acoustics, gas dynamics, hydrodynamics, electrical engineering, meteorology, environmental science, plasma physics, ... These fields are different in physics, but the dynamic interaction problems are in common from a point of view of the mathematical mechanics. In the early stage up to 80s, studies on dynamic interaction are conducted separately without sufficient communication between specialists working in different fields. But this situation is changed in recent years, and the interaction among specialists in different fields is indeed helpful to promote the studies. So at the very beginning, I would like to advise you that it is better to pay attention to the related studies in other fields, when you study the dynamic interaction problems in a particular field.

Two groups of examples in earthquake engineering may be used to illustrate importance of the studies on dynamic interaction. The first group of examples is the earthquake resistant designs in critical engineering projects. In China, we are now designing and going to construct arch dams of a height up to about 300 meters in the area with high seismicity. Seismic safety has become one of the critical problems for hydropower construction. The dams are built in deep valley between mountains. To estimate the dynamic response of a dam to earthquake correctly, we have to consider the interaction between the dam and the surrounding irregular topography and geological structures. Since non-linearities appear in the joints of the dam and in the neighboring geological structures, this is a complicated 3-dimensional problem of dynamic interaction. For earthquake resistant design of other important man-made structures. Such as nuclear power plants, large bridges, high-rise buildings and so on, we also need to consider the soil-structure dynamic interaction. This group of examples belongs to the forward problem of dynamic interaction, that is, we estimate the dynamic response of the structure as the underground medium and seismic excitation are known. The second group of examples is inverse problems of dynamic interaction. For instance, finding out a local underground geological structure of a depth about tens of meters for engineering purpose using data of observed ground motions excited by an artificial wave source. Since these inverse problems are non-linear and so complicated that they have to be solved generally by the trial-and-error method. Suppose that a model of the local structure, for instance, is characterized by 20 parameters, and that the value of each parameter is selected from 50 possible values, the model space would contain 50^{20} models. Although the extremely large model space may be reduced significantly using some optimization method, thousands of models, at least, must be considered, that is, thousands of forward computations of dynamic interaction between the local structure and the surrounding medium have to be performed. Thus we need the dynamic interaction analysis to be performed not only accurately but also very efficiently.

1.2 Statement of Problem

A concrete statement of the dynamic interaction problem is presented in this section via an example. Figure 1 shows a model of the nuclear power plant for dynamic interaction analysis. The reactor building, as a finite actual structure is embedded in the earth medium, which is infinite. The infinite earth medium is divided into two parts: the irregular bounded medium adjacent to the structure and the remaining unbounded medium extending to infinity. The structure and the irregular bounded medium form the generalized structure, which can exhibit non-linear behavior. The regular unbounded medium must remain linear. The generalized structure interacts with the regular unbounded medium at the interface between the structure and the medium. For simplicity, the generalized structure is called structure, the regular unbounded medium, the medium, and the interface between the generalized structure and the regular unbounded medium, the structure-medium interface.

The dynamic loading may be introduced into the structure-medium system through the medium. For example, incident waves from earthquakes, underground explosions. Alternatively, the loading may act directly on the structure, arising from machines, impacts, or vehicles moving within the structure, etc. The dynamic interaction problem may be stated as how to interpret the response of the above structure-medium system to a dynamic loading.

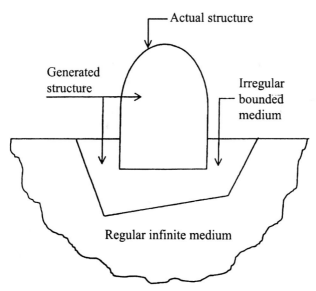

Fig. 1. Structure-medium dynamic interaction model

1.3 Methodology

General remarks on the methodology for studying the dynamic interaction problem are addressed below. The first remark is related to application of the domain discretization techniques to the dynamic interaction analysis. An actual structure is usually so complicated in geometrical configuration and in material properties that the dynamic response of the structure are almost impossible to be dealt with by the analytical methods except in extremely simple cases. For this reason, we will put the dynamic interaction analysis on the basis of the domain discretization techniques, because it is well known that the latter can be used to model any structure of finite dimension. The widely used domain discretization techniques are the finite element method and the finite difference method. Both of them are powerful in dealing with problems of wave motion in complicated situations. Researchers in the engineering circle usually use finite elements while physicists and geophysicists generally prefer finite differences. The selection of a particular domain discretization technique is mainly a custom. For definiteness, the finite element method is used for the domain discretization in this lecture. Note that most discussions below are also suitable for the finite difference method.

The second remark is related to a special and important notion in the dynamic interaction analysis, that is, the so-called artificial boundary. Although the domain discretization technique can be used to model any system of finite dimension, it cannot be used to model a system of infinite dimension. For example, the system that we concern now consists of a bounded structure and an unbounded medium. To study the dynamic response of the structure in the infinite system using the domain discretization technique, we have to introduce a fictitious surface, which encloses the structure and possibly part of the neighboring medium. A boundary condition must be imposed on the boundary to account for effects of the unbounded medium. A number of such boundary conditions have been proposed by specialists in different fields from various, physical and mathematical considerations. Thus the boundary condition has tens of names, for example, absorbing, non-reflecting, transmitting, silent, anechoic, radiating, transparent, open, free-space, one-way boundary condition, and the force-motion relationship, etc. Since the boundary does not exist in the realistic model and is artificially introduced merely for conducting the domain dicretization computation, we will use the general name, the Artificial Boundary Condition (in brief, ABC) in this lecture notes. The ABC problem is fundamental to a complete analysis of the dynamic interaction. In fact, significant advances made in the dynamic interaction field in the last 30 years are due to, to a large extent, the studies on ABC. It would be helpful for you to catch the essence of the dynamic interaction analysis if you keep the ABC problem in your mind throughout discussions in this lecture.

The third remark is related to the basic governing equations of the dynamic interaction analysis. Having discretized the region bounded by the artificial boundary using a domain discretization technique, the discrete nodes are divided into two groups: the boundary nodes which are on the artificial boundary, and the interior nodes which include all the others. The solution of dynamic interaction problem is governed by two groups of equations: the first group governing the motion of the interior nodes and the second group governing the motion of the boundary nodes. The first group of equations is the domain discretization equations, which have been well studied; the second group of equations is the ABC, which has not been studied as much as the first. So the focus of this lecture is on the ABC issues

The fourth remark is related to classification of major methods for the interaction analysis developed so far. The general framework to the analysis may be summarized as the domain discretization technique plus ABC. Various methods have been developed in realizing the general framework, and they may be classified into two major groups: the substructure method and the direct method. The classification may be simply made according to the relative location of the artificial boundary with respect to the structure-medium interface. If the artificial boundary coincides with the structure-medium interface, as shown in Fig. 2a, it leads to the substructure method; if the artificial boundary encloses part of the unbounded medium adjacent to the structure-medium interface as well as the structure, as shown in Fig. 2b, it leads to the direct method.

An outline of the remainder of the lecture is as follows. In Chapter 2, the state-of-the-art of the substructure method is discussed. In Chapters 3, the direct method is discussed, particularly, a decoupling technique is introduced and demonstrated in detail. In Chapter 4, the focus of the direct method-the ABC problem, is further addressed. In Chapter 5, the governing equations of the interior nodes are clarified via studying the wave motions in discrete grids. In the final Chapter, applications of the dynamic interaction analysis to the forward problems and a challenge to cope with the related inverse problems in the next century are briefly addressed.

2. SUBSTRUCTURE METHOD

The substructure method is the predominant technique in the dynamic interaction studies conducted so far, which has been significantly developed and widely applied to analysis of the related problems, particularly in the linear case. The fundamentals, major results, advantages and disadvantages of the method will be briefly discussed in this Chapter.

2.1 Fundamentals of Substructure Method

Since the artificial boundary coincides with the structure-medium interface in the substructure method, a rigorous ABC has to be imposed on the artificial boundary to achieve sufficient accuracy in computation of the structure response. The ABCs are usually expressed by the force-displacement relationship on the structure-medium interface. The relationship in the frequency domain is written as:

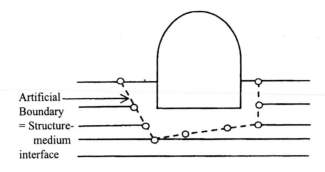

(a) Substructure method

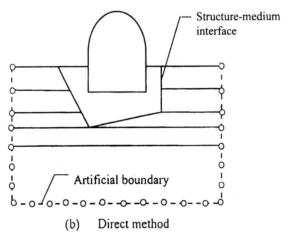

(b)   Direct method

Fig. 2. Classification of methods for structure-medium dynamic interaction analysis

$$\mathbf{R}_b = \mathbf{S}_{bb}^{\infty} \mathbf{U}_{sb} \tag{1}$$

where the vectors $\mathbf{R}_b$ and $\mathbf{U}_{sb}$ represent the interaction forces and displacements, respectively. The dimension of $\mathbf{R}_b$ and $\mathbf{U}_{sb}$ is equal to the total number of all degrees of freedom of the nodes on the artificial boundary. $\mathbf{S}_{bb}^{\infty}$ is the dynamic stiffness matrix of the infinite medium (Wolf and Song, 1996). For the linear cases, the equation of motion of the structure, using the standard finite element method, may be written:

$$\begin{bmatrix} \mathbf{S}_{ii} & \mathbf{S}_{ib} \\ \mathbf{S}_{bi} & \mathbf{S}_{bb} \end{bmatrix} \begin{pmatrix} \mathbf{U}_i \\ \mathbf{U}_b \end{pmatrix} = \begin{pmatrix} \mathbf{F}_i \\ -\mathbf{R}_b \end{pmatrix} \tag{2}$$

$$\mathbf{U}_{sb} = \mathbf{U}_b - \mathbf{U}_{gb} \tag{3}$$

where S_{jk} are the dynamic stiffness submatrices of the structure, U_j are the displacement vectors, the subscripts $j, k = i, b$, and i and b indicate the interior node and the boundary node, respectively, F_i is the external load vector acting on the interior nodes, U_{gb} represents the free-field displacements of all nodes on the structure-medium interface in the infinite medium under earthquake excitation with the structure being removed, the interaction displacement vector U_{sb} denotes the scattering displacements of the nodes on the structure medium interface, that is, the total displacements minus the free-field displacements on the interface, as shown by Eq. (3). In the context of the finite element method, S_{ik} are submatrices of the structure dynamic stiffness matrix S,

$$\mathbf{S} = \begin{bmatrix} \mathbf{S}_{ii} & \mathbf{S}_{ib} \\ \mathbf{S}_{bi} & \mathbf{S}_{bb} \end{bmatrix} = (1 + 2\mathbf{di}) \, \mathbf{K} - \omega^2 \mathbf{M} \tag{4}$$

where K is the static stiffness matrix of the structure, M is the mass matrix of the structure, d is the linear damping ratio, ω is the circular frequency. If the earthquake excitation U_{gb} and the external load F_i are known, the dynamic response of the structure in the interaction system may be solved from the simultaneous equations (1 -3).

Three remarks on the above results are as follows: 1. The dynamic response of the structure thus obtained in the frequency domain is accurate in the finite element sense, because the ABC is rigorous. 2. The dynamic response of the structure in the time domain may be obtained from that in the frequency domain via FFT. 3. Since the finite element formulation and programming are routine operations for the structure dynamic stiffness matrix S, the difficult problem in realizing the substructure method is related to the rigorous ABC, Eq. (1). This problem has become the focus of studies on the substructure method, and will be discussed in the next section.

2.2 Rigorous Modeling of the Infinite Medium

The state-of-the-art of the rigorous modeling of the infinite medium will be briefly reviewed in this section. Not all methods developed so far for formulation and implementation of the force-motion relationship can be discussed, and a historical review of the development lies outside the scope of this lecture. Preference is given to those methods which are widely applied in Earthquake Engineering. Studies on the force-motion relationship will be first discussed in the frequency-domain and then in the time-domain.

2.2.1 Force-motion relationship in the frequency domain

In the frequency-domain, the force-motion relationship Eq. (1) is rewritten as

$$\mathbf{R}(\omega) = \mathbf{S}^{\infty}(\omega)\mathbf{U}(\omega) \qquad (5)$$

where the subscripts s and b in Eq. (1) have been removed for simplicity, and the quantities as functions of the frequency ω are indicated for clarity. It is worth mentioning that, as a result of the rigorous modeling of the infinite medium, S^{∞}(ω) is a full matrix. Therefore, the ABC (Eq. (5)) is global in space, that is, the interaction force of a specific degree of freedom depends on the displacements of all degrees of freedom of the nodes on the structure-medium interface. Three major approaches to the formulation of Eq. (5) and their advantages and disadvantages are briefly discussed below.

1. Boundary-element method: The boundary-element method is based on the discretized form of the boundary-integral equation. The latter is formulated by using the fundamental solution, which satisfies all governing differential equations and physical boundary conditions in the infinite medium. Therefore, the force-motion relationship of the boundary nodes derived from the boundary element method provides a rigorous modeling of the unbounded medium in the discretization sense. The method is also general because it is suitable to any linear infinite medium once the fundamental solution is available. Unfortunately, the fundamental solution is not always available. For instance, for anisotropic or two-phase media, the fundamental solution of even the homogeneous full-space is generally not available. As for isotropic elastic media, except for the homogeneous full-space, the fundamental solution is so complicated that even for the homogeneous half-space the fundamental solution involves infinite integrals of Bessel functions. Thus the elements in the matrix S^{∞}(ω) are difficult to evaluate because of the special functions and singularities in the integrands. In addition, the evaluation is also impossible to be accurate because only a finite part of the boundaries, which extend to infinity such as the free surface and the interfaces between two different materials, can be discretized, and the truncation results in errors.

2. Thin layer method: This method is only applicable to a horizontally layered unbounded medium with its base fixed. Having divided the medium into thin layers in the finite element sense, the lateral ABC in the form of Eq. (5) may be set up based on the solution of an eigenvalue problem resulting from the radiation condition and the discretization in the vertical direction. The ABC is called the consistent-boundary condition, which is exact in the finite element sense for the particular model of infinite media. Since this layered model was once acceptable in the engineering circle, the ABC was widely used in earthquake engineering up to 80s. The advantage of the method is the computational effort for a layered medium being the same as that for a homogeneous medium.

3. *Procedure based on similarity:* The third important approach to setting up

the force-motion relationship is a procedure based on similarity. This procedure was originally proposed by Dasgupta (1982). To explain the concept, a two-dimensional homogeneous half-plane is considered (Fig. 3). Suppose the similarity center *o* is outside of the unbounded medium and the free surface passes through it. The origin of the coordinates *x, y* coincides with the similarity center. To construct another similar structure-medium interface, all coordinates of the nodes on the interface are multiplied by a positive constant. Thus the structure-medium interface can be determined by a characteristic length *r*, as shown in Fig. 3. A relationship linking the two dynamic stiffness matrices at the two similar interfaces of the same unbounded medium can be set up via a dimensional analysis based on the geometrical similarity. Another relationship between the two matrices simply provided by the equation of motion of finite element nodes between the two similar interfaces. The dynamic stiffness matrix can then be obtained from these two relationships. Significant developments from the original idea have been made recently by Wolf and Song (1996), so that the procedure is now applicable to a wide range of problems of wave motion in the compressible and incompressible elastic media and to the static and diffusion problems. Compared the procedure with the thin-layer method, the latter may be regarded as a special case of the former as the similarity center approaches infinity. Compared it with the boundary-element method, the distinguishing feature of the procedure is that it is formulated without using the fundamental solution and thus avoids all drawbacks of the boundary-element method. The only limitations arise from the geometrical requirements imposed to obtain similarity.

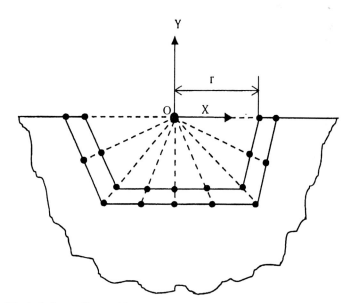

Fig. 3. The infinite medium with structure-medium interface defiened by characteristic length *r* and similarity center *o*

For further information refer to the works by Beskos (1987), Brebbia et al. (1984), Manolis et al. (1988), Dominguez (1933), Banerjee (1994) for the boundary-element method, Lysmer (1970), Lysmer and Waas (1972), Waas (I 972) for the thin layer method, and Dasgupta (I 982), Wolf and Song (l 996) for the procedure based on similarity.

2.2.2 Force-motion relationship in the time domain

The force-motion relationship in the time domain must be considered if non-linearities appear in the dynamic response of the structure. The relationship may be derived from the Fourier transform of Eq. (5) for the infinite medium of a quiescent past

$$\mathbf{r}(t) = \int_0^t \mathbf{s}^\infty (t - \tau)\mathbf{u}(\tau)d\tau \tag{6}$$

where $r(t)$, $s^\infty (t)$ and $u(t)$ are the Fourier transforms of $R(\omega)$, $S^\infty(\omega)$ and $U(\omega)$, respectively.

$$\mathbf{s}^\infty (t) = \frac{1}{2\pi} \int_{-\infty}^{\infty} \mathbf{S}^\infty (\omega) e^{i\omega t} d\omega \tag{7}$$

$s^\infty (t)$ is called the unit-impulse response matrix of the infinite medium. Since the matrix $s^\infty (t)$ is full, as a result of the Fourier transform of the full matrix $S^\infty (\omega)$, the distinguishing feature of Eq. (6) is global, that is, the interaction force of a specific degree of freedom at a specific time depends on the displacements of all degrees of freedom of nodes on the structure-medium interface at all previous times from the start of the excitation onwards. Eqs. (6) and (7) are used merely to provide a general idea of how to derive $r(t)$ and $s^\infty (t)$, not the actual procedure to perform the computation. For example, Eq. (7) is merely a formal expression because $S^\infty (\omega)$ is not square integrable. The actual procedure to compute the interaction force from the motion (displacement, velocity and acceleration) on the structure-medium interface have been discussed in detail in Chapter 2 for compressible elastic media and Chapter 6 for incompressible elastic media in the book by Wolf and Song (l996).

Since the pioneering contributions more than sixty years ago by Reissner (1936), who studied the dynamic response of a circular foundation supported on an elastic half-space, and by Sezawa and Kanai (1935, 1936), who studied the seismic response of a cylindrical elastic structure supported on a hemispherical foundation resting on a elastic half-space, significant advances have been made in the field of dynamic interaction analysis based on the rigorous consideration of effects of the infinite medium. The advances have led to a powerful and complete framework for the analysis, that is, the domain discretization plus the rigorous ABC, and to a series of valuable methods to realize the framework, particularly, the development of the similarity-based procedure has made implementation of the framework practical in many cases. However, future studies are still needed in this area because the similarity-based method is not general enough though it is practical and the boundary-element method is not practical in many cases though it is general. Furthermore, essential feature of the rigorous force-motion relationship in the substructure method is global in space and time. This fact leads to considerable computing costs in analysis of the dynamic interaction. Thus this kind of rigorous modeling of infinite media is less attractive in dealing with forward dynamic interaction problems of large-scale and nonlinearity, and particularly, in dealing with the inverse problems, which require a number of dynamic interaction forward computations. These considerations have led to searching for other solutions, which result in development of the direct method.

3. A DECOUPLING DIRECT METHOD

In contrast to the substructure method, the artificial boundary is apart from the structure-medium interface in the direct method. This fact results in a computational region in the direct method larger than that in the substructure method. However, it will be shown that the same fact also results in that the ABC in the former is much simpler than that in the latter, thus the computational cost may be considerably reduced by the direct method, particularly, for nonlinear analysis of large-scale dynamic interaction system. On the other hand, it can be shown that an appropriate direct method may have accuracy consistent with that of the domain discretization. Therefore, the direct method is attractive for applications to a wide range of forward and inverse problems of dynamic interaction. A very simple direct method, which is called a decoupling numerical simulation of wave motion, will be introduced in this Chapter. This procedure is based on an simple idea, namely, the local feature of wave motion. A dynamic disturbance at a spatial point cannot be transmitted to another spatial point instantaneously: it takes time to reach there because the wave speed in any realistic medium is finite. In other words, the motion of a specific spatial point at the next moment is determined completely by the motions of its neighboring points at the present and previous times within a short time window. Therefore, the natural form of governing equations of wave motion should be local in space and time either for the interior nodes or the boundary nodes of the computational region. Our introduction to the simple direct method will be started with formulating the local, discretized governing equations for the two groups of nodes.

3.1 Governing Equations

The governing equations of the interior nodes and the boundary nodes may be formulated for the decoupling numerical technique as follows.

3.1.1 Governing equation of the interior nodes

The governing equation of the interior nodes may be set up using the standard finite element technique, and some software for this purpose is already available. One note that the lumped-mass formulation is suggested for the spatial discretization in this study. This is because not only the lumped-mass formulation is much more practical than the consistent-mass for large-scale computations, but also it is the simplest manner to reflect the local feature of wave motion. The governing equation of the interior nodes are written in the following form

$$\mathbf{M\ddot{u}} = \mathbf{P}_c + \mathbf{P}_e \tag{8}$$

where M is the diagonal mass matrix for the lumped-mass formulation, $\ddot{u}$ is the acceleration vector, P_e is a given external force vector, and P_c is the constitutive force vector. In the linear case,

$$\mathbf{P}_c = -\mathbf{C}\dot{u} - \mathbf{K}u \tag{9}$$

where C is the damping band matrix, K is the stiffness band matrix, $\dot{u}$ and u are the velocity and displacement vector, respectively. The explicit numerical integration of Eq. (8) is preferable if it is possible. The displacement may be used as a single independent variable in performing the integration for the linearly elastic case, namely, C= 0. In the case of $C \neq$ 0 or in the non-linear cases, two independent variables, namely, the velocity and displacement may be used to perform the integration. In the two independent variables case, an additional equation must be provided by defining $v = \dot{u}$. It is worth mentioning that the governing equations expressed in terms of velocity and stress may be used as well to form an integration scheme similar to the well-known staggered scheme of the finite difference computation (Liao, 1996).

3.1.2 Governing equation of the boundary nodes

As mentioned earlier, the governing equations of the boundary nodes are the local ABCs. Since the local ABCs are the most difficult part in formulation of the direct method, the related studies will be discussed in Chapter 5. Merely the local ABC adopted in the suggested decoupling direct method is presented here for practical use. The ABC is called the Multi-Transmitting Formula (MTF), which is written (Liao, 1996)

$$u((p+1)\Delta t,0) = \sum_{j=1}^{N}(-1)^{j+1}C_j^N u((p+1-j)\Delta t,-jc_a\Delta t) \tag{10}$$

where $u(t, x)$ is a function of time $t = p\Delta t$ and $x = -jc_a\Delta t$, Δt is the time step, $j =$ 0, 1,…, N, p is an integer. The coordinates $x = -jc_a\Delta t$ indicate the sampling points on the x-axis, which is perpendicular to the artificial boundary at a boundary point 0 under consideration (Fig. 4), C_j^N are the binomial coefficients, c_a is the artificial speed, N is the approximation order of MTF. The most practical form of the local ABC is the second-order MTF, which is written as

$$u((p+1)\Delta t, 0) = 2u(p\Delta t, -c_a\Delta t) - u((p-1)\Delta t, -2c_a\Delta t) \qquad (11)$$

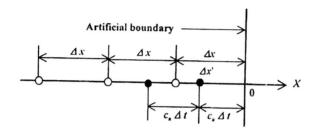

Fig. 4. Sampling points • involved in MTF and the nodal points o

Two notes on Eq. (10) are as follows:

1. Major features of MTF: The first is the locality of Eq. (10), that is, the governing equation of a boundary node under consideration in each time step integration is decoupled from all the nodes except those in its neighborhood and from all the times except those within a short time window. In fact, Eq. (10) shows that the displacement of the boundary point 0 at the time $t = (p+1)\Delta t$ is determined by the displacements of the sampling points adjacent to the point 0 and at the times $t = p\Delta t$, …, $(p+1-N)\Delta t$. The second is the generality of MTF. It will be shown later in this Chapter and Chapter 5 that Eq. (10) is applicable not only to the scalar waves, such as sound waves and SH waves, but also to the vector waves, such as in-plane and three-dimensional waves in a wide range of infinite media. In the latter case, u(x, t) should be understood as one of the components of a vector wave field concerned. And u(x, t) in Eq. (10) may denote any wave motion quantities, such as velocity and stress, etc. besides displacement. In addition, Eq. (10) is applicable to all the boundary nodes including those on corners of the computational region and on the interfaces between different materials because only the nodes on the normal to the boundary are involved in Eq. (10). The third feature of MTF is related to accuracy of MTF. It will be shown in Chapter 5 that within a desired frequency band $0 < \omega \leq \omega_c$ Eq. (10) can be made accurate enough to meet the practical needs, where the upper limit ω_c may be regarded as the frequency limit, beyond which the wave motions cannot be modeled by the finite elements. In other words, the accuracy of MTF can be made consistent with that of numerical simulation of wave motion by using the finite elements.

2. Interpolation schemes: Since $u(t, x)$ in Eq. (10) are sampled at points $x = -jc_a\Delta t$, which do not generally coincide with the discrete nodes $x = -n\Delta x$, Δx is the space step (Fig. 4). In order to implement Eq. (10), an interpolation scheme is required to express $u(t, -jc_a\Delta t)$ in terms of $u(t, -n\Delta x)$. The interpolation may be realized in a number of ways. Two of the interpolation schemes are as follows:

Interpolation scheme 1: The displacements at the nodal points are denoted by

$$u_n^p = y(p\Delta t, -n\Delta x) \tag{12}$$

Suppose that a quadratic interpolation is used, the displacements at the computational points may be written:

$$u(p\Delta t, -jc_a\Delta t) = \sum_{n=0}^{2j} t_{j,n} u_n^p \tag{13}$$

$t_{1,0} = T_1, \ t_{1,1} = T_2, \ t_{1,2} = T_3,$

$t_{2,0} = T_1^2 \ t_{2,1} = 2T_1T_2, \ t_{2,2} = 2T_1T_3 + T_2^2, \ t_{2,3} = 2T_2T_3, \ t_{2,4} = T_3^2,$

$\vdots$ $\tag{14}$

$t_{j,n} = \sum T_{k_1} T_{k_2} \cdots T_{k_j} \qquad \text{for} \quad n = 0, 1, 2, \cdots 2j,$

where the notation $\sum$ in Eq. (14) indicates that the summation includes all the terms in which the subscripts k_m satisfy the following condition,

$$k_1 + k_2 + \cdots k_j = n + j, \quad k_m = 1, 2, 3 \quad \text{for} \quad m = 1, \cdots, j \tag{15}$$

and

$$T_1 = \frac{1}{2}(2 - S)(1 - S), \quad T_2 = S(2 - S) \quad T_3 = \frac{1}{2}S(S - 1), \tag{16}$$

$$S = c_a\Delta t / \Delta x \tag{17}$$

Interpolation scheme 2: Another type of interpolation schemes is for implementation of MTF of $N=2$. Let the space step Δx of the finite element adjacent to the boundary point 0 be denoted by $\Delta x'$, and $u(p\Delta t, x)$ for

$(\Delta x' + \Delta x) < x < 0$ be interpolated in terms of the nodal displacements u_0^p, u_1^p

and u_2^p at $x = 0$, $\Delta x'$ and $-(\Delta x' + \Delta x)$, respectively, by a quadratic interpolation,

$$
\left.
\begin{aligned}
u(p\Delta, x) &= a_2(x)a_2^p + a_1(x)u_1^p + a_0(x)u_0^p \\
a_2(x) &= \frac{(x/\Delta x + \Delta x'/\Delta x)x/\Delta x}{1 + \Delta x'/\Delta x} \\
a_1(x) &= \frac{(x/\Delta x(1 + x/\Delta x))/\Delta x'/\Delta x}{\Delta x'/\Delta x} \\
a_0(x) &= \frac{(x/\Delta x + 2(\Delta x'/\Delta x) + 1)/x/\Delta x}{(\Delta x'/\Delta x + 1)\Delta x'/\Delta x} + 1
\end{aligned}
\right\}
\tag{18}
$$

Having assumed that $c_a \leq c_{max}$ and that $c_{max}\Delta t \leq \min(\Delta x, \Delta x')$, Eq. (18) yields the sampling displacements $u(p\Delta t, -C_a\Delta t)$ and $u((p-1)\Delta t, -2c_a\Delta t)$. Substituting the expressions of the sampling displacements into Eq. (11), the final result is

$$
u_0^{p+1} = \sum_{n=0}^{2}\sum_{j=0}^{1}\gamma_n^j u_n^{p-j}
\tag{19}
$$

where

$$
\left.
\begin{aligned}
\gamma_n^0 &= 2a_n(-c_a\Delta t) \\
\gamma_n^1 &= -a_n(-2c_a\Delta t)
\end{aligned}
\right\}
\quad \text{for } n = 0, 1, 2
\tag{20}
$$

Two special cases are as follows:
Case 1: $\Delta x' = \Delta x$

$$
\left.
\begin{aligned}
\gamma_2^0 &= S(S-1), & \gamma_1^0 &= 2S(2-S), & \gamma_0^0 &= (S-2)(S-1), \\
\gamma_2^1 &= -S(2S-1), & \gamma_1^1 &= -4S(1-S), & \gamma_0^1 &= -(S-1)(2S-1).
\end{aligned}
\right\}
\tag{21}
$$

Case 2: $\Delta x' = 2C_a\Delta t$

$$\left.\begin{array}{lll} \gamma_2^0 = -\dfrac{2S^2}{1+2S}, & \gamma_1^0 = 1+S, & \gamma_0^0 = \dfrac{S+1}{2S-1} \\[2ex] \gamma_2^1 = 0, & \gamma_1^1 = -1, & \gamma_0^1 = 0 \end{array}\right\} \qquad (22)$$

Equation (19) together with Eq. (21) or Eq. (22) provide two simple but practical schemes to implement MTF of N=2. For example, substituting Eq. (22) into Eq. (19) leads to:

$$u_0^{p+1} = \sum_{n=0}^{2} \gamma_n^0 u_n^p - u_1^{p-1} \qquad (23)$$

which is identical to the interpolation scheme initially suggested by Shao and Lan (1995).

The two groups of governing equations of the interior and boundary nodes are all local in both the space and time, thus, they provide a practical procedure for the decoupling numerical simulation of wave motion. This procedure explicitly and locally updates the wave field one grid point at a time for each time step. Once the entire grid is updated, the method is repeated to advance the wave field to the next time step. The key problem in analysis of the structure-medium dynamic interaction is understanding of the waves scattered from the structure, thus the dynamic interaction problems are similar to the source problems of wave motion in essence. Numerical solutions to the two types of problems are used to illustrate the performance of the decoupling procedure. It has been shown that the procedure with MTF of $N = 2$ is practical for use in earthquake engineering and seismic prospecting (Liao (1984, 1990, 1996), Lan (1997), Shao and Lan (1995, 1997), Wang and Yang (1994), Chen et al. (1997)). The performance of the method will be demonstrated in the next section.

3.2 Demonstration

Examples presented in this Section are divided into two groups. The first group is related to the numerical simulation of scalar wave motions in isotropic elastic media. Since the scalar wave motions are comparatively simple, they will be used to demonstrate the major features of the decoupling direct method. The second group is related to the vector wave motions in comparatively complicated media, including the anisotropic medium, the two-phase medium as well as the isotropic elastic medium. These examples will be used to demonstrate further the applicable scope of the decoupling procedure.

3.2.1 Numerical simulation of scalar waves

The examples involve the propagation of SH-waves in an isotropic and linearly elastic half space (Fig. 5). The half-space consists a homogeneous top layer of thickness h and an underlying homogeneous half-space. The wave source embedded at a depth h_s below the free surface is described by

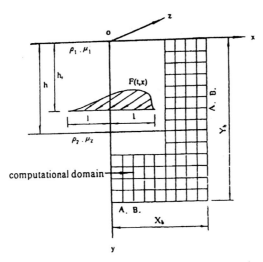

Fig. 5. A layered half-space with an embedded loading

a time dependent load,

$$F_i(t,x) = T_i(t)S(x) \tag{24}$$

where the time function $T_i(t)$ is assumed to be either a triangular impulse function ($i=1$) with duration T, or a semi-infinite sine function ($i=2$) with period T defined as $T_2(t)=\sin(2\pi/T)$ for $t>0$. The spatial load distribution, $S(x)$ in Eq. (24), is assumed to be an approximate Delta function defined as:

$$
\begin{aligned}
s(x) &= 1+6(x^3-x^2) & 0 \le x \le 0.5 \\
&= -2(x-1)^3 & 0.5 < x \le 1 \\
&= 0 & x > 1 \\
&= S(-x) & x < 0
\end{aligned}
\tag{25}
$$

Since the load is symmetrical about the y-axis, symmetry conditions will be used. The only component of interest for SH-waves, the antiplane component, will be denoted as $u(x, y, t)$ in the following examples. The lumped-mass finite element equation with an explicit integration scheme is used to simulate the wave propagation inside the artificial boundary while Eq. (10) and Eq. (13) are used to estimate the boundary nodal values. The effectiveness of the direct method with MTF will be tested by comparing the related numerical results with the corresponding exact solution, which is either an analytical one if available or a numerical one for an extended mesh. Conclusions drawn from this group of numerical experiments are as follows.

1. *Accurate for the entire computational region :* Finite element solutions using MTF of $N=2$ are presented for the entire computational region up to the artificial boundary. The response to the impulse time function $F_1(t, x)$ and the harmonic time function $F_2(t, x)$ at ten distinct locations are shown in Figs. 6 and 7, respectively. The dimensions of the models (Fig. 5) are $X_b = 2$ and $Y_b = 2$. Numerical results of the response of a homogeneous half space (h = 0) to the load function $F_1(t, x)$ are shown in Fig. 6 for five locations along the free surface ($x = 0, 0.5, 1, 1.5, 2$ with $y = 0$) and five locations down the symmetry axis ($y = 0, 0.5, 1, 1.5, 2$ with $x = 0$). The difference between these solutions and the exact ones is undetectable on the scale of this particular graph. No reflection from the artificial boundary is observed although the model size is only two times the source dimension. To aid the readers, pointers marked AR (Artificial Reflections) are placed along the time scale to indicate the time that artificial reflections would normally return if the effects of the artificial boundary were not removed. These arrival times can be estimated as the shortest path (from the source to the nearest artificial boundary and back to the observer) divided by the shear wave speed.

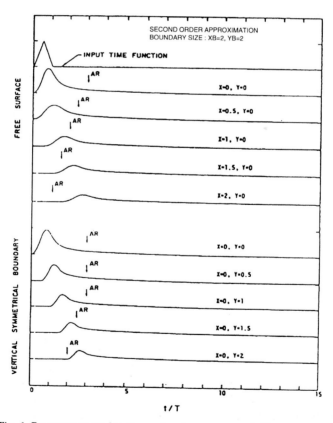

Fig. 6. Response at ten locations of an homogeneous half space subjected to a triangular time function loading. MTF of $N=2$

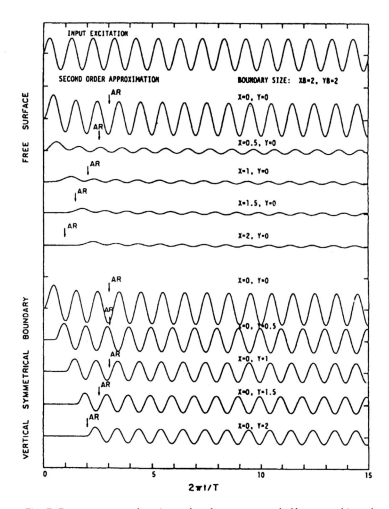

Fig. 7. Response at ten locations of an homogeneous half space subjected to a harmonic loading. MTF of $N = 2$

To show that the transmitting formula does not accumulate error over an extended period of time with non-decreasing excitations, a semi-infinite harmonic load function, $F_2(t, x)$, is applied. The numerical results are shown in Fig. 7; the locations chosen are the same as those used for Fig. 6. After the transient initial responses, the response at each location settles to a harmonic wave of constant and stable amplitude, which is identical to the steady-state solution at the corresponding frequency; no artificial reflections can be detected. The AR pointers serve the same purpose as those in Fig. 6.

2. Large angles of incidence : For the examples shown above, the maximum angle at which the outgoing waves approach the artificial boundary is $45°$ because of the model geometry. In this section, the case of a layered half space with surface loading is considered. The reflection and refraction at the layer interface of this model can produce a wider range of incident angle toward the vertical artificial boundary. The results of the calculation are shown in Fig. 8 for three special locations of the layered half space subjected to the impulse loading $F_1(t, x)$. The shear wave speeds of the layers are $c_1 = 1$ and $c_2 = 2$ respectively; the mass densities, ρ_1 and ρ_2, are equal to 1. The thickness of the top layer was specified as $h{=}1$. The artificial wave speed $c_a = c_1$ in the top layer, $c_a = c_2$ in the underlying half space, $c_a = (c_1 + c_2)/2$ along the layer interface. It is worth mentioning that the numerical results are not sensitive to the choice of the value of c_a, for example, changes of the numerical results are not observed in the figure if $c_a = c_1$ for the entire vertical artificial boundary. The results also indicate that the first order MTF is incapable of a complete transmission of the outgoing waves, but the second order solution yields an excellent approximation to the exact solution. By all indications, the transmitting boundary performs equally well for a layered media as for the homogeneous media. The exact solution for this particular example was obtained using a model with an extended mesh of $X_b = Y_b = 7$ to avoid reflections in the duration concerned.

To further demonstrate the capacity of MTF in dealing with the waves of large angles of incidence, a convincing example is shown in Fig. 9. This figure

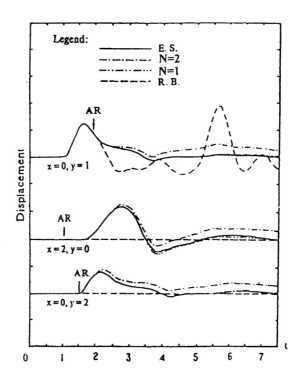

Fig. 8. Performance of MTF of $N{=}1, 2$ for transmiting SH waves in a layered half-space with a surface loading

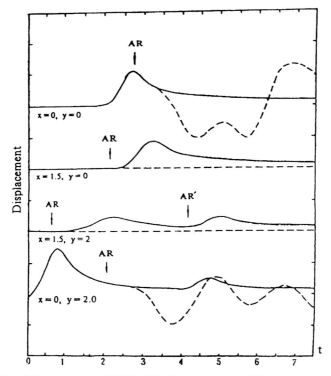

Fig. 9. Performance of MTF of $N=3$ for transmitting SH wave at a large incident angles in a half-space with an embedded loading

shows the responses at four locations $((x, y) =(0, 0), (1.5, 0.0), (1.5, 2.0)$ and $(0.0, 2.0))$ for a buried load $(h_s = 2.0)$ in a homogeneous half-space $(h = 0)$ and $X_b=1.5$, $Y_b =3.5, \rho_2 =1, c_2 =1$ and the time increment $\Delta t= 0.05$. The numerical results for MTF of $N= 3$ and $c_a= c_2$ are shown by the dashed lines separated by dots in Fig. 9. The exact solutions in the figure are calculated using the analytical solution (Liao, 1996). The response at the location $(x, y)=(1.5, 2.0)$, that is, the third group of the curves from top in Fig. 9 is most valuable to explain the capacity of MTF in dealing with the large angle of incidence. Besides the stated mark AR, another mark AR' is placed over the time scale of this group of curves. The mark AR' indicates the time at which the waves reflected from the free surface just arrive at the location under consideration. The geometrical relation of the rays indicated in Fig.10 shows that the incident angles of the predominant portion of the mentioned waves are about 69.4° while the maximum reaches 82.9°. The excellent consistence between the dashed line separated by dots and the solid line, particularly, after the pointer AR', speaks the point.

3.2.2 Numerical simulation of vector waves

The decoupling direct method is applied to the vector wave cases in the same way as to the scalar wave cases. The governing equations of interior nodes are

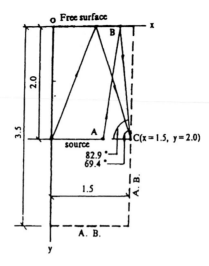

Fig. 10. The computational region for Fig. 5 and the geometrical relation of the rays reflected from the free surface and then impinging upon the boundary point of *x*=1.5 and *y*=2.0

obtained by using the lumped-mass finite elements and an explicit scheme of time-step integration. The governing equations of the boundary nodes are given by MTF via an interpolation scheme, as shown in Chapter 3, for each component of the vector waves, independently. All the numerical results presented here are obtained using the decoupling procedure with MTF of $N=2$.

1. Scattering of 3-dimensional vector waves: Consider a cubical-shaped depression of size 1 on the ground, which is assumed to be the free surface of a homogeneous, isotropic and elastic half-space (Fig. 11). The rectangular

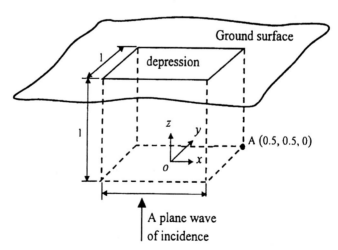

Fig. 11. A cubical depression on the ground surface subject to a vertically incident plane wave

coordinate system *oxyz* is shown in Fig. 11 with the *z*-axis up vertically and the origin *O* coincides with the center of the bottom of the depression. Suppose that a shear plane wave of incidence is vertically impinging upon the depression. The displacement component of the incident wave is written

$$\left. \begin{aligned} u_y^{(i)} &= \hat{\delta}\left(t - z/c_s\right) \\ u_x^{(i)} &= u_z^{(i)} = 0 \end{aligned} \right\} \tag{26}$$

where the subscripts *x*, *y* and *z* indicate the directions of the components, respectively, c_s is the shear speed of the medium. The impulse function $\hat{\delta}$ (*t*) of duration 1 is defined by

$$\left. \begin{aligned} \hat{\delta}(t) &= 16\left[t^3 H(t) - 4\left(t - \frac{1}{4}\right)^3 H\left(t - \frac{1}{4}\right) \right], &t \le 1/2 \\ &= \hat{\delta}(1 - t) &t > 1/2 \end{aligned} \right\} \tag{27}$$

where *H*(*t*) is the unit step function.

The scattering from the depression subject to the plane wave of incidence is addressed below. Introduce a cubical-shaped box whose edges (namely, the artificial boundaries) are specified by *x*=±(1/2+*L*), *y*=±(1/2+*L*) , y=±(1/2+L) and z=−L where *L* is a constant length describing the size of the computational region. The computational region is bounded by the artificial boundaries and the free surface, which are specified by

$$x = \pm 1/2,\ y = \pm 1/2,\ z = 0 \text{ and } z = 1$$

Having considered symmetry of the model, the computational region may be reduced to fourth of the above region, in which $x \ge 0$ and $y \ge 0$. Define the free-field by

$$\left. \begin{aligned} u_y^{(f)} &= \hat{\delta}(t - z/c_s) + \hat{\delta}(t + (z - 2)/c_s) \\ u_x^{(f)} &= u_z^{(f)} = 0 \end{aligned} \right\} \tag{28}$$

The numerical integration of the governing equations of all nodes in the computational region for each time step may be performed by the following computations. The total displacements *u* at the time level *p*+1 are first computed in terms of the data at the time levels p and *p*–1 for each interior node one after another. The total displacements *u* of the boundary nodes at the time level *p*+1 are then computed by

$$u = u^{(s)} + u^{(f)} \tag{29}$$

where the scattering displacements $u^{(s)}$ of the boundary nodes at time level *p*+1 are computed by Eqs. (21) and (23) in terms of the data at time levels *p* and *p*–1.

Suppose that the computational region is discretized by uniform cubic finite elements of size Δ. The spatial discretization parameter Δ is determined by $\Delta = c_s/(10f_{max})$, f_{max} is the upper limit of a desired frequency band, $f_{max} = 8$ in this example. Implication of the selection of f_{max} in accuracy of the numerical simulation may be revealed by analysis of the Fourier spectrum of $\hat{\delta}(t)$, which is written

$$\left| F(f) \right| = \left| \int_{-\infty}^{\infty} \hat{\delta}(t) e^{-i2\pi ft} dt \right| = \frac{12}{\pi^4} \frac{1}{f^4} \left| \cos(\pi f) + 3 - 4\cos\left(\frac{\pi}{2} f\right) \right| \tag{30}$$

Equation (30) yields that

$$\left. \begin{array}{l} \left| F(f) \right|_{max} = \underset{f \to 0}{Lim} \left| F(f) \right| = 0.375 \\ \left| F(f) \right| = 0 \quad \text{for} \quad f = 4, 8, 12, \cdots \end{array} \right\} \tag{31}$$

$f_{max} = 8$ implies that the amplitudes of f > 8 neglected are about 10^{-4} of the maximum value of $\left| F(f) \right|$. The time step Δt is determined by the stability condition $\Delta t \leq \Delta/c_p$, c_p is the p-wave speed.

The vector waves in the scattering problem are numerically simulated for $c_s = 4$ and Poisson ratio $v = 0.25$, and the parameters for the discrete computation are $\Delta = 0.05$, $\Delta t = 0.005$ and $L = 1$. Typical numerical results of the dynamic response are shown by dash lines in Fig. 12 for the three components of the displacement vector at the point A(1/2, 1/2, 0). The numerical exact solution obtained by using the extended mesh is shown in Fig. 12 by the solid lines for comparison.

2. Vector waves in anisotropic and 2-phase media : The lumped-mass finite elements plus MTF of *N*=2 have been successfully applied to modeling vector waves in anisotropic and 2-phase media by Lan (1995), Shao and Lan (1995) and Shao (1997). They proposed a simple scheme for implementation of MTF, as shown by Eq. (23) in the present paper. The 3-dimensional wave fields at a depth of z = 2000 m and at successive times in an anisotropic half-space excited by a concentrated force horizontally acting on the free surface are shown

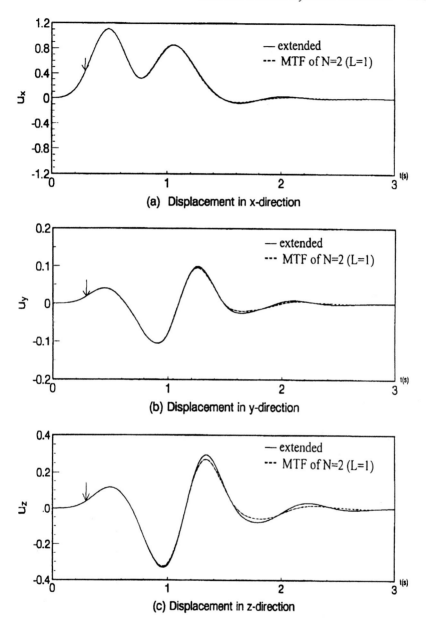

Fig. 12. Three components of the response at point A

in Fig. 13 for the vertical component. The synthetic seismograms on the ground surface of a two-phase medium half-space excited by an embedded explosion center are shown in Fig. 14(a) and (b) for the solid-phase and the fluid-phase, respectively. These results provided by Shao and Lan (1995, 1997) show excellent performances of the decoupling direct method of MTF of $N=2$ in the complicated

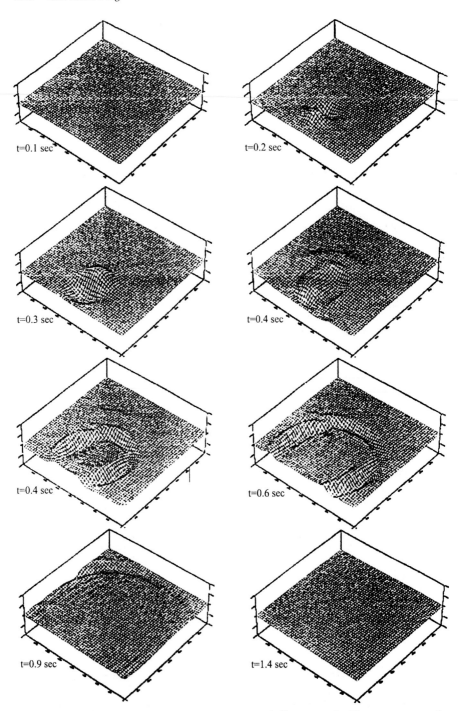

Fig. 13. 3-D anisotropic waves in a homogeneous, half-space excited by a concentrated horizontal faces acting on the free surface (vertical component received at Z=2000 m, Lan (1995))

trace

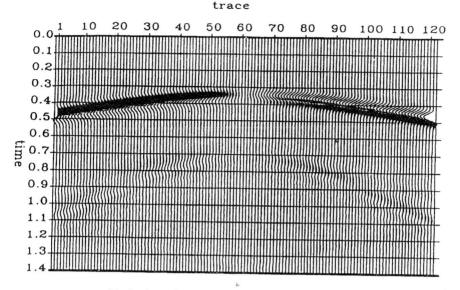

(a)   horizontal components of the solid-phase

trace

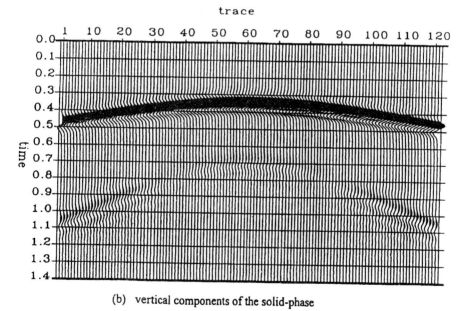

(b)   vertical components of the solid-phase

Fig. 14a. Synthetic seismograms of the solid-phase on the ground surface for waves in a two-phase medium excited by an embedded explosion center

media. In addition, Shao and Lan (1995) have applied successfully the decoupling direct method to modeling of the dynamic interaction between a number of irregular thin layers and the surrounding earth medium.

trace

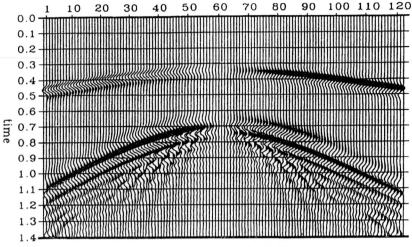

(a) horizontal components of the fluid -phase

trace

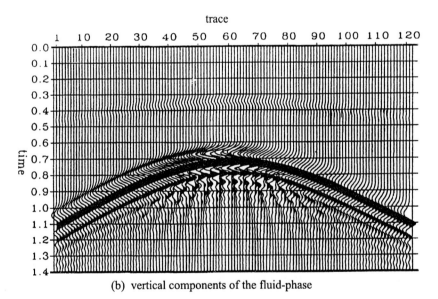

(b) vertical components of the fluid-phase

Fig. 14b. Synthetic seismorgan of the fluid-phase on the ground surface for waves in a two-phase medium excited by an embedded explosion center

The decoupling direct method has been successfully tested by numerical experiments in various cases, as demonstrated in the above. In order to understand the performance of the simple and practical method, accuracy and numerical stability of the method will be further clarified in association with studies on the governing equations of the interior nodes and the boundary nodes in the following two chapters, respectively.

4. WAVE MOTIONS IN DISCRETE GRIDS

Errors of the governing equations of interior nodes in replacing the corresponding partial differential equations may be clarified by the laws of wave motion in the discrete grids and their differences from those in the corresponding continuous media. Fundamental features of waves in the discrete grids have been studied by Liao and Liu (1980), Liu and Liao (1989, 1990, 1992a, 1992b) and by Guan and Liao (1995) for one-, two- and three-dimensional models. Major results in association with the decoupling direct method are summarized below.

4.1 Applicable Scope of Finite Element Modeling

It is conditional that finite elements can be used to simulate waves in the corresponding continuous media. One important condition is that an accurate simulation can be realized only within a frequency band specified by a high frequency limit ω_c. This conclusion may be first illustrated by using a simple one-dimensional model. The partial differential equation of wave motion is written

$$\frac{\partial^2 u}{\partial t^2} = c_0^2 \frac{\partial^2 u}{\partial x^2} \tag{32}$$

where the constant c_0 is the wave speed in the continuous model. It is well-known that a harmonic wave with wave number ω / c_0 satisfies Eq. (32) for any frequency ω. If Eq. (32) is discretized in space by finite elements of space step Δx and in time by central finite differences of time step Δt, the equation of motion of the discrete model in the frequency domain is written

$$U_{j+1} + 2bU_j + U_{j-1} = 0 \tag{33}$$

where U_j represents the amplitudes at $x = j\Delta x$, $j = 0, \pm 1, \pm 2 \ldots$, b is a function of ω, $\Delta \tau = c_0 \Delta t / \Delta x$, $\omega_N = \pi / \Delta t$ and the mass matrix, which may be either lumped or consistent. Analysis of Eq. (33) yields that wave motions satisfying the equation are strongly frequency-dependent. The frequency dependence may be clarified via two characteristic frequencies: Nyquist frequency ω_N and cut-off frequency ω_u. The time discretization leads to ω_N, and wave motions in the discrete model at any frequency $\omega > \omega_N$ would appear at a lower frequency within the band $(0, \omega_N)$. This frequency aliasing implies that the discrete model cannot be used to simulate wave motions at any frequency $\omega > \omega_N$. The time and space discretization leads to another characteristic frequency ω_u which is less than ω_N. Wave motions are dispersed in $0 \le \omega \le \omega_u$. The dispersion error increases with ω increasing from $\omega = 0$ at which the dispersion does not exist. As $\omega > \omega_u$ the waves are reduced to the parasitic oscillation in the frequency range of $\omega_u \leq \omega \leq \omega_N$. The above conclusion will be explained later in some detail. Thus wave motions in a desired band of $0 \leq \omega \leq \omega_c$, can be simulated with a given accuracy if ω_u is properly larger than ω_c. The last condition may be satisfied by refining the grid to enlarge ω_u for a desired ω_c or by restricting ω_c for a given grid.

In the multi-dimensional cases, the similar condition holds. That is wave motions in a continuous model can and only can be simulated with a desired accuracy at frequencies lower than a high frequency limit ω_c. Differences between the one- and multi-dimensional discrete models mainly lie in the undesired wave motions at frequencies higher than ω_c. For example, besides dispersion and parasitic oscillation the anisotropy and the polarization drift appear in wave motions at the high frequencies in the multi-dimensional models, whose corresponding continuous models are even isotropic, homogeneous and linearly elastic. The anisotropy is shown by the direction-dependent dispersion curves even for the anti-plane case. Besides the anisotropy, the polarization drift of the waves may appear for the P- and SV-waves in the in-plane case. The drift may be described by the deviation of the angle of the wave propagating direction with respect to the particle vibrating direction in the discrete grid from that in the corresponding continuum. The drift for the in-plane case results in that the direction of the particle vibration in the discrete model is no longer parallel or perpendicular to the direction of wave propagation for the P- or SV-plane wave, respectively. The polarization drift might also appear in the SH type of waves in the 3-D discrete model. However, all the undesired wave motions do not affect the desired accuracy of the numerical simulation in the required frequency band of $\omega < \omega_c$. And the undesired high frequency components can be easily filtered out if necessary.

4.2 Remarks on Lumped-mass Finite Elements

The governing equations of interior nodes may be formulated by the lumped-mass or the consistent-mass finite elements. The former has been suggested for use since the decoupling direct method was first proposed by Liao (1984). The suggestion was originally based on two considerations: (1) the lumped-mass model yields a spatial decoupling scheme, which is substantially more efficient than the spatial coupling scheme resulting from the consistent-mass model; (2) the suggested model is more reasonable in the sense of physics that the wave speed must remain finite. This suggestion has then been further justified by comparing accuracies and numerical stabilities of the two types of finite elements in the numerical simulation of wave motion. The results of the comparisons are summarized below. The conclusion drawn from the studies of accuracy comparison is that the lumped-mass models are not less accurate than the corresponding consistent-mass models as far as the numerical simulations of wave motion in a desired frequency band are concerned. In fact, the numerical errors resulting from all sorts of discretization errors including dispersion, anisotropic, polarization drift, etc. within the desired frequency band for the lumped-mass models are not larger than those for the corresponding consistent-mass models. The stability criterion in the step-by-step numerical integration of the governing equations of interior nodes may be set up in the frequency domain according to the dispersion equation. The criterion is stated as that the numerical stability is guaranteed by the frequency in the dispersion equation being real, if the wave number is independent and real. The stability criteria thus derived for the lumped-mass and consistent-mass models in a few typical cases are as follows:

in the one-dimensional case

$$
\left.
\begin{aligned}
c_0 \Delta t &\leq \Delta x && \text{for lumped - mass} \\
c_0 \Delta t &\leq \Delta x / \sqrt{3} && \text{for consistent - mass}
\end{aligned}
\right\}
\tag{34}
$$

in the anti-plane case

$$
\left.
\begin{aligned}
c_s \Delta t &\leq \Delta x && \text{for lumped - mass} \\
c_s \Delta t &\leq \Delta x / \sqrt{6} && \text{for consistent - mass}
\end{aligned}
\right\}
\tag{35}
$$

in the in-plane case

$$
\left.
\begin{aligned}
c_p \Delta t &\leq \Delta x && \text{for lumped - mass} \\
c_p \Delta t &\leq \left[3\left(1 + c_x^2 / c_p^2\right)\right]^{-1/2} \Delta x && \text{for consistent - mass}
\end{aligned}
\right\}
\tag{36}
$$

Equations (35) and (36) correspond the homogeneous, isotropic and linearly elastic medium, which is discretized by uniform square finite elements, c_s and c_p are the S-wave and P-wave speed, respectively, Δx is the size of the finite elements. It is observed that in all the cases the restrictions imposed on the time step Δt by the criteria for the consistent-mass models are harsher than those for the lumped-mass ones, respectively.

4.3 Forms of Wave Motion in Discrete Grids

Understanding general forms of wave motion in discrete grids is important not only for analyzing accuracies of the discrete simulations of wave motion but also for studying numerical stabilities of the local ABC coupled with the governing equations of interior nodes.

The one-dimensional case is first addressed. The central difference approximation of Eq. (32) is written

$$u_j^{p+1} - 2u_j^p + u_j^{p-1} - \Delta\tau2(u_{j+1}^p - 2u_j^p + u_{j-1}^p) = 0 \qquad (37)$$

where $\Delta\tau=c_0\Delta t/\Delta x$.The general form of solution of Eq. (37) is written

$$u_j^p = (e^{s\Delta t})^p (e^{\gamma\Delta x})^j \qquad (38)$$

Substituting Eq. (38) into Eq. (37) leads to the following dispersion relation,

$$sh^2\left(\frac{\gamma\Delta x}{2}\right) = \frac{1}{\Delta\tau^2} sh^2\left(\frac{s\Delta t}{2}\right) \qquad (39)$$

If $\Delta x \to 0$ and $\Delta t \to 0$, Eq. (39) yields

$$\gamma^2 = \frac{1}{c_0^2} s^2 \qquad (40)$$

Equation (40) is the dispersion equation in the corresponding continuous case. The general form of solution of Eq. (32) is written

$$u = e^{st} e^{\gamma x} \qquad (41)$$

For understanding similarities and differences between the forms of wave motion in the discrete and continuous models, we describe the manner in which g varies as s varies with Re $s \geq 0$ for the discrete and continuous cases. First consider the case where s lies in the closure of the first quadrant, then sh^2 (sDt/2) and s^2/c_0^2 both lie in the upper half-plane, and g^2 and sh^2 (gDx/2) must also lie in the upper half-plane. One solution of g lies in the third quadrant, and the other lies in the first quadrant for both Eq. (39) and Eq. (40). As for the case where s lies in the fourth quadrant, the two solutions of g lies in the fourth and second quadrants for both Eq. (39) and Eq. (40). Thus the general forms of wave motion in the discrete and continuous models are similar. Particularly, the following two basic forms are in common: the form with Re $s > 0$ corresponds to solutions exponentially increasing with time and the form with Re $s = 0$ corresponds to solutions of purely oscillatory waves. However, substantial differences do exist in the wave motion forms of the two models. We denote the complex numbers s and g by

$$s = r + \omega i \quad \text{and} \quad \gamma = \beta + ki \qquad (42)$$

where ω and k are the circular frequency and wave number, respectively. Having substituted Eq. (42) into Eq. (39) and noticed the periodic property of the hyperbolic functions with respect to k or ω, it is observed that any wave form having wave number $k > 2\,k_N$ would appear as one with smaller k' in the wave number band

$$0 \le k' \le 2k_N \tag{43}$$

where $k_N = \tau / \Delta x$ is called the Nyquist wave number and that any wave form having frequency $\omega > 2\omega_N$ would appear as one with ω' in the frequency band

$$0 \le \omega' \le 2\omega_N \tag{44}$$

where $\omega_N = \pi / \Delta t$ is the Nyquist frequency. The wave number aliasing introduced by the space discretization and the frequency aliasing introduced by the time discretization imply that the high frequency errors in the discrete models are unavoidable, but that the errors can be reduced to an allowable extent by properly reducing Δx and Δt. Consider the purely oscillatory wave of Re $s = 0$, which is the most important case from a physical point of view, Eq. (39) is written

$$\sin^2 \frac{\omega \Delta t}{2} = \Delta \tau^2 \sin^2 \frac{k \Delta x}{2} \tag{45}$$

Triangle functions of k and ω in Eq. (45) imply that the fundamental branches of wave number and frequency may be written

$$\left. \begin{array}{l} 0 \le k \le k_N \\ 0 \le \omega \le \omega_N \end{array} \right\} \tag{46}$$

because the values of $\sin(\omega \Delta t / 2)$ at $\omega = \omega_N \pm \omega'$ are equal. Furthermore, the numerical stability requires ω must be real for any real number of k, thus

$$\Delta \tau^2 \le 1 \tag{47}$$

Under the condition of (47), Eq. (45) yields the cut-off frequency

$$\omega_u = \frac{2}{\pi} \omega_N \arcsin \Delta \tau \tag{48}$$

and the related conclusion on accuracy of the discrete simulation of wave motion as pointed out earlier. For late reference, the solutions of the governing equations of the interior nodes for the multi-dimensional, isotropic and linearly elastic media have the following general form for a rectangular coordinate system,

$$u^p_{j,m_1} = (e^{\gamma \Delta x})^j (e^{il_y \Delta y})^m (e^{s \Delta t})^p \mathbf{q} \tag{49}$$

for the uniform rectangular finite elements in the two-dimensional case, where q is a complex vector with two components, and

$$u^p_{j,m_1,m_2} = (e^{\gamma \Delta x})^j (e^{il_y \Delta y})^{m_1} (e^{il_z \Delta z})^{m_2} (e^{s \Delta t})^p \mathbf{q} \tag{50}$$

for the uniform cubic finite elements in the three-dimensional case, where q is a complex vector with three components. The dispersion relations and the corresponding various forms of wave motion in the multi-dimensional cases are more complicated than those in the one-dimensional case. However, the basic relationship between solutions of the discrete and the corresponding continuous models in the multi-dimensional cases is similar to that in the one-dimensional case. That is, the wave motions in a continuous model can always be simulated to a required accuracy in a required frequency band with a upper limit ω_c by an appropriate discrete model. Errors in the discrete simulation resulting from various wave motions at undesired frequencies higher than ω_c are unavoidable, but can be reduced to an extent acceptable for the accuracy required by a reasonably designed discretization.

5. LOCAL ARTIFICIAL BOUNDARY CONDITIONS

Although local ABCs have been proposed with resource to different mathematical or physical principles Givoli (1991), Kausel (1988), Wolf (1986, 1992), Liao (1997), the methodologies of their formulations may be classified into two major types. The first one is prevalent, which is started with a system of partial differential equations of wave motion and the local ABCs are then derived from the particular equation system. For example, those proposed by Clayton and Engquist (1977, 1980), Halpern and Trefethen (1988), Higdon (1986, 1987, 1991, 1992). Although this methodology has led to valuable results, which must be developed separately for different differential equation systems. In addition, efforts are required occasionally in dealing with special boundary points, which are also located on the actual boundaries, such as the free surface and interfaces in a layered medium, because the actual boundary effects are not included in the differential equation system. Another methodology suggested and developed by Liao and his co-workers(1982-1996) is based solely on wave propagation and is not geared to any particular system of differential equations. A concise comparison between the two methodologies is listed in Table 1.

Table 1. A comparison of two methodologies for formulating local ABCs

Methodology	Prevalent	Suggested
Starting point	A particular set of differential equations for a specific infinite- linear- exterior-medium and for a specific motion mode in the medium.	A general expression of the one-way waves for any infinite-linear-exterior-medium and for all sorts of motion modes in the medium.
Formulation	A particular set of one-way wave equations in the continuous form is derived from the particular set of differential equations, and the analytical local ABC is then discretized by finite differences.	A unified transmitting formula in the discrete form is derived from a direct replacement of the general expression, and the discrete local ABC is expressed by the nodal values via an interpolation scheme.
Applicability	With modification as the local ABC applies to different cases and to special boundary points each case.	Without modification as the local ABC applies to all the cases and to all the boundary points in each case

The remainder of the Chapter is organized as follows. MTF is first formulated using the suggested methodology, relationship of MTF with other major local ABCs derived by the prevalent methodology is then discussed in Section 5-1. In Section 5-2, accuracy of MTF is discussed. The stable implementation of MTF is discussed in Section 5-3.

5.1 Multi-transmitting Foumula

5.1.1 Formulation of MTF

Since ABCs are used to prevent the wave reflection at the edge of the computational box, a natural approach to set up the boundary conditions is a direct simulation of the one-way wave motion, which are passing through the edge from the interior to the exterior of the box. The starting point to perform the simulation is a general expression of the outgoing waves in all linear exterior media and for any sorts of motion modes in the media. Having considered the local characteristic of wave motion, the expression may be presented in the local sense, that is, it describes the one-way wave motion merely in the neighborhood of a boundary point under consideration and within a short time interval. Suppose that a point o on the boundary is considered (Fig. 15). Let the x-axis pointing to the exterior of the box pass this point and be perpendicular to the boundary. The outgoing waves impinging upon the boundary point in the local sense may be generally expressed by their apparent propagation along the x-axis,

$$u(t,x) = \sum_{j=1} f_j(c_{xj}t - x) \tag{51}$$

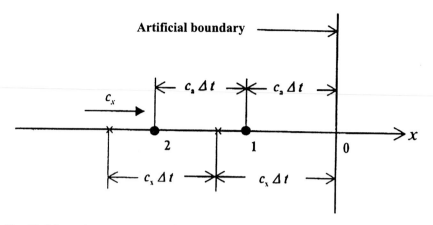

Fig. 15. Schematic representation of the apparent propagation of a one-way wave along the x-axis approaching the artificial boundary

where $u(t, x)$ is a function of t and x and consists of arbitrary number of one-way waves $f_j(c_{xj} t - x)$, each of which propagates along the x-axis with an apparent speed c_{xj}, f_j are arbitrary functions. It can be shown below that a system of local ABCs can be derived via a direct simulation of Eq. (51) using the space-time extrapolation without knowing any detail of Eq. (51). The apparent propagation of one of the one-way waves in Eq. (51) is written as (neglecting the subscript j),

$$u(t,x) = f(c_x t - x), \tag{52}$$

where f is an arbitrary and unknown function and the apparent speed c_x is an unknown real positive number. Using the forward characteristic, $c_x t - x$, it is easy to see that

$$u(t + \Delta t, x) = u(t, x - c_x \Delta t), \tag{53}$$

where Δt is the time step. Although this type of space-time extrapolation in association with Kirchhoff's formula may be used to set up a rigorous ABC in some simple cases (Liu and Miksis, 1986), it is impossible to derive ABCs directly from Eq. (53) in general, because c_x and f both are unknown. However, if an artificial wave speed c_a is introduced to replace the unknown physical speed c_x in Eq. (53), local ABCs can be derived as follows. First, Eq. (53) is replaced by

$$u(t + \Delta t, x) = u(t, x - c_a \Delta t) + \Delta u(t + \Delta t, x), \tag{54}$$

where $\Delta u(t + \Delta t, x)$ expresses the error caused by c_a replacing c_x, and Eq. (54) yields

$$\Delta u(t + \Delta t, x) = u(t + \Delta t, x) - u(t, x - c_a \Delta t) \tag{55}$$

Substituting Eq. (52) into Eq. (55) leads to that $\Delta u(t + \Delta t, x)$ would be a function of c_x, c_a, Δt and $c_x t - x$. If c_x, c_a and Δt are fixed, we recognize

$$\Delta u(t + \Delta t, x) = f_1(c_x t - x) \tag{56}$$

where f_1 is a function of the forward characteristic. Eq. (56) means that the error $\Delta u(t + \Delta t, x)$ is also a wave propagating along the x-axis with the same speed c_x. Recognition of $\Delta u(t + \Delta t, x)$ being a wave is the point in formulation of MTF. Following Eq. (54) the error wave may be written as

$$\Delta u(t + \Delta t, x) = \Delta u(t, x - c_a \Delta t) + \Delta^2 u(t + \Delta t, x) \tag{57}$$

Having replaced t and x in Eq. (55) by $t - \Delta t$ and $x - c_a \Delta t$, respectively, the first term in the right-hand side of Eq. (57) is written as

$$\Delta u(t, x + c_a \Delta t) = u(t, x - c_a \Delta t) - u(t - \Delta t, x - 2c_a \Delta t) \tag{58}$$

and the error term in Eq. (57) is written:

$$\Delta^2 u(t + \Delta t, x) = \Delta u(t + \Delta t, x) - \Delta u(t, x - c_a \Delta t) \tag{59}$$

Substitution of Eq. (57) into Eq. (54) yields

$$u(t + \Delta t, x) = u(t, x - c_a \Delta t) + \Delta u(t, x - c_a \Delta t) + \Delta^2 u(t + \Delta t, x) \tag{60}$$

It is easy to see that $\Delta^2 (t + \Delta t, x)$ and the higher-order error terms similarly introduced are all functions of the forward characteristic $(c_x t - x)$. Following the above discussion we obtain

$$u(t + \Delta t, x) = u(t, x - c_a \Delta t) + \sum_{n=1}^{N-1} \Delta^n u(t, x - c_a \Delta t) + \Delta^N u(t + \Delta t, x) \tag{61}$$

where

$$\Delta^n u(t, x - c_a \Delta t) = \Delta^{n-1} u(t, x - c_a \Delta t) - \Delta^{n-1} u(t - \Delta t, x - 2c_a \Delta t) \tag{62}$$

$$\Delta^N u(t + \Delta t, x) = \Delta^{N-1} u(t + \Delta t, x) - \Delta^{N-1} u(t, x - c_a \Delta t) \tag{63}$$

Next, suppose that the origin of the x-axis coincides with the boundary point o under consideration. Substituting $x = 0$ and $t = p\Delta t$ into Eq. (61) and neglecting the higher-order term $\Delta^N u(t + \Delta t, 0)$ yield

$$u((p+1)\Delta t, 0) = \sum_{n=1}^{N} \Delta^{n-1} u(p\Delta t - c_a \Delta t) \tag{64}$$

where $\Delta^0 u = u$, the integer p stands for the time level. If c_a used for simulating the error wave $\Delta^{n-1} u(t + \Delta t, x)$ is replaced by c_{an}, Eq. (64) may be written

$$u((p+1)\Delta t, 0) = \sum_{n=1}^{N} \Delta^{n-1} u(p\Delta t, -c_{an} \Delta t) \tag{65}$$

Since Eq. (64) or (65) is valid for each one-way wave in Eq. (51), it remains valid for the wave system expressed by Eq. (51) if a common artificial speed c_{an} is used for transmitting the error wave $\Delta^{n-1} u(t + \Delta t, x)$ of each term in Eq. (51). The discrete quantities of $u(t, x)$ in Eq. (64) or (65) stand now for the total displacement;. contributed by all the incident one-way waves. Although Eq. (65) seems more general than Eq. (64), the latter is more practical for use. If a single artificial speed is used, it is obtained using the Eq. (62) that

$$\Delta^n u(p\Delta t, - c_a \Delta t) = \sum_{j=1}^{n+1} (-1)^{j+1} C_{j-1}^n u((p+1-j)\Delta t, - jc_a \Delta t) \tag{66}$$

where C_j^n are the binomial coefficients. Substituting Eq. (66) into Eq. (64) yields

$$u((p+1)\Delta t, 0) = \sum_{j=1}^{N} A_{N,j} u((p+1-j)\Delta t, - jc_a \Delta t) \tag{67}$$

$$A_{N,j} = (-1)^{j+1} C_j^N \tag{68}$$

Equation (67) is the MTF, which provides a family of local ABCs with variable approximation orders. The formula was derived in the early 80s for the incidence of plane waves, therefore, the applicable scope of MTF was not fully revealed (Liao, 1996a). The present derivation is solely based on Eq. (51), the general expression of linear one-way waves, thus the applicable scope of MTF is clear: it can be applied to all cases where Eq. (51) is valid in principle.

Having coupled a space-extrapolation to the space-time extrapolation, MTF may be generalized as (Liao, 1996a, 1997)

$$u_0^{p+1} = \sum_{j=1}^{N} A_{N,j} u_j^{p+1-j} + \sum_{q=1}^{M} A_{M,q} u_q^{p+1} \sum_{j=1}^{N} \sum_{q=1}^{M} A_{N,j} A_{M,q} u_{q+j}^{p+1-j} \qquad (69)$$

The approximation order of Eq. (69) is given by $N + M$, A_{Nj} and $A_{m,q}$ are determined by Eq. (68). Eq. (69) is reduced to MTF as the integer $M = 0$.

5.1.2 Relationships between MTF and other local ABCs

Since all local ABCs are essentially one-way wave equations, a relationship must exist between the boundary conditions derived by the two methodologies. To clarify the relationship, the differential form of MTF is first derived from Eq. (67) as Δt approaches zero (Chen et al., 1995, Liao, 1996),

$$\left(\frac{\partial}{\partial t} + c_a \frac{\partial}{\partial x} \right)^N u = 0 \qquad (70)$$

It can be shown that Eq. (65) as Δt approaches zero reduces to the following equation

$$\left[\prod_{j=1}^{N} \left(\frac{\partial}{\partial t} + c_{a\partial} \frac{\partial}{\partial x} \right)^N \right] u = 0 \qquad (71)$$

Similarities and differences between MTF in the above differential forms and the major types of local ABCs derived by the prevalent methodology, that is, the Clayton-Engquist's and Higdon's ABCs, are discussed as follows.

1. MTF and Clayton-Engquist ABC: It has been shown that Eq. (70) is equivalent to the Clayton-Engquist ABC in the acoustic case if c_a is equal to the acoustic wave speed (Kausel, 1988). However, the artificial speed c_a in MTF can be adjusted so that perfect transmission may be achieved by MTF for a particular angle of incidence. Thus the Clayton-Engquist ABC is a special case of MTF as the artificial speed is equal to the acoustic wave speed. A more important difference between the two types of ABCs is in their applicability. The Clayton-Engquist ABC in the acoustic case is derived from the paraxial rational expansion of the dispersion relation of the acoustic equation. The derivation cannot be directly extended to even the case of vector waves in homogeneous, isotropic and elastic media because the vector wave equations are not uniquely specified from their dispersion relations. In order to generalize the Clayton-Engquist ABC to the vector wave case, a general form of the paraxial approximation must be assumed using the hint provided in the acoustic case and the general form is then specified by matching it to the full elastic wave equations (Clayton and Engquist, 1977). The ABCs thus derived for the elastic case are more complicated than that for the acoustic case. On the contrary, MTF applies without modification to the vector wave case and all cases in which Eq. (51) can express in principle the one-way waves concerned.

2. MTF and Higdon ABC: Started with the acoustic equation and the plane wave assumption, Higdon (1986) presented a set of local ABCs:

$$\left[\prod_{j=1}^{N}\left(\cos\theta_j \frac{\partial}{\partial t} + c \frac{\partial}{\partial x}\right)^N\right] u = 0 \tag{72}$$

where θ_j is the angle of incidence, $|\theta_j| < \pi/2$, c is the acoustic speed. Since Eq. (72) is derived by the prevalent methodology, the physical speed c and the incident angle θ_j are inherent in Eq. (72). The ABC in the acoustic case, Eq. (72), has been then generalized to the vector waves in the isotropic and elastic media via some modifications of the physical wave speed c and the parameters in the equation (Higdon, 1987, 1991, 1992). It is interesting to notice that the Higdon's generalization is still following the prevalent methodology, but the results are much simpler than the Clayton-Engquist's. This may be explained by a relation of the Higdon's theory with Eq. (71) under the p Lane wave assumption. This relation is easily seen by substituting $c_{aj} = c/\cos\theta_j$ into Eq. (71) for the acoustic case. As regards the Higdon's generalization of Eq. (72) to the vector wave cases under the plane wave assumption, it may be greatly simplified by the following consideration. If the physical speed c is replaced by the artificial speed c_a and the one-dimensional apparent wave propagation of the form $u=f(ct-x)$ satisfies the first-order differential relation

$$\left(\frac{\partial}{\partial t} + c_a \frac{\partial}{\partial x}\right) u = 0, \tag{73}$$

a combination of the first-order differential relations is identical to Eq. (69) based on the Eq. (51).

5.2 Spurious Reflection Analysis

The spurious reflection from an artificial boundary may be described by a reflection coefficient R defined by

$$R = \left| U_0^R / U_0^I \right| \tag{74}$$

where U_0^I and U_0^R are the amplitudes at the boundary point $x=0$ of a harmonic plane wave of incidence and the wave reflected from the boundary, respectively. The reflection coefficient R is usually defined in the steady-state sense. This definition is also correct for the transient-state computation if it is performed by the Fourier transform from the frequency domain to the time domain. However, if the transient computation is performed by the step-by-step integration in the time-space domain, justification of R defined in the steady-state case is no longer obvious. For this reason, the definition of R in the transient-state case has been suggested to account for the spurious reflection in the step-by-step integration (Liao,1996a). In this case, it is assumed that the wave field merely consists of the incident wave in implementation of the local ABCs. Suppose that a plane acoustic wave with a speed c impinges upon the boundary at an angle of incidence θ with respect to the x-axis, the explicit formulas of R have been obtained (Liao, 1996a),

$$
R = \left| 2\sin\left(\pi \frac{\Delta t}{T} \left(\frac{c_a}{c} \cos\theta - 1 \right) \right) \right|^N \left| 2\sin\left(\pi \frac{\Delta t}{T} \frac{c_a}{c} \cos\theta \right) \right|^M , \text{ in the transient-state} \tag{75}
$$

$$
R = \left| \frac{\sin(\pi(\Delta t/T)(1-(c_a/c)\cos\theta))}{\sin(\pi(\Delta t/T)(1+(c_a/c)\cos\theta))} \right|, \text{ in the steadystate} \tag{76}
$$

where T stands for the vibration period. Numerical results of R versus θ in the entire range of $0 \le \theta \le 90°$ in the two extreme cases are shown in Fig. 16 for Eq. (69), $c_a = c$ and $\Delta t/T = 0.1$. Conclusions drawn from the figure are as follows. The first, a coupling of the space extrapolation to MTF does reduce substantially R in the transient case for the large incident angles. However, the coupling has no effect on R in the steady case. The second, considerable differences between R values in the two cases appear only for the large incident angles. The differences become more obvious as $\Delta t/T$ approaches zero. In fact, R in the transient-state case approaches zero with an order of magnitude of $(\Delta t/T)^{N+M}$ within the entire range $|\theta| \le \pi/2$, as shown by Eq. (75) with $\Delta t/T \to 0$; while R in the steady-state case is reduced to

$$
R = \left| \frac{c - c_a \cos\theta}{c + c_a \cos\theta} \right|^N \tag{77}
$$

which means that R in this case would not be further reduced as $\Delta t/T \to 0$, as shown by the solid line in Fig. 16(b). Since the differences of R values in the two extreme cases are significant for the large angles of incidence, a question arises: which R would be more appropriate for describing the spurious reflection from the boundary in the step-by-step integration? To answer this question, a realistic

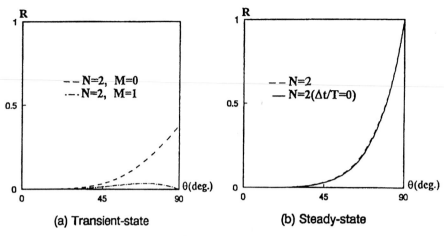

(a) Transient-state   (b) Steady-state

Fig. 16. R in the transient-state and steady-state cases for
Eq. (69), $c_a = c$ and $\Delta t/T = 0.1$

reflection analysis has been performed via a numerical experiment of the reflection for incidence of a plane acoustic wave. The numerical experiment is summarized as follows. Having discretized a rectangular computational domain with uniform square finite elements, the MTF is imposed on its one edge, the artificial boundary, whose center is the boundary point under consideration (Fig. 17). The displacements of the nodal points on the other edges are assigned by the given plane acoustic wave of incidence. Suppose that the incident wave is of an impulse form, the numerical simulation is started with $t = 0$, at which the wave front of the incident wave just arrives at the boundary point. Having assigned the initial values of the entire grid in terms of

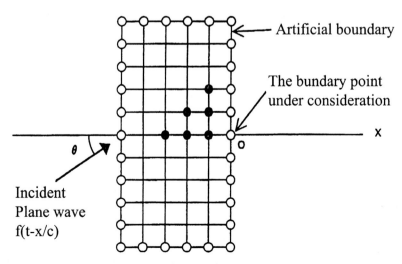

Fig. 17. Model for definition of R in the realistic case

the incident wave, the numerical simulation is performed by the step-by-step integration. If the computed displacement of the boundary point is denoted by $u(t, \theta)$, which is a function of time t and the angle θ of incidence, the total error of the computed displacement is

$$e(t,\theta) = f(t) - u(t,\theta) \tag{78}$$

where $f(t)$ is the exact displacement of the incident wave of the boundary point. Since all sorts of numerical errors, except that caused by the ABC, have been reduced to a negligible extent by using a sufficiently large size of the computational region, refining the grid and selecting the computing duration and the time increment, $e(t, \theta)$ may represent the realistic spurious reflection mainly caused by the artificial boundary. Substituting the Fourier amplitude spectra of $e(t, \theta)$ and the incident wave into Eq. (74) leads to the realistic reflection coefficients. Fig. 18 presents a comparison of the values of R in the realistic case with those of R in the two extreme cases for Eq. (69), $c_a = c$ and $\Delta t / T = 0.025$. The close relation between R values in the transient and realistic cases shown in Fig. 18 speaks that R in the transient case does play a more important role than R in the steady case in governing accuracy of the local ABC in the step-by-step integration. The conclusion has been demonstrated by a number of numerical experiments, for example, Fig. 9 in this lecture notes, where the excellent results is well explained by R in the transient case, and impossible to be explained by R in the steady case. This conclusion also greatly simplifies analysis of the spurious reflection, because the formulation of R in the transient case does not involve the reflection waves, which would make formulation of R in the steady-

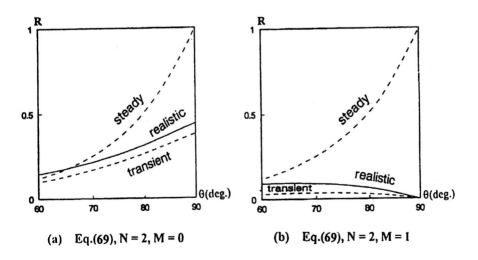

(a)   Eq.(69), $N = 2$, $M = 0$ (b)   Eq.(69), $N = 2$, $M = 1$

Fig. 18. Comparison of R in the realistic case with R in two extreme case for Eq. (69), $c_a = c$ and $\Delta t / T = 1/40$

state case quite involved in the vector wave cases and other comparatively complicated situations. Therefore, Eq. (75) is correct for all sorts of one-way waves of incidence though it was originally derived merely for the acoustic case.

5.3 Stable Implementation of Multi-transmitting Formula

The numerical instability problems are occasionally encountered in implementation of the local ABC in the direct numerical simulation. Studies on the problems aim at providing a stable implementation of the ABC without significant effects on accuracy of the numerical simulation in the desired frequency band. In this section, the mechanism of numerical instabilities in implementation of MTF are first clarified by GKS criterion and the high-frequency stability analysis using a simple one-D model, and stable implementation schemes are then presented for use.

5.3.1 Interpretation of numerical instabilities of local ABCs

1. Interpretation drawn from GKS criterion: The numerical stability of the local ABCs is commonly interpreted by GKS criterion developed by Gustafsson, Kreiss, and Sundstrom (1972). GKS criterion for a discrete initial-boundary value problem may be stated by that a discrete local ABC satisfies the GKS criterion if and only if the boundary condition is not satisfied by any nonzero solution of the governing equation of the interior nodes, which belongs to either of the following categories:

    (a) Solutions (38) ((49) or (50)) for which Re $s > 0$, and Re $\gamma < 0$ as $j \to \infty$,

    (b) Solutions that are limits of the solutions in (a) as Re s approaches zero through positive value (Higdon (1990)).

If solutions of (a) were allowed, then the initial-boundary valve problem would admit solutions that have exponential rates of growth with respect to t. Thus solutions of type (a) means the catastrophic exponential instabilities. If solutions of (b) were allowed, then the solutions may contain traveling waves, which carry energy propagating into the spatial domain $x > 0$. Thus existence of such waves would lead to a form of instability though that is less explosive than the kind associated with the solutions of type (a).

It can be shown that MTF satisfies GKS stability criterion except at zero frequency and zero wave number. This exceptional case can be coped with by adding a small positive number into the formula, as discussed later on. Since GKS criterion is related and only related with the local ABC at the boundary of a semi-infinite domain $x > 0$, it cannot explain completely the instabilities, which take place within a finite computational domain of several boundaries. New type of numerical instabilities of local ABCs would appear in the latter case, as addressed below.

2. Interpretation of high-frequency instabilities : Consider the one-dimensional wave propagation in the spatial domain $x \in = [0, L]$, which is excited by a given vibration at $x = L$ under the zero-initial condition. The wave motion is simulated by the decoupling numerical approach suggested in Chapter 3. The instability phenomenon encountered in the simulation is shown in Fig. 19(a). It can be seen that the instability is of the form of oscillation at high frequencies close to but less than the cut-off frequency ω_u, and that it does not always occur (Fig. 19(b)). In order to understand the phenomena, the exact solution of the discrete model has been derived in the frequency domain. It is found that the necessary and sufficient condition for the exact solution being finite is (Liao, 1996b)

$$|R| < 1, \quad \omega < \omega_u \tag{79}$$

where R is the reflection coefficient in the steady state. Fig. 20 shows the relation of the reflection factor $|R|^{1/N}$ versus ω for $\omega \le \omega_u$. This figure plus the stability condition (79) provide a full explanation of major features of the stability phenomena. First, the stability criterion derived from the condition (79) for the simple 1-D model is

$$\Delta \tau c_a / c_0 < 1.5 \tag{80}$$

The criterion (80) explains why does the high-frequency instability occur in Fig. 19(a), but not in Fig. 19(b). This is because $\Delta \tau = 1$, $c_a/c_0 = 2.0$ and $\Delta \tau = 1$, $c_a/c_0 = 1.2$ for Fig. 19(a) and Fig. 19(b), respectively, and the former does not satisfy (80), but the latter does. Second, the frequencies, at which the unstable oscillation takes place in Fig. 19(a), are within the range of $0.5 \omega_n < \omega < \omega_u$, as counted directly from the time domain numerical results; and the unstable frequency range coincides with those frequencies at which $|R| > 1$, as shown by the curve of $\alpha = c_a/c_0 = 2.0$ in Fig. 20. In short, the instability comes from the amplification ($|R| > 1$) at the artificial boundary and the multi-reflection of the high frequency waves in the finite computational region.

5.3.2 Stable implementation of MTF

Two types of measures for stable implementation of MTF in the numerical computations are addressed for satisfying GKS criterion and eliminating the undesired high frequency components, respectively.

1. *Technique for satisfying GKS criterion:* In order to guarantee GKS criterion at the zero frequency and zero wave number, Eq. (70) may be modified as

$$\left(\frac{\partial}{\partial t} + c_a \frac{\partial}{\partial x} + \gamma \right)^N u = 0 \tag{81}$$

where the small quantity $\gamma > 0$. Effects of this modification on accuracy of MTF

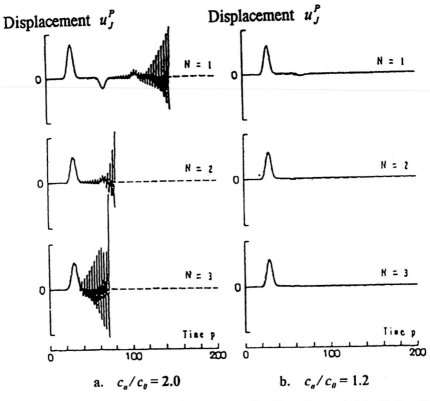

Fig. 19. Numerical solutions of the motion at the artificial boundary node ($J = 20$, $\Delta \tau = 1$)

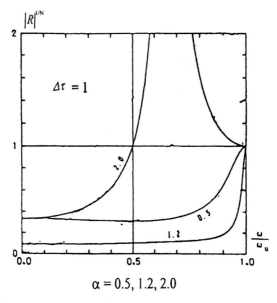

$$\alpha = 0.5, 1.2, 2.0$$

Fig. 20. The reflection factor in the steady case for the one-dimensional model

can be observed via the following solution of Eq. (81),

$$u = f(t - x/c_a)e^{-\gamma t} \tag{82}$$

where $u = f(t-x/c_a)$ satisfies Eq. (81) for $\gamma = 0$, that is, MTF in the analytical form. Suppose that T is the desired computational length, if γ is sufficiently small so that γT is a small number, then effects of the modification on the original solution are not significant in the time period $t \leq T$. In the discrete form, the corresponding modification of MTF can be obtained similarly.

2. Technique for eliminating the high-frequency instabilities: Because amplification of the undesired high frequency waves at the artificial boundary and the multiple reflection of the waves in the finite computational region interpret the high frequency instability, it follows several schemes to filter out the undesired high frequency components before implementing the local ABC. Several alternatives for the filtering are addressed by Liao(1966b). The most practical scheme for this purpose is introducing a material damping into to the equations of the interior nodes, which is proportional to the derivative of the strain with respect to time, that is, the damping coefficient matrix is proportional to the stiffness matrix. The damping force may be introduced into the computations naturally by consideration of the damping effect in realistic materials. However, this scheme may be applied to problems of wave motion in the perfectly elastic media as well. In this case, such damping effects are artificially introduced into the vibration modes of a system under consideration to eliminate the undesired high-frequency components but with negligible effects on the response of the desired low-frequency modes. A detailed demonstration of this scheme will be presented in another report.

6. CONCLUSIONS

Structure-medium dynamic interaction problems have been intensively studied in various fields in the 20th century. It is not only in the frequency domain but also in the time domain that the techniques to cope with these problems have been significantly developed. The techniques have been successfully applied to dealing with many forward problems in these fields, including some complicated nonlinear dynamic interaction problems of large-scale. For example, in the earthquake resistant design of Xiaowan arch dam of a height 292 m, which is located in an area of high seismicity in the southwest of China, the following factors are required to be considered in estimation of the seismic response of the dam: (1) effects of the infinite rock medium on the dam response; (2) the complicated geometrical configuration of the dam and the irregular geological structures adjacent to the dam; (3) the nonlinearities taken place in the joints of the dam and the faults in the adjacent rock medium; (4) effects of the water in the reservoir on the dam response. The three-dimensional nonlinear dynamic interaction of the arch damfoundation-reservoir system has been analyzed systematically by Chen etc. (1997) using the decoupling direct method with MTF of $N=2$. Having considered all the factors just mentioned the authors have provided scientific data for the engineering design of the dam and convinced that the time domain technique is not only reliable but also operable for dealing with the dynamic interaction problems, even for extremely complicated cases of large scale, such as Xiaowan arch dam. However, future work is still needed to improve further the computational efficiency of the dynamic interaction analyses as well as their accuracies. This is because the inverse problems of dynamic interaction will become a real challenge in many engineering fields in the coming century, and each inverse problem generally requires performing a number of forward computations of the dynamic interaction, as mentioned earlier in this paper. Thus the dynamic interaction analysis must be performed very efficiently besides accurately. For this reason, more attention should be paid to studies on those techniques, which are promising to meet this need, such as the decoupling direct method. The potentiality of the decoupling direct method in this aspect is also implied by the interpretation of a complicated geological profile in an oil field using the decoupling direct method with MTF of $N=2$ (Shao and Lan, 1995, 1997).

Acknowledgements

This study has been supported by China Joint Seismological Science Foundation.

REFERENCES

Boore D. M., 1972. Finite difference methods for seismic wave propagation in heterogeneous materials. In Vol. 11, Methods in computational physics (ed. Bolt B. A.), Academic Press, New York.

Beskos D. E., 1987, Boundary element methods in dynamic analysis. Applied Mechanics Reviews, Vol. 40, 1-23.

Brebbia C. A., Telles J. C. F. and Wrobel L. C., 1984, Boundary element techniques. Springer. Berlin.

Banerjee P. K., 1994, The boundary element methods in engineering. McGraw-Hill, London.

Chen HQ. Du XL and Hou SZ, 1997, Application of transmitting boundaries to non-linear dynamic analysis of an arch dam-foundation-reservoir system. Dynamic Soil-Structure Interaction (ed. Zhang CH and J. P. Wolf), International Academic Publishers, Beijing, China, 115-124.

Chen N. and C. H. Chen, 1995, Relationship between Liao and Clayton-Engquist absorbing boundary conditions: Acoustic case, Bull. *Seisml. Soc. Amer.*, 85, 954-956.

Clayton, R. and B. Engquist, 1977, Absorbing boundary conditions for acoustic and elastic wave equations, Bull. *Seismol. Soc. Amer.* 67, 1529-1540.

Clayton, R. and B. Engquist, 1980, Absorbing boundary conditions for wave-equation migration, *Geophysics* 45, 895-904.

Dominguez J., 1993, Boundary elements in dynamics, Computational Mechanics Publications, Southampton.

Dasgupta, G., 1982, A finite element formulation for unbounded homogeneous media, *Journal of Applied Mechanics*, ASME, 49,136-140.

Engquist B. and A. Majda, 1979, Absorbing boundary conditions for the numerical simulation of waves, *Math Comput.* 31, 629-652.

Guan HM and Liao ZP, 1995, Elastic waves in 3-D discrete grids, *Acta, Mechanica Solida*

Sinica, Vol. 8, No. 4, 283-293.

Gustapsson, B., H. O. Kreiss, and A. Sundstrom, 1972, Stability theory of difference approximations for mixed initial boundary value problems, II, *Math. Comp.*, 26: 649-686.

Halpern. L. and L. N. Trefethen, 1988, Wide-angle one-way wave equations, *J. Acoust. Soc. Amer.* 84,1397-1404.

Higdon, R. L. 1986, Absorbing boundary conditions for difference approximations to the multi-dimensional wave equation, *Math. Comput.* 47,437-459.

Higdon, R. L. 1987, Numerical absorbing boundary conditions for the wave equation, *Math. Comput.* 49, 65-90.

Higdon, R. L., 1990, Radiation boundary conditions for elastic wave propagation: SIAM J. *Numerical Analysis,* 27, 831-870.

Higdon, R. L. 1991, Absorbing boundary conditions for elastic waves, *Geophysics*, 56(2), 231-241.

Higdon, R. L. 1992, Absorbing boundary conditions for acoustic and elastic waves in stratified media, *J. Comput. Phys.* 101, 386-418.

Kausel, E., 1988. Local transmitting boundaries. *Journal of Engineering Mechanics*, ASCE, 114, 1011-1027.

Kausel E., Roesset J. M. And Waas, G., 1975. Dynamic analysis of footings on layered media, *Journal of Engineering Mechanics*, ASCE, 101, 679-693.

Lan ZL, 1997, Finite element solutions of several equations of wave motion and their application to seismic prospecting, Ph.D Thesis, Institute of Mathematics, Chinese Academy of Sciences, (in Chinese)

Liao ZP, 1996a. Extrapolation non-reflecting boundary conditions, Wave Motion, 24, 117-138.

Liao ZP, 1996b. An introduction to wave motion theories in engineering. Academic Press, Beijing, China (in Chinese)

Liao ZP, 1984, A finite element method for near-field problems of wave motions in heterogeneous materials. *Earth. Eng. Eng. Vib.* 4(2), 1-14 (in Chinese)

Liao ZP, 1990, Seismic Microzonation, Seismological Press, Beijing, China (in Chinese)

Liao ZP, 1997, A decoupling numerical simulation of wave motion, Dynamic Soil-Structure Interaction (ed. Zhang CH and J. P. Wolf), International Academic Publishers, Beijing, China, 115-124.

Liao ZP and Liu JB, 1986, Elastic wave motion in discrete grids I, *Earthq. Eng. Eng. Vib.*, Vol. 6, No.2, 1-16 (in Chinese)

Liu JB and Liao ZP, 1989, Elastic wave motion in discrete grids II, *Earthq. Eng. Eng. Vib.*, Vol. 9, No. 2, 1-14 (in Chinese)

Liu JB and Liao ZP, 1990, Elastic wave motion in discrete grids III, *Earthq. Eng. Eng. Vib.*, Vol. 10, No. 2, 1-10 (in Chinese)

Liu JB and Liao ZP, 1992a, In-plane wave motion in finite element models, *Acta Mechanica Sinica*, Vol. 8 No. 1, 80-87.

Liu JB and Liao ZP, 1992b, Antiplane wave motion in finite element models, *Acta Mechanica Sinica*, Vol. 24, No. 2, 207-214 (in Chinese)

Luco J. E., 1986. Linear soil-structure interaction: a review, Applied mechanics division, ASME, 53:41-57.

Lu Ting and M. J. Miksis, 1984, Exact boundary conditions for scattering problems, *J. Acoust. Soc. Amer.* 80, 1825-1827.

Lysmer J. and Was G., 1972. Shear waves in plane infinite structures. *Journal of Engineering Mechanics Division.* ASCE, 98, 85-105.

Manolis G. D. and Beskos, D. E., 1988, Boundary element methods in elastodynamics. Unwin Hyman. London.

Shao XM, Lan ZL, 1995, Numerical simulation of the seismic wave propagation in inhomogeneous isotropic elastic media, *Acta Geophysica Sinica*, Vol. 38, Supplement 1, 55-72 (in Chinese)

Shao XM, Lan ZL, 1995. Absorbing boundary conditions for anisotropic elastic wave equations. *Acta Geophysica Sinica.* Vol. 38. Supplement 1, 73-79 (in Chinese)

Shao XM, 1997, Non-reflecting boundary conditions for wave propagation problems and their stability analysts, Dynamic Soil-Structure interaction (ed. Zhang CH and J. P. Wolf), International Academic Publishers. Beijing, China, 161-174.

Waas G., 1972, Linear two-dimensional analysis of soil dynamics problems in semi-infinite layered media, Ph.D. dissertation, University of California, Berkeley, CA, USA.

Wang LJ and Yang BP, 1994, A two-dimensional analysis of ground motion response at a standard test site—Turkey flat, California, Proceedings of 5th National Conference on Earthquake Engineering, Chicago, USA.

Wolf J. P., 1985, Dynamic soil-structure interaction. Prentice-hall, Englewood-Cliffs, N J.

Wolf J. P., 1988, Soil-structure-interaction analysis in time domain. Prentice-Hall, Englewood-Cliffs, N J.

Wolf J. P and Song Ch, 1996. Finite-element modeling of unbounded media. John Wiley & Sons Ltd, N. Y.

Yuan, YF. B. Yang and S. Huang, 1992. Damage distribution and estimation of ground motion in Shidian Basin, China, Proc. Int. Symp. On the Effects of Surface Geology on Seismic Motion (Odawara, Japan), 281-286.

5

Deep Compaction of Granular Soils

Dr. K. Rainer Massarsch

GEO Engineering AB, Bromma, Sweden

1. INTRODUCTION

Foundation engineering plays an increasingly important role in social and industrial development, for example in connection with major construction projects such as power stations or the expansion of infrastructure systems (railways, harbors, airports etc.). However, challenging foundation problems can also be encountered on smaller projects, for example when working in the vicinity of sensitive structures or installations. In many countries, limited availability of space, especially in urbanized areas, has led to the need to reclaim land from the sea or to build on ground with marginal or poor foundation conditions. Until recently, most major structures to be constructed on loose, compressible soils were founded on piles or other types of deep foundations. This has been an acceptable but costly solution to most foundation problems. However, soil improvement methods are being used increasingly in many countries, as they can be more cost-effective in many cases and may be adapted more easily to variable ground conditions and design requirements. Therefore, soil improvement methods have been among the most rapidly developing specialty areas of geotechnical engineering over the past 15 years and will continue to do so in the foreseeable future. Soil compaction can be used to improve the geotechnical properties of natural or man-made soil deposits, consisting primarily of granular materials, such as gravel, sand and silt. The objective of the present paper is to discuss the possible applications and limitations of deep compaction methods, to illustrate their practical applications and to outline new developments.

1.1 Objectives of Deep Compaction

The objective of deep compaction is to improve the strength and deformation characteristics of granular soils to such a degree, that structures can be supported safely and economically on or below the improved ground surface. Several important developments in foundation engineering have taken place during the past decade, which have made modem soil improvement methods very competitive:

- Availability of powerful construction equipment (e.g. vibrators and cranes), which can achieve high compaction effects to large depths,
- More reliable and accurate geotechnical field investigation methods, such as penetration tests and dynamic (seismic) tests, as well as improved concepts of data interpretation,
- Better understanding of the static and dynamic stress-strain behavior of soils, which has made it possible to model more accurately their deformation characteristics,
- More advanced analytical tools for geotechnical analysis and design, e.g. for predicting settlements and the effect of soil-structure interaction.
- Increasing awareness of seismic effects on foundations, especially in water-saturated granular soils (liquefaction hazard), and
- Reliable electronic measuring systems, which can be used in rough site conditions, making it possible to monitor on-site the compaction process and its effectiveness.

1.2 The Active Design Concept

Another important development has been the "Active Design" (AD) concept, which was originally applied to soil and rock tunneling projects (e.g. NATM, New Austrian Tunneling Method). However, the AD concept can also be used as a powerful tool for the planning and implementation of soil compaction projects, Massarsch and Westerberg (1995). The planning and implementation of a deep soil compaction project according to the AD concept is based on the following steps:

1. Identification of loading situation (static and dynamic forces) and specification of structural requirements (allowable deformations),
2. At an early stage of the project, determination of the geotechnical conditions by detailed field and laboratory investigations (traditionally, this aspect is not part of the planning phase of a project),
3. Analysis of the foundation problem with the aim to identify critical parameters for the short-term (construction) situation and the long-term (operating) performance of the structure,
4. Evaluation of different foundation alternatives, considering the site-specific requirements (e.g. environmental aspects, availability

of equipment and know-how etc.), cost/benefit analysis of alternatives and selection of optimal solution(s),

5. Specification of compaction criteria which are relevant for the actual situation and which can be verified on site, as well as statement of field control methods to be used (e.g. penetration testing),

6. Detailed planning of compaction procedure (trial compaction can be useful on larger projects in order to optimize compaction process) and method statement including specification of required competence, equipment specifications, verification of project execution and quality control measures,

7. Implementation of compaction project and optional trail compaction (for optimization of equipment and procedure), monitoring of production process and on-site review as well as interpretation of field data (geotechnical investigations)

8. Comparison of achieved results with initial design assumptions; if the requirements are not fulfilled, modification of compaction process (over-compaction should be avoided as this can cause unnecessary costs).

Optimization of compaction process to variations of ground conditions or modification of the compaction process/method (if the required compaction results are not achieved).

Performance monitoring of structure after completion of construction, including instrumentation of important elements and verification of performance after completion (detailed documentation of field measurements can provide valuable information for future projects and makes it possible to build up data base).

The above listed steps indicate that soil compaction projects require significantly more effort during the planning phase of the project. The involvement of knowledgeable geotechnical engineers at an early stage of a project and the execution of the work by an experienced foundation contractor are important factors, which are not always taken into consideration. However, soil compaction projects offer to the project planner significant possibilities to optimize the time for project execution and to achieve important cost savings.

In the recent past, several large soil compaction projects have been carried out according to the above described concept, which have lead to improvements of soil compaction methods and helped to understand fundamental aspects, such as the factors influencing the densification effect.

1.3 Soil Placement

Deep soil compaction if frequently used to improve the geotechnical properties of hydraulic fill in connection with land reclamation projects. The placement method has great influence on the geotechnical properties of the landfill. Careful planning and monitoring of the placement operation should be an essential part of the ground improvement project. The method of soil placement during land reclamation is of importance for the geotechnical characteristics of the soil deposit and does influence the selection of the compaction method and the required degree of compaction. Therefore, these aspects are discussed briefly.

A common and economical placement method for dredged sand is bottom dumping from large barges. This method requires a minimum water depth of at least 5 -7 in. Special vessels have been developed for efficient bottom dumping, called "split barge". A split barge has the ability to open the bottom of its hull, thereby releasing the entire load of fill. When the fill material is dumped, only little segregation between fine and coarse material does occur. Since the sand is placed by impact, it achieves a relatively dense state. However, soft soil below the fill can be disturbed and remolded significantly. It is important for sand barges to work in one area at a time and to fill the sand to the construction level before they move on to a new location. The sand barges often operate in pairs or even three or more at one time to make the shore line as wide as possible.

In the case of shallow water depth in the reclamation area, split barges can not be used and the fill must be placed by pumping, cf. Fig 1. The water-sand slurry is pumped along pipelines over distances of several hundred meters from the dredger to the reclamation area. The pipelines form a grid covering the reclamation area and are moved to other locations when the fill has reached the required level. Pumped fill is usually in a very loose state and can only support light structures without further measures.

A major disadvantage of sprayed sand fill is the risk of particle segregation, which can result in the formation of intermittent layers and/or seams of fine, silty and clayey material. These less permeable zones can negatively affect the efficiency of the compaction process.

Fig. 1. Spraying of pumped fill from barge

Fill above the water level is usually placed mechanically by excavators, dumpers or bulldozers, Fig. 2. Movement of heavy construction equipment along certain zones of a construction site densities the superficial soil layers and can create heterogeneous soil conditions. The densification caused by the construction equipment is limited to a depth of about 2-5 in. If the fill has been placed in two or more phases then the movement of heavy construction equipment can create several well compacted layers with less compacted zones in between. Dense surface layers, especially in the partially water-saturated zone above the ground water table can cause problems for compaction equipment to penetrate. This aspect should be taken into account when planning and interpreting the soil investigations for a compaction site.

Figure 2. Mechanical placement by bulldozer

1.4 Objectives of Deep Soil Compaction

Deep compaction of loose, granular soils can be used for the solution of a variety of foundation problems, such as:

- improvement of soil strength in order to increase the bearing capacity of foundations, or the stability of slopes and excavations with respect to static and/or dynamic loading,
- increase of soil stiffness for reduction of total and differential deformations and settlements which may be caused by static, cyclic or dynamic loading,
- reduction of lateral earth pressure against retaining structures (provided that compaction is carried out prior to their installation),
- mitigation of liquefaction hazard in loose, granular soils below the ground water level as a result of dynamic and cyclic loading (e.g. caused by earthquakes),
- lowering of soil permeability in, below or adjacent to dams and water front structures and general improvement of geotechnical properties in heterogeneous soil deposits.

The two most common applications of deep compaction are the improvement of reclaimed land for infrastructure projects (e.g. ports and airports) and the mitigation of liquefaction risk in seismic areas. Experience from a large number of case histories in different parts of the world confirms the satisfactory performance of improved ground and supported structures in recent earthquakes. Available ground improvement methods provide the means for effectively treating large undeveloped sites, as well as smaller, constrained areas, Mitchell et al. (1998).

2. SOIL INVESTIGATION METHODS

In-situ tests are an essential part of all ground improvement projects and may be needed during several phases of a compaction project, as discussed above:
- During the planning phase of a project for establishing of geotechnical conditions,
- During the design phase for selection of appropriate compaction method,
- During a trial phase for evaluation of compaction procedure and verification of compaction effect,
- During the production phase for production control and verification of compaction effect, and
- After completion of the project for documentation of long-term performance.

A variety of techniques, tests and observations can yield useful information regarding the effectiveness of deep soil compaction, Mitchell (1986). Visual observations of the compaction process, of soil response and of the performance of compaction equipment on site can provide a valuable background for the planning and implementation of more detailed geotechnical investigations. The compaction effectiveness can also be monitored indirectly by recording machine performance data, using electronic sensors attached to different components of the compaction equipment. These aspects will be discussed in more detail below.

However, different types of penetrometers are most frequently used for compaction projects. During the exploratory phase of a soil investigation it is possible to determine the soil conditions in general such as the depth, thickness and lateral extent of the various strata. They can also be used to check the density of fills and to estimate the compressibility and shear strength of primarily granular soils. A variety of different static and dynamic penetrometers are used today. The most advanced sounding method today is the Cone Penetration Test (CPT), which has become very popular in Europe and in the Far East and in some parts of North America. In the United States and in many other parts of the world the Standard Penetration Test (SPT) is still used extensively, primarily due to familiarity and experience of test interpretation with this method. The CPT and SPT have been standardized by the International Society for Soil Mechanics and Foundation Engineering (ISSM FE). National standards also exist in different countries.

Deformation properties (modulus values) can be determined in the laboratory on disturbed, reconstituted samples. The main problem in granular soils is the difficulty and high cost to obtain undisturbed soil samples. Therefore, different types of field investigation methods are used at most compaction projects. Empirical correlation of deformation characteristics and stress conditions with in situ tests, such as the Menard pressure-meter (PMT), the Machetti flat dilatometer (DMT) or seismic tests have been developed, Belotti et al. (1986). In the following section, the most important field investigation methods for compaction control will be discussed.

2.1 Standard Penetration Test (SPT)

The SPT is a well-established and unsophisticated method, which was developed in the United States around 1925. It has since undergone refinements with respect to equipment and testing procedure. The testing procedure varies in different parts of the world. Therefore, standardisation of SPT was essential in order to facilitate the comparison of results from different investigations. The equipment is simple, relatively inexpensive and rugged. Another advantage is that representative but disturbed soil samples are obtained. The reliability of the method and the accuracy of the result depend largely on the experience and care of the engineer on site.

A split-barrel sampler is driven from the bottom of a pre-bored hole into the soil by means of a 63.5 kg hammer, dropped freely from a height of 0.76 m. The diameter of the pre-bored hole varies normally between 60 and 200 mm. If the hole does not stay open by itself, casing or drilling mud should be used. The sampler is first driven to a depth of 15 cm below the bottom of the pre-bored hole, then the number of blows required to drive the sampler another 30 cm into the soil, the so-called N_{30} count, is recorded. The rods used for driving the sampler should have sufficient stiffness. Normally, when sampling is carried out to depths greater than around 15 m, 54 mm rods are used.

The quality of test results depends on several factors, such as actual energy delivered to the head of the drill rod, the dynamic properties (impedance) of the drill rod, the method of drilling and borehole stabilisation. The actually delivered energy can vary between 50-80% of the theoretical free-fall energy. Therefore, correction factors for rod energy (60%) are commonly used, Seed and De Alba (1986). The SPT can be difficult to perform in loose sands and silts below the ground water level (typical for land reclamation projects), as the borehole can collapse and disturb the soil to be tested. The following factors can affect the test results: nature of the drilling fluid in the borehole, diameter of the borehole, the configuration of the sampling spoon and the frequency of delivery of the hammer blows. Therefore, it should be noted that drilling and stabilisation of the borehole must be carried out with care. The measured N-value (blows/0.3 m) is the so-called standard penetration resistance of the soil. The penetration resistance is influenced by the stress conditions at the depth of the test.

Peck et al. (1974) proposed, based on settlement observations of footings, the following relationship for correction of confinement pressure. The measured N-value is to be multiplied by a correction factor C_N to obtain a reference value, N_1, corresponding to an effective overburden stress of 1 t/ft² (approximately 107 kPa),

$$N_1 = N \cdot C_N \tag{1}$$

where C_N is a stress correction factor and p' is the effective vertical overburden pressure.

$$C_N = 0.77 \cdot \log_{10}(20/p') \tag{2}$$

Seed (1976) proposed a similar correction factor for the assessment of liquefaction problems in loose saturated sands. This relationship was developed for earthquake problems and is based on extensive laboratory tests on mainly loose to medium dense sands,

$$C_N = 1 - 1.25 \cdot \log_{10}(\sigma_0'/\sigma_1') \tag{3}$$

where σ_0' is the effective overburden pressure (in t/ft²) and σ_1' is the reference stress (1 t/ft²). The correction of SPT results with respect to the effective overburden pressure is of importance for the evaluation of compaction results. Therefore, consideration should be given to this aspect when compaction criteria are to be based on N-values. Unfortunately, this fact is not always appreciated.

The resistance (N_{30}) has been correlated with the relative density of granular soils. Sand and gravel can be classified as shown in Table 1, Broms (1986).

The Standard Penetration Test is mainly used to estimate the relative stiffness and strength (bearing capacity) of soils. Deformation characteristics of granular soils can be estimated from empirical correlations, Peck et al. (1974). It is also possible to get some indications from SPT of the shear strength in

Table 1. Classification of sand and gravel after Broms (1986)

Relative density	Standard penetration resistance (N_{30},blows/0.3m)
Loose	≤ 10
Medium dense	10-30
Dense	≥ 30

cohesive soils. The SPT used frequently for the evaluation of the liquefaction potential of water-saturated, loose sands and silts in seismic areas, Seed and De Alba (1986).

2.2 Cone Penetration Test (CPT)

The CPT was invented and developed in Europe but has gained increasing importance in other parts of the world, especially in connection with soil compaction projects. Different types of mechanical and electric cone penetrometers exist but the electric cone is most widely used. A steel rod with a conical tip (apex angle of $60°$ and a diameter of 35.7 mm) is pushed at a rate of 2 cm/s into the soil. The steel rod has the same diameter as the cone. The penetration resistance at the tip and along a section of the shaft (friction sleeve) is measured. The friction sleeve is located immediately above the cone and has a surface area of 150 cm^2. The electric CPT is provided with transducers to record the cone resistance and the local friction sleeve.

A CPT probe, equipped with a porewater pressure sensor is called CPTU. It is important to assure complete saturation of the filter ring of the porewater (piezo) element. Otherwise, the response of the piezo-transducer, which registers the variation of pore water pressure during penetration, will be slow and may give erroneous results. The CPTU offers the possibility to determine hydraulic soil properties (such as hydraulic conductivity-permeability) but is most widely used for identification of soil type and soil stratification. The CPT can also be equipped with other types of sensors, for example vibration sensors (accelerometer or geophone) for determination of vibration acceleration or velocity. The "seismic cone" is not yet used on a routine basis but has, because of the relative simplicity of the test, potential for wider application especially on soil compaction projects.

The CPT is standardised and the measurements are less operator-dependent than the SPT, thus giving more reproduceable results. The recent geotechnical literature contains comprehensive information about different types of cone penetration tests, detailed descriptions of the test procedures and data evaluation/interpretation, Lunne et al. (1998). Therefore, only some aspects of importance for soil compaction projects will be discussed below.

The CPT measures the cone resistance q_c and the sleeve friction f_s from which the friction ratio, FR can be determined. FR is the ratio between the local sleeve friction and the cone resistance, expressed in percent (f_s/q_c). In spite of the limited accuracy of sleeve friction measurements, the valuable information, which can be obtained in connection with compaction projects, has not yet been fully appreciated. As will be discussed below, the sleeve friction measurement reflects the variation of lateral earth pressure in the ground, and can be used to investigate the effect of soil compaction on the state of stress, as will be discussed later. Cone and sleeve friction measurements are also strongly affected by the effective overburden pressure. It is necessary to take this effect into account, similar to the SPT. A correction factor C_M for the cone resistance was proposed by Massarsch (1994),

$$C_M = (100/\sigma'_m)^{0.5} \tag{4}$$

where σ_m' is the mean effective stress. It should be noted that for SPT correction the overburden pressure is used, which does not take into consideration the effect of lateral earth pressure. The mean effective stress can be determined from

$$\sigma'_m = \sigma'_v(1 + 2K_0)/3 \tag{5}$$

where σ_v' is the vertical effective stress and K_0 is the coefficient of lateral earth pressure at rest. Figure 3 compares the different correction factors for SPT and CPT, respectively. It should be noted that for Massarsch uses the mean effective stress while the SPT correction factor is based on the vertical effective stress. The measured cone penetration resistance, q_c can be corrected for the effect of the mean effective stress σ_m' at any given depth

$$q_{co} = q_c C_M = q_c\left(100/\sigma'_m\right)^{0.5} \tag{6}$$

where q_{co} is the normalised cone penetration value. When considering the difference between the vertical and the mean effective stress, there is good agreement between the different correction factors for the SPT and CPT, respectively. However, as will be shown later, soil compaction can significantly increase the horizontal stress and this effect should be taken into consideration when evaluating the densification effect. It is recommended to limit the correction

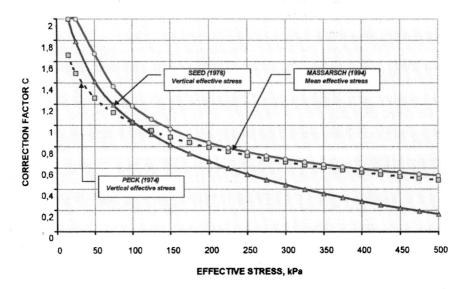

Fig. 3. Stress correction factors C_N (SPT) and C_M (CPT) with $K_0 = 0.57$, cf. equations (1, 2 and 3)

factor C_M to a value of 2.5. It is suggested that the corrected cone resistance q_{co} be used for specification of compaction criteria, as this will assure more homogeneous soil layers and avoid unnecessary overcompaction close to the ground surface.

One important objective of the CPT investigations in connection with soil compaction is to obtain information concerning soil stratification and variation in soil properties both in horizontal and vertical direction. The friction ratio is often used as an indicator of soil type (grain size) and can provide valuable information when evaluating alternative compaction methods.

Measurement of the excess pore water pressure with the CPTU can detect layers and seams of fine-grained material (silt and clay). It is also possible to obtain more detailed data information concerning soil permeability and thus soil stratification.

2.3 Comparison between SPT and CPT

In areas like North America, many geotechnical engineers have developed considerable design experience, based on local correlation between the SPT and performance of foundations. Therefore, engineers may feel more comfortable by converting CPT data to equivalent SPT N-values and then comparing these with their SPT-based design methods. A considerable number of studies have been carried out in the past to quantify the relationship between SPT N-value and CPT cone bearing resistance (q_c), Robertson and Campanella (1986). The author has compiled data from various case histories, which are compared in Fig. 4 with the data published by Robertson and Campanella (1986).

In spite of some scatter between individual test points and different soil types, there exists a clear correlation between the q_c/N ratio and the mean particle size.

2.4 Pressure Meter Test (PMT)

The PMT was invented by the Menard in 1962-1963 in France, where this test is widely used. National standards exist and geotechnical design is based almost exclusively on this type of test. Over the years the PMT has been further developed in France, the United Kingdom and Japan, and has found increasing acceptance in several countries. However, the PMT is still a specialist tool, which requires experience in test performance and data interpretation.

The standard pressuremeter is either inserted into a pre-bored hole or directly jacked or driven into the ground. A slotted tube protects the measuring cell, which consists of a cylindrical rubber membrane. In order to reduce the influence of soil disturbance during probe insertion, the self-boring pressuremeter was developed. This type of pressuremeter is, however, limited to fine-grained soils, while the standard pressuremeter can be used in most soil types. A detailed description of the PMT is beyond the scope of this paper. However, guidelines for data evaluation and interpretation as well as design

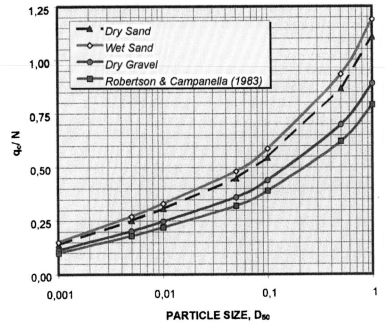

Fig. 4. Variation of q_c/N ratio with mean grain size (note that the cone resistance in MPa)

recommendations were published by Baguelin et al. (1986). The PMT is an intermittent test and can thus not provide a continuous profile. The test is comparatively time-consuming and thus expensive.

From the PMT, a stress-deformation curve (applied pressure vs. volumetric strain) can be obtained in situ. From this curve, a deformation modulus and a value of the limit strength can be obtained. Also the "at rest" lateral earth pressure can be estimated, which is of considerable interest for soil compaction project. Few correlations exist between the PMT and other in-situ tests. Because of the necessity to drill borehole, the quality of the test results may be suspect in loose sand below the ground water table.

2.5 Dilatometer Test (DMT)

The dilatometer test (DMT), which is a simple and reliable in-situ testing tool, was developed in Italy and later introduced in Europe and North America, Marchetti (1980). The dilatometer consists of a flat, 15 mm thick and 95 mm wide blade and has a length of 220 mm. A flexible, stainless steel membrane, 60 mm in diameter, is located on one face of the blade. Inside the steel membrane there is a pressure chamber and a distance gauge for measurement of the movements of the membrane when the pressure inside is changed. The probe is pushed into the soil with the aid of hollow sounding rods and does not require drilling of a hole. When the membrane is inflated, the pressure required to just lift the membrane off the sensing device (p_0) and to cause 1.10 mm deflection (p_1) are recorded. As the pressure is released and the membrane returns to its initial lift-off position, another reading can be taken. The pressure values p_0 and p_1 can be used to define three index parameters. Marchetti (1980) calls these parameters the material index (I_D), the horizontal stress index (K_0) and the dilatometer modulus (E_D), respectively.

$$I_D = (p_1 - p_0) / (p_0 - u_0) \tag{7}$$

where u_0 = pore water pressure at rest (not excess pore pressure). The I_D value varies from about 0.6 to 1.8 for silt and is about 1.8 for sand. The DMT is especially suited for monitoring of compaction projects as it can be used to assess the deformation characteristics of soils. From these index values, empirical relationships have been developed to determine geotechnical parameters. For instance, assuming that the soil behaves elastically, the dilatometer modulus can be deduced from the relation:

$$E_D = 48.1 (p_1 - p_0) \tag{8}$$

The dilatometer modulus is commonly used to assess the compression (oedometer) modulus M of sand, silt and clayey silt. Experience has shown that the following relation obtains a good estimation of M

$$M = 1.1 R_m E_D \tag{9}$$

where R_m varies depending of the soil. Schmertman (1986) has suggested design procedures for settlement estimates based on the DMT.

2.6 Seismic Tests

Conventional seismic tests, such as wave refraction measurements, have been used in the past primarily for soil and rock layer identification. However, during the past decade, several new seismic in situ tests have been developed and applied successfully on a variety of soil compaction projects, Massarsch and Westerberg (1995). The "seismic" cone penetrometer, which has been briefly discussed above, incorporates a small rugged velocity sensor in an electronic penetrometer. Woods (1986) has published a detailed description of different seismic field testing methods.

The most common seismic test for compaction control is the down-hole test. A vibration sensor is installed in a borehole, or by pushing the sensor into the ground (cf. seismic CPT). A polarised shear (and/or compression) wave is generated at the ground surface and the time required for the wave to travel across the soil layers to a receiver is measured. Different methods of signal interpretation can be used to determine the first arrival time of the signal. From the known distance the wave propagation velocity (shear wave or compression wave) can be calculated. Down-hole tests are relatively easy to perform, as only one sensor must be installed in the ground. The down-hole test is suitable for compaction control as it measures the average properties of a relatively large soil volume, compared to penetration tests. Signal interpretation is basically simple (determination of first arrival time at the two sensor locations), but more complex evaluation concepts (e.g. signal cross-correlation) are used. An important advantage of the shear wave velocity is that the ground water level does not affect the measurements.

In the case of a cross-hole test, two sensors are installed in the ground and the wave is generated in a borehole at the same level. The distance between the vibration source and the sensors must be determined accurately in order to obtain sufficient accuracy. The distance between the sensors is typically 3-6 m. The cross-hole test is more cumbersome to perform than the down-hole test and is mainly used for research purposes.

Another, increasingly popular seismic testing method is the Spectral Analysis of Surface Wave Technique, SASW, Woods (1986). It uses a seismic source (impact or vibration generator) at the ground surface and at least two vibration transducers at the ground surface. The vertical transducers record the propagation of surface (Rayleigh) waves. By analysing the phase information for each frequency contained in the wave train, the Rayleigh and shear wave velocity, can be determined. The evaluation of SASW measurements is relatively complex and requires specially developed computer software. However, user-friendly hardware and software has been developed which simplify the application of SASW also for compaction control. SASW measurements can determine wave velocity profiles to depth exceeding 20m, which is sufficient for most foundation projects. The main advantage of SASW is that large soil volume can be investigated relatively rapidly.

From the calculated shear wave velocity C_s, the shear modulus G can be calculated from the following relationship

$$G = C_s^2 \cdot \rho \tag{10}$$

where ρ is the bulk density of the soil mass. As the strain level of the propagating shear wave is low ($< 10^{-4}\%$) and the elastic wave velocity is measured. It should be noted that the "dynamic" (small-strain) shear modulus decreases with increasing strain level and can thus not be directly converted into a "static" (large-strain) modulus value. A modulus reduction factor can be used to estimate the static modulus G_{stat} from the dynamic modulus, G_{max}. Approximate values of the modulus reduction factor R can be obtained from Table 2. Semi-empirical correlations can be used to estimate the equivalent static shear modulus (secant modulus), G_{stat}

$$G_{stat} = G_{max} \cdot R \tag{11}$$

Table 2. Modulus reduction factor R for determination of the static (secant) modulus from dynamic tests, Massarsch (1984)

Soil type	Reduction factor, R
Gravel	0.20
Sandy gravel	0.19
Loose sand	0.18
Medium dense sand	0.15
Dense sand	0.12

where R is a reduction factor, which takes into account the strain-softening effect of soils (at strains of approximately 0.1 % strain).

It is interesting to note that the static modulus in sand is only about 10 to 20% of the dynamic modulus. The Young's modulus, E_{stat} and the constrained (oedometer) modulus M_{stat} can be readily calculated from the shear modulus G_{stat} (Poisson's ratio of v: 0.3)

$$G_{stat} = 2.6 \cdot G_{stat} \tag{12}$$

$$M_{stat} = 3.5 \cdot G_{stat} \tag{13}$$

From the above relationships it is obvious that the Young's modulus and the constrained modulus are significantly larger than the shear modulus. This aspect must be taken into account when evaluating the results of dynamic soil tests.

The damping (attenuation) characteristics of the soil can also be determined using seismic techniques, Woods (1986). However, the practical application of these measurements for compaction projects is not yet well understood.

3 DESIGN ASPECTS OF SOIL COMPACTION

The geotechnical engineer must at an early stage of the project assess whether the ground needs to be improved. In most cases, settlement criteria govern the design for static loading conditions. Therefore, as a first step an estimate of settlements of the untreated soil must be made. The most important parameters for soil compaction projects are the determination of relative density, of settlements and shear strength.

3.1   Relative Density

Soil density, or more commonly, the relative density, D_R is often used as an intermediate soil parameter to specify compaction criteria. In the present paper, the cone penetration test will be used for assessing relative density. Recent work in large calibration chambers has provided numerous correlations of cone resistance with relative soil density. Most investigations have shown that no single relationship exists for all granular soils between relative density, effective stress and cone resistance. In Fig. 5, results from extensive pressure chamber tests are summarised, Robertson and Campanella (1983).

The cone penetration resistance, the vertical and the horizontal effective stress affect the relative density D_R. Jamiolkowski et al. (1988) have pointed out that a survey of the literature shows surprisingly poor correlations between relative density and the (uncorrected) cone penetration resistance. Thus the correlation in Fig. 5 should be interpreted with caution. A range, rather than specific values of cone penetration resistance should be used in practice to specify compaction requirements, based on relative density.

In order to facilitate the estimation of relative density from cone penetration

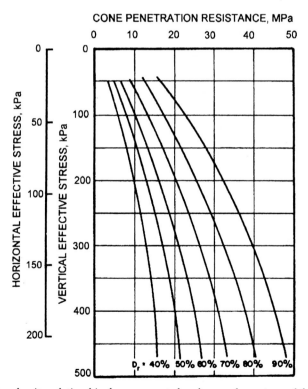

Fig. 5. Relative density relationship for uncemented and unaged quartz sand for a coefficient of lateral earth pressure K_0 = 0.45 (after Robertson and Campanella, 1986)

tests, it is proposed to correct the measured cone penetration resistance with respect to the mean effective stress of 100, cf. equation (6) and Fig. 3.

3.2 Settlements Estimates

Many geotechnical problems in granular soils require the assessment of settlements. Schmertmann (1970) has proposed a method of settlement analysis below isolated footings in sand, based on results from cone penetration tests. In this case, the deformation characteristics of the soil are governed by the modulus of elasticity, E_s. Since the factor of safety against bearing capacity failure is usually high for footings on sand, the designer is usually interested in a modulus for an average mobilised stress around 25% of the failure stress. Based on a review of calibration chamber tests, Robertson and Campanella (1986) have proposed a relationship between the elastic modulus E_{25} at a stress level of 25% and the cone penetration resistance q_c which for normally consolidated sand varies between $E_{25}/q = 1.5$ to 3.0. This value is in agreement with the recommendations by Schmertmann (1970). The Canadian Foundation Engineering Manual (I 985) proposes the following coefficient E_{25}/q_c, which is based on soil type and compactness, Table 3.

Table 3. Estimation of E_{25}/q_c based on results from Static Cone Penetration Test, Canadian Foundation Engineering Manual (1985)

Soil type	E_{25}/q_c
Silt and sand	1.5
Compact sand	2
Dense sand	3
Sand and gravel	4

The ratio E_{25}/q_c for overconsolidated sand appears to be higher, on the order of 3 to 6 times the value for normally consolidated sand. It is obvious that this simplified method of assessing the soil modulus value is not accurate and should only be used for preliminary settlement estimates. In the following section, a new method is proposed for the determination of the soil modulus, which uses the results from cone penetration tests in a rational way, Massarsch (1994).

3.2.1 Tangent modulus approach

Settlements in cohesive and cohesionless soils can be analysed using the tangent modulus method, which takes into account the non-linearity of the load-deformation relationship of most soils. Janbu (1963) has shown that the tangent modulus M_t can be defined by the following relationship,

$$M_t = m \cdot \sigma_r (\sigma'/\sigma_r)^{(1-j)}$$

where m is a dimensionless modulus number, σ_r is an arbitrarily chosen reference (14) stress (100 kPa), σ' is the vertical effective stress and j is a stress exponent. The strain increase of a soil layer $\Delta\varepsilon$ caused by an increase of the vertical effective stress $\Delta\sigma'$ can be calculated from

$$\Delta\varepsilon = \{ [(\sigma_0' + \Delta\sigma')/\sigma_r]^j - [\sigma_0'/\sigma_r]^j \}/(m \cdot j) \qquad (15)$$

where σ_0' is the initial vertical effective overburden stress and $\Delta\sigma'$ is the increase of the vertical effective stress. For cohesive soils, values of the modulus number m and the stress exponent j can be determined by conventional laboratory tests. For cohesionless soils, however, it is often difficult to obtain undisturbed soil samples and the results from laboratory tests are therefore uncertain. Thus, empirical values are often used to estimate the modulus number and the stress exponent, cf. Canadian Foundation Engineering Manual (1985), Table 4.

Table 4. Typical values for the stress exponent *j* and the modulus number *m* for granular soils (after Canadian Foundation Engineering Manual, 1985)

Soil type	Stress exponent, j	Modulus number, m
Gravel	0.5	40-400
Dense sand	0.5	250-400
Compact sand	0.5	150-250
Loose sand	0.5	100-150
Dense silt	0.5	80-200
Compact silt	0.5	60-80
Loose silt	0.5	40-60

3.2.2 Determination of soil modulus from CPT

The cone penetration test (CPT) is widely used to investigate granular soils and several authors have proposed empirical correlations between the cone penetration resistance and the tangent modulus (Robertson and Campanella 1983, 1986). However, the results from cone penetration tests are influenced by many factors, the two most important are the relative density and the stress level, especially the effective horizontal stresses (Jamiolkowski et al., 1988). Attempts have been made to correlate directly the tangent modulus M_t to the cone penetration resistance q_c, but with limited success (Robertson and Campanella, 1986; Jamiolkowski et al., 1988). The normalised cone penetration resistance q_{co} is mainly a measure of the strength properties of the soil. Janbu (1974) has proposed the following theoretical relationship between the cone resistance and the tangent modulus M_t

$$M_t = m_p(q_p \cdot \sigma_r)^{0.5} \qquad (16)$$

where m_p is a dimensionless number and q_p is the net cone resistance (q_c–σ_v). However, the cone penetration resistance is generally not corrected with respect to the effective confining pressure. Janbu also suggested that m_p would not change substantially with porosity. Massarsch (1994) has proposed the following correlation, which is similar to that introduced by Janbu (1974), but the normalised cone resistance q_{co} is used instead of the net cone resistance q_p,

$$m = a \cdot (q_{co}/\sigma_r)^{0.5} \qquad (17)$$

In this equation, *a* is an empirical modulus factor, which mainly depends on soil type and varies within a relatively narrow range. Experience from several case histories has shown that the values of the modulus number, as proposed initially by Massarsch (1994), are conservative and that the settlements are over-predicted by 30 to 50%. Revised values for the modulus parameter *a*, which are used in the present paper, are given in Table 5.

Several comparisons between predicted and observed settlements of

Table 5. Modulus factor *a* for different soil types, Massarsch et al. (1997)

Soil type	Modulus parameter, *a*
Silt, organic soft	7
Silt, loose	12
Silt, compact	15
Silt, dense	20
Sand, silty loose	20
Sand, loose	22
Sand, compact	28
Sand, dense	35
Gravel, loose	35
Gravel, dense	45

compacted ground suggest that the above proposed concept gives reliable results, Massarsch et al. (1997)

3.3 Shear Strength

The friction angle of sand φ cannot be determined directly in the field but only by laboratory tests on reconstituted samples. Thus all correlations with field tests are at best approximations. No unique relationship does exist between the friction angle of sands and cone resistance. However, the CPT is still one of the more reliable field methods which can be used to assess approximately the friction angle and the change of the friction angle as a result of densification. Robertson and Campanella (1983) have proposed a semiempirical relationship for estimation of the friction angle from cone penetration resistance, Fig. 6. The diagram is applicable for soils with a friction ratio lower than 1%.

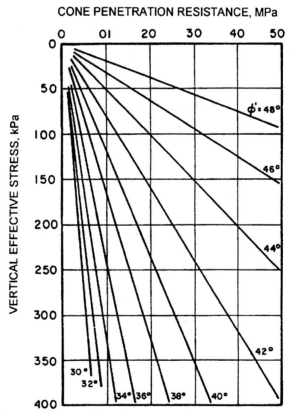

Fig. 6. Correlation between cone resistance and peak friction angle for uncemented sand (after Robertson and Campanella, 1983)

3.4 Liquefaction

Another important application of compaction projects is the mitigation of the liquefaction hazard in seismic areas. Simplified design procedures have been developed for preliminary design purposes, Seed and De Alba (1996). These are based on empirical correlations between the SPT or CPT resistance and the observation of liquefaction during earthquakes. In the case of complex or large projects, more sophisticated analytical tools can be used to investigate the need and degree of compaction. Experience from recent earthquakes suggests that even moderate compaction by various methods has had a beneficial effect on the performance of natural and man-made soil deposits, Mitchell (1998).

3.5 Compactability of Soils

Another important question for the geotechnical engineer is the determination, whether the soil can be improved by deep compaction. Mitchell (1982) identified soils according to grain size distribution and suggests that most granular soils with a fines content (particles < 0.064 mm) lower than 10% can be compacted by vibratory and impact methods. The disadvantage of compaction criteria based on grain size curves is that soil samples have to be taken. It is preferable to use the results of penetration tests for assessment of soil compactability. Massarsch (1991) has proposed compaction criteria based on CPT cone resistance and friction ratio values, Fig. 7. It should be noted that the above diagram assumes homogeneous soil conditions. In the case of thin layers of silt and clay, the effectiveness of soil compaction will be reduced. The CPTU with excess porewater pressure measurements can also be used to determine the occurrence of less permeable silt and clay layers.

4 DEEP SOIL COMPACTION METHODS

For a general overview of compaction methods, reference is made to the comprehensive paper by Mitchell (1981). In this paper, emphasis is placed on the description of recent developments and the application of new compaction system. As has been discussed in the preceding section, the planning of soil compaction requires geotechnical competence and careful planning on the part of the design engineer. Similarly, only contractors with demonstrated experience and suitable equipment should be entrusted to carry out deep soil compaction with the method to be used.

Each of the below discussed compaction methods has its optimal applications. The selection of one or several suitable methods is controlled by a variety of factors, such as: soil conditions, required degree of compaction, type of structure, maximum depth of compaction, site-specific considerations such as sensitivity of adjacent structures or installations, available time for completion of project, competence of contractor, access to equipment and material etc. It is common practice to award the soil compaction project to the lowest bidder.

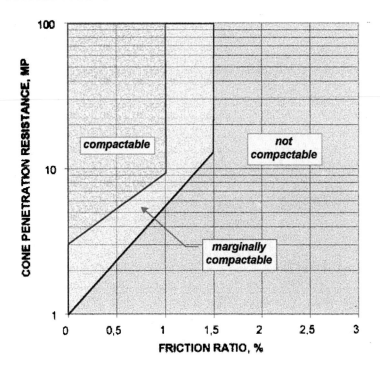

Fig. 7. Soil classification for deep compaction based on CPT, from Massarsch (1991)

However, at the end of completion of a project, this may not always turn out to be the optimal solution, if the required compaction result is not achieved, or the duration of work is significantly exceeded. Therefore, it is paramount for all types of soil compaction projects that a high degree of quality control and site supervision is achieved.

In contrast to pile foundations, soil compaction is a repetitive process and much can be gained from properly planned and executed compaction trials. The most important factors, which should be established and verified at the start of the project, are:
- required compaction energy at each compaction point,
- spacing between compaction points,
- duration of compaction in each point,
- ground settlements due to compaction (in compaction point and overall settlements),
- time interval between compaction passes (time for reconsolidation of soil),
- verification of the achieved compaction effect by field measurements and penetration tests,
- potential increase of compaction effect with time after compaction,
- ground vibrations in the vicinity (effects on adjacent structures and installations),

- effect on stability of nearby slopes or excavations,
- monitoring of equipment performance and review of safety aspects.

Soil compaction methods can be classified according to the following categories:

a) **Energy transfer from ground surface:**

Type of energy source : *impact:* Dynamic Compaction, DC and impact roller

vibratory: vibratory plate

Energy transfer below ground surface:

Type of energy source: *impact:* driven probe, pile, Franki stone column, explosives

vibratory: vibroflotation, vibratory probes, MRC compaction

4.1 Energy Transfer from Ground Surface

The compaction energy can be applied to the soil at the ground surface either by impact (falling weight) or by steady state vibrations (vibratory action). Thus, the highest compaction effect will be achieved close to the ground surface and decreases with depth. The effective depth of compaction is difficult to assess and is influenced by a variety of factors, such as the geotechnical conditions, the type and quality of equipment, compaction procedure etc.

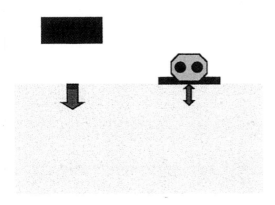

a) Dynamic compaction   b) Vibratory plate compaction

Fig. 8. Deep compaction methods applying energy at the ground surface

4.1.1 Dynamic compaction

Soil densification by dynamic compaction (DC), also called "heavy tamping is a well-known compaction method. The method was "rediscovered" by Menard, who transformed the crude tamping method into a rational compaction procedure. Soil is compacted by repeated, systematic application of high energy using a heavy weight (pounder). The imparted energy is transmitted from the ground surface to the deeper soil layers by propagating shear and compression waves types, which force the soil particles into a denser state. In order to assure effective transfer of the applied energy, a 1 to 2 m thick stiff layer usually covers the ground surface. Pounders can be square or circular in shape and made of steel or concrete. Their weights normally range from 5 to 25 tons and drop heights of up to 25 m have been used. Heavier weights and larger drop heights have been used for compaction of deep soil deposits, but are not very common.

Dynamic compaction is carried out in several passes. During each pass, the weight is dropped repeatedly in a predetermined grid pattern. The distance between the compaction points is normally decreased in the subsequent passes and compaction is carried out in-between the previously compacted points. The final pass, also called "Ironing pass", usually performed with low compaction energy, is carried out with a reduced drop height. The objective is to densify the superficial soil layers without remoulding the already densified deeper layers. Mayne (1984) presented a detailed description of the dynamic compaction method. Choa (1996) has summarised the experience from dynamic compaction work a major project, the Changi Airport in Singapore, Fig. 9.

Although the dynamic compaction method appears to be very simple, it requires careful design of the compaction process. The densification effect is strongly influenced by the dynamic response characteristics of the soil to be compacted, but also by the underlying soil layers. Usually, extensive compaction trials are needed to optimise the compaction process with respect to the required energy for achieving specified densification criteria. A major limitation of dynamic compaction is the lack of monitoring and quality control during the production phase. However, for research purposes, the pounder can be equipped with sensors to monitor the applied energy and to record the dynamic response of the soil layer.

The maximum depth which can be achieved by dynamic compaction depends on several factors, such as the geotechnical properties of the soil layer to be compacted, the dynamic soil properties in and below the layer to be compacted (e.g. a soft clay layer below the layer to be densified can significantly reduce the compaction effect), the ground water level, the compaction grid, the number of compaction passes and the time interval between passes. As a general rule, the maximum depth d_{max} to which a soil deposit can be estimated from the following relationship

$$d_{max} = \alpha \sqrt{H \cdot M} \qquad (18)$$

Fig. 9. Dynamic compaction carried out in the trial area of Changi East Reclamation
Project, Phase 1B (mass 25 tons, drop height 25 m)

where H is the average drop height and M the mass of the pounder. The factor
a should be determined for each site, but varies between 0.3-0.5. The typical
depth of compaction for drop height of 15 m and a pounder mass of 15 tons is 7-
8 m.

4.1.2 Impact roller

A simple, but in some cases surprisingly effective surface compaction method
is the impact roller, cf. Fig. 10. A conventional tractor pulls a heavy prism-
shaped mass, consisting of steel or concrete. The impact generated by the rotation
of the heavy mass (up to 50 tons) transfers sufficient energy to achieve medium
compaction to a depth of several meters. The compaction process is usually
based on site-specific correlation and little documented evidence about actually
achieved compaction effects is available. The impact roller can be used on
granular soils including coarse-grained material such as gravel and rock fill.

4.1.3 Vibratory compaction plate

Deep soil compaction can be carried out using a heavy steel plate, which is excited by strong vertical vibrations, cf. Fig. 11. This compaction method has been made possible by the development of powerful hydraulic vibrators. The first vibrators used for pile driving were developed some 60 years ago in Russia and have since been used extensively on foundation projects worldwide. During the past decade, very powerful vibrators have been developed for foundation applications, such as pile and sheet pile driving and soil compaction. These vibrators are hydraulically driven, which allows continuous variation of the vibrator frequency during operation.

Fig. 10. Impact roller for deep compaction

Fig. 11. Heavy vibratory compaction plate

Modern vibrators can generate a centrifugal force of up to 4,000 kN (400 tons). The maximum displacement amplitude can exceed 30 mm. These enhancements in vibrator performance have opened new applications to the vibratory driving technique. Recently, vibrators with variable frequency and variable static moment (displacement amplitude) have been introduced. Figure 12 shows a vibrator with eccentric masses, arranged at separate rotation levels. During any stage of vibrator operation the position of the lower row of masses can be changed relative to that of the upper row, thereby affecting the static moment and the displacement amplitude.

The maximum depth of compaction depends primarily on the size (geometry) of the plate and on the applied energy (force and number of compaction cycles). Extensive investigations have been performed in connection with submarine soil compaction projects, Nelissen (1983). It was found that the compaction effect depended on the vibration frequency and the dynamic interaction of the plate-soil system. Based on field trials on land and on the seabed, the optimal compaction parameters could be established.

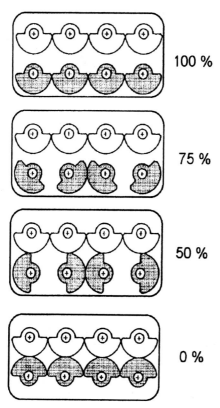

Fig. 12. Operating principle of vibrator with dual rows of eccentric masses, allowing variation of the static moment (displacement amplitude)

4.2 Energy Transfer Below Ground Surface

The most efficient way to densify deep deposits of granular material is to introduce the compaction energy at depth. The energy can either be applied by impact or vibratory action, Fig. 13.

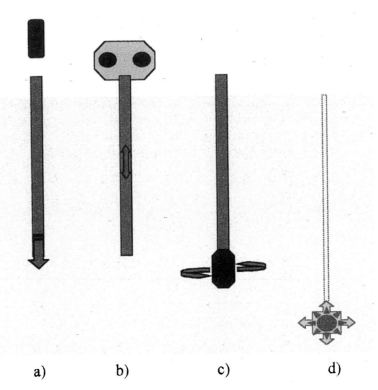

a)   b)   c)   d)

Fig. 13. Deep compaction methods applying energy at depth

4.2.1 Displacement pile method

Driving of timber or concrete piles into deposits of granular soil is probably the oldest deep compaction method (Fig. 13a). However, because of the high cost and the limited efficiency, this method is not used extensively on larger projects. The Franki stone column method is based on the same concept, but the energy is transmitted at the bottom of an open steel tube. Figure 14 shows the basic principle of the Franki system. A heavy impact hammer with a mass of up to 10 tons is dropped on a compacted soil plug at the bottom of a thick-walled steel tube. The steel tube is driven into the ground and the surrounding soil is displaced and at the same time compacted. After the maximum depth has been reached, the soil plug is expelled and the tube is withdrawn. The soil below the tube is compacted during the extraction phase, thereby further increasing the densification effect. Very high compaction effects can be achieved to depth

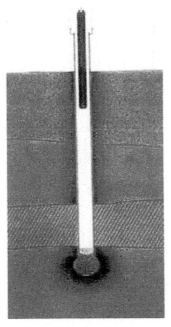

Fig. 14. Franki compaction method

exceeding 20 m. However, the method is time-consuming and therefore costly.

4.22 Vibro-probe and MRC method

The vibro-rod principle is employed by several deep compaction methods (Fig. 13b). A compaction probe or rod is inserted in the ground a repeated number of times with the aid of a heavy vibrator, thereby gradually compacting the surrounding soil. Different types of compaction probes have been developed, ranging from conventional tube or sheet pile profiles to more sophisticated, purpose-designed tools. The vibro-probe method was initially developed in Japan and used a probe with short wings, attached to a vibrating pile driving hammer. The so-called Vibro-Wing was developed in Sweden and uses an up to 15 m long steel rod, which is provided with 0.8 m long wings, spaced 0.5 m apart.

The vibratory hammer is usually operated from a piling rig, Fig. 15. The frequency of the vibrator can be varied to fit the conditions at a particular site. The duration of vibration and rate of withdrawal of the probe depend mainly on the permeability of the soil, the depth of the soil deposit and the spacing between compaction points. The duration of compaction, the grid spacing and number of probe insertions are chosen empirically or are determined by field tests. The maximum depth of compaction depends on the capacity of the vibrator and size of the piling rig and is on the order of 10 to 15 m.

Fig. 15. Vibro wing method

4.2.3 MRC compaction system

The MRC system is a recently developed method for deep vibratory soil compaction, Massarsch (1991). The MRC concept uses the resonance effect in soil layers to increase the efficiency of vibratory soil densification, cf. Fig. 16.

A heavy vibrator with variable frequency is attached to the upper end of a flexible compaction probe. The probe is inserted into the ground at a high frequency in order to reduce the soil resistance along the shaft and the toe. When the probe reaches the required depth, the frequency is adjusted to the resonance frequency of the soil layer, thereby amplifying the ground response. The probe is excited in the vertical direction and the vibration energy is transmitted to the surrounding soil along the probe surface. When resonance is achieved, the whole soil layer will oscillate simultaneously and this is an important advantage, compared to other vibratory methods.

The compaction probe is an essential component of the MRC system and is designed to achieve optimal transfer of compaction energy from the vibrator to the soil. The probe profile has a double Y-shape, which increases the compaction influence area. Reducing the stiffness of the probe further increases the transfer of energy to the surrounding soil. This is achieved by openings in the probe, cf. Figure 16b. The openings also have the advantage of making the probe lighter and thereby providing larger displacement amplitude during vibration, compared to a massive probe of the same size.

a) MRC compaction equipment   b) MRC compaction probe

Fig. 16. MRC compaction system using variable frequency vibrator and flexible compaction probe

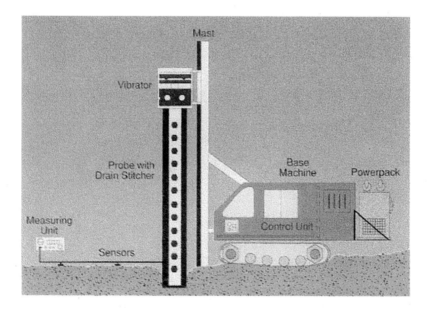

Fig. 17. MRC electronic monitoring system for compaction control

An important feature of the MRC system is the electronic monitoring system, which continuously records all parameters of importance for the compaction process, such as vibrator frequency, probe penetration depth, power supply (hydraulic pressure) and ground vibration velocities, cf. Fig. 17.

The recorded information can be printed out on site, providing an accurate documentation of the actually performed compaction process. The documentation can be used for quality control as well as for optimisation of the continuing compaction execution. The recorded information is also transmitted to an electronic control unit, mounted in the cabin of the compaction machine. It collects and evaluates important parameters and provides the machine operator with instructions on a display, Fig. 18. Thus, the machine operator can continuously adapt the compaction work to achieve optimal densification throughout the whole compaction process.

Figure 19 shows as an example the variation of the vertical vibration velocity on the ground as a function of the vibration frequency. During probe penetration and extraction, a high vibration frequency (around 30 Hz) is used, which does not cause significant ground vibrations. During compaction, the speed of the vibrator is reduced to the resonance frequency of the probe-soil system. At resonance, ground vibrations are strongly amplified. The probe and the surrounding soil vibrate in phase, resulting in an efficient transfer of compaction energy.

The MRC system uses state-of-the-art electronic process control, the information of which can be used to evaluate the soil conditions during each compaction pass. By recording the penetration speed of the compaction probe during insertion at a given vibration frequency, a record of the soil resistance is obtained in each compaction point. This concept can be used during trial compaction in order to determine the optimal spacing of compaction points.

The compaction duration depends on the soil properties and on the required degree of densification, to be achieved. Compaction is usually carried out in a

a) MRC field computer   b) electronic process control system

Fig. 18. MRC electronic control unit for adaptation of compaction process

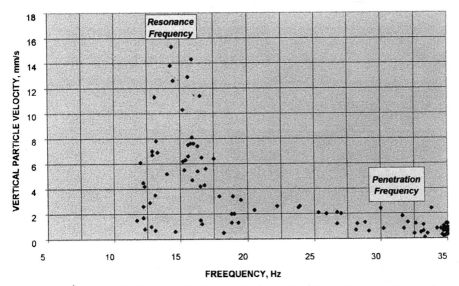

Fig. 19. Vertical ground vibration velocity at a distance of 4 m from the compaction probe penetration and resonance compaction

grid pattern, in two or more passes. The grid spacing ranges typically between 3.5 to 4.5 m. During the first compaction pass, which is carried out at an assumed grid spacing, the time for full penetration of the compaction probe is measured. Thereafter, soil compaction is carried out according to the MRC process. During the following compaction pass, the probe is inserted in the diagonal point of the compaction grid and the time required for the probe to penetrate the soil layer is again recorded. If the time is the same as during the first compaction pass, then a closer compaction grid spacing must be chosen. However, if the probe cannot penetrate during the second pass, then the chosen grid was too close. As a result of such field trials it is possible to establish the optimal compaction parameters for a specific site.

The dynamic response of the soil deposit during compaction can also be used to monitor the compaction effect. With increasing densification of the soil layers, the resonance compaction frequency rises. Also the ground vibration velocity increases and soil damping is reduced. With the aid of vibration sensors placed on the ground surface, the change in wave propagation velocity can be determined, which reflects the change of soil stiffness and soil strength, Massarsch (1995).

4.2.4 Vibroflotation

This method was invented in Germany almost 60 years ago, and its development has continued mainly there and in North America, where it was introduced in the 1940s. The equipment consists of three main parts: the vibrator, extension tubes and a supporting crane, Fig. 20. Vibroflotation is the most widely used deep compaction equipment and extensive experience has been accumulated

Fig. 20. Vibroflotation equipment with water jetting

over the past 20 years. The vibrator is incorporated in the lower end of a steel probe. The vibrator rotates around the vertical axis to generate horizontal vibration amplitude. Vibrator diameters are in the range of 350 to 450 mm and the length is about 3-5 m, including a special flexible coupling, which connects the vibrator with the extension tube.

Units developing centrifugal forces up to 160 kN and variable vibration amplitudes up to 25 mm are available. Most usual vibroflot probes are operating at frequencies between 30 and 50 Hz. The extension tubes have a slightly smaller diameter than the vibrator and a length dependent on the depth of required penetration.

The vibroflot is slowly lowered to the bottom of the soil layer and then gradually withdrawn in 0.5-1.0 m stages. The length of time spent at each compaction level depends on the soil type and the required degree of compaction. Generally, the finer the soil, the longer the time required achieving the same degree of compaction. In order to facilitate penetration and withdrawal of the equipment, water jetting is utilised with a water pressure of up to 0.8 MPa and flow rates of up to 3000 l/min. The water jetting transports the fine soil particles to the ground surface and by replacing the fines with coarse material, well-compacted soil columns are obtained.

4.2.5 Compaction by explosives

Compaction using explosives has been used in Europe and North America for a variety of projects. This technique is very attractive from an economic viewpoint, but is limited to large projects with deep deposits of water-saturated sands and gravels. The psychological effect of detonating explosive limits the application of this method to unpopulated areas.

Small explosive charges are installed in pre-bored holes at depth of approximately 2/3 of the depth of the zone to be densified. The shock of the explosive liquefies the soil, which starts to reconsolidate due to the total overburden pressure.

5. HYPOTHESIS FOR COMPACTION MECHANISM IN SAND

Although extensive information is available in the literature concerning the application of different soil compaction methods, little is mentioned about the mechanism, which causes the rearrangement of soil particles and densification. In order to obtain a better understanding of the compaction process, the stress-strain behavior of granular soils will be discussed first. Then the most important factors controlling energy transfer from source to the soil and in the soil will be briefly addressed.

5.1 Energy Transfer from Source to Soil

Different types of energy sources can be used for soil densification. However, the basic mechanism, governing the energy transfer from the source to the soil will be illustrated by the MRC compaction probe. A fundamental criterion is that the maximum energy, which can be transmitted from the probe to the surrounding soil in the plastic zone depends on the strength τ_f (or in the case of the CPT, f_s)

$$\tau_f = v_{max} \cdot Z_s = v_{max} \cdot C_s^* \cdot \rho \tag{19}$$

where v_{max} is the maximum particle vibration velocity, Z_s is the soil impedance and ρ is the bulk density. Similar relationships do apply to the energy transfer at the base of the probe. From equation (19), the maximum particle vibration velocity can be estimated.

$$v_{max} = \gamma_f/(C_s^* \cdot \rho) \tag{20}$$

It should be noted that the shear wave velocity C_s^* corresponds to a wave propagating at large strains and is significantly lower than the elastic shear wave velocity, which is determined by seismic field tests. Assuming a medium dense sand with a sleeve friction f_s of about 100 kN/m², a shear wave velocity C_s^* of 50 m/s and a bulk density of 2 t/m³, a maximum particle vibration velocity of 1 m/s can be transmitted to the soil. The shear strain level γ can be estimated from the following relationship

$$\gamma = v_{max}/C_s^*$$ (21)

Thus, for the above assumed case, the shear strain level adjacent to the compaction probe is on the order of 2% (1/50). Clearly, the soil is in the plastic state and sets an upper limit on the propagating vibration energy. Figure 21 shows a typical stress strain relationship for a sandy soil, as determined in the laboratory by a resonant column tests, Massarsch (1983). The vertical axis shows the reduction of the maximum shear modulus as a function of shear strain. As mentioned above, the shear modulus is related to the shear wave velocity according to the following relationship

$$G = C_s^2 \cdot \rho$$ (10)

Thus the shear wave reduction factor is the square root of the shear modulus reduction factor.

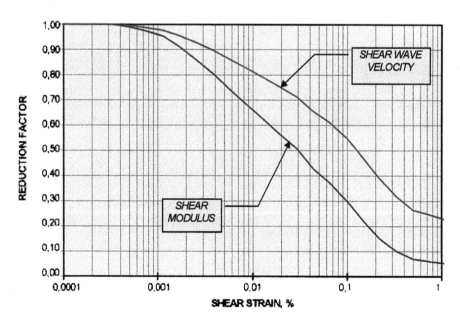

Fig. 21. Reduction of shear modulus and shear wave velocity as a function of shear strain in a saturated sand

5.2 Vibration Propagation from the Source to the Surrounding Soil

It is apparent that at a shear strain level exceeding 1%, the shear modulus has decreased to a value on the order of 5% of the maximum value, and the shear wave velocity to about 20%. Figure 21 also suggests that when the shear strain level is lower than about 10^{-3}% then the strain-softening effect is negligible. With this information, three zones can be identified around the vibration source:

- elastic zone: strain level is below 10^{-3}%, where no permanent deformations can be expected,
- elasto-plastic zone: strain level between 10^{-3} and 10^{-1}%, where some permanent deformations will occur, and
- plastic zone: where the soil is in a failure condition and experiences large deformations.

These three zones are indicated schematically in Fig. 22. Also shown is the

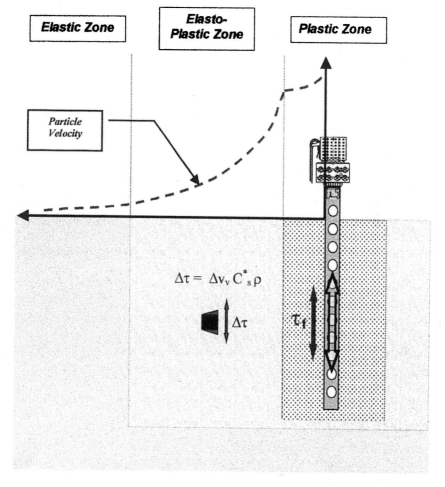

Fig. 22. Transfer of energy from the compaction probe to the surrounding soil

attenuation of the vibration velocity (particle velocity) in the ground. In the plastic zone the vibration velocity is relatively constant and limited by the shear strength of the soil. The vibration amplitude attenuates rapidly in the elasto-plastic zone. In the plastic, and the elasto-plastic zone, the wave propagation velocity is strain dependent and increases with distance from the energy source. In the elastic zone, the wave propagation velocity is constant.

Although the above presented picture is based on several simplifications, it provides a rational basis for the fundamental understanding of the parameters governing the compaction process.

5.3 Compaction Criteria Based on Vibration Velocity

From Figure 21 and assumed or measured values of the "elastic" shear wave velocity it is possible to establish ground vibration levels which are related to the densification process. It can be assumed that when sandy soil is subjected to repeated vibration cycles with a strain level exceeding $10^{-2}\%$, permanent shear deformations (soil compaction) will occur. The "elastic" shear wave velocity in most medium dense, sandy soils is on the order of 150 m/s and the "elasto-plastic" shear wave velocity is on the order of $C_s = 0.8 \cdot 150 = 120$ m/s. From equation (21), the particle velocity corresponding to this strain level can be estimated, and is on the order of $12 \cdot 10^{-3}$ m/s ($120 \cdot 10^{-2}/100$). This order of magnitude is in good agreement with experience form vibratory compaction projects. The compaction effect in the soil can thus be assessed and monitored by measuring the ground vibration velocity in the vicinity of the compaction probe. The above-described concept has been incorporated in the electronic process control system of the MRC method.

6. EFFECTS OF COMPACTION ON GEOTECHNICAL PROPERTIES

6.1 Change of Lateral Effective Stress

Man-made sand fills are usually normally consolidated prior to compaction. However, a pre-consolidation effect is obtained as a result of vibratory soil compaction, which causes the fill to become overconsolidated. Leonards and Frost (1988) have pointed out the importance of the stress history in order to assess realistically the settlements of compacted granular fills. Figure 23 shows the stress path for a soil element before, during and after compaction.

The initial state of stress for a normally consolidated sand fill is indicated by point A. During vibratory compaction, high centrifugal forces are generated (up to 4.000 kN) and thus the vertical and the horizontal effective stress are increased temporarily along the K_0-line to point B. However, as a result of soil densification, the inclination of the K_0-line is changed and the stress point C is

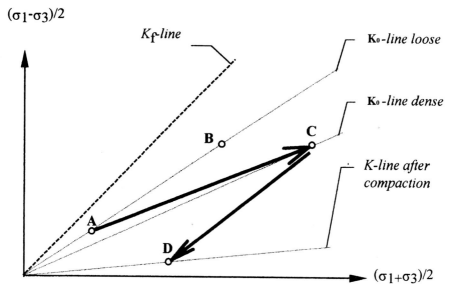

Fig. 23. Stress path for a soil element before (A), during (C) and after compaction (D)

actually reached instead of B. After compaction, the vertical stress is reduced (at zero lateral strain), and the stress conditions correspond to D. It is important to note that the soil has not only been densified but also been pre-loaded, resulting in a permanent increase of lateral effective stress.

The degree of pre-consolidation depends mainly on the compaction method. At predominantly horizontal vibrations, the compaction effect will thus be lower than if the soil is subjected to both vertical and horizontal stress cycles.

The modulus is increased as a result of soil densification (reduced void ratio), but also the stress conditions have changed (from A to D), cf. Fig. 23. As point D is located further away·from the failure line, the settlements for a load increase at D will be smaller than for a similar load increase at A.

An increase of lateral stress by compaction has been observed on many soil compaction projects, e.g. Frost and Leonards (1988) and Massarsch (1991). This change of the lateral effective stress, which is presently not taken into account, is especially important for compacted, man-made sand fills.

It should be noted that the sleeve resistance of CPT can provide quantitative information on the increase of lateral effective stresses as proposed by Massarsch (1994) and that the sleeve friction, f_s is affected by changes of the horizontal effective stress. If the sleeve friction value before compaction f_{s1} and after compaction f_{s2} are compared, an improvement value n can be determined, which reflects the change in lateral stress,

$$n = f_{s2}/f_{s1} \cong \sigma' hD/\sigma' hA \qquad (22)$$

where σ'_{hA} and σ'_{hD} correspond to the stress conditions at Points A and D in Fig. 23, respectively. Massarsch (1997) has proposed a method to estimate the overconsolidation effect from the friction sleeve measurements. Alternatively, the DMT or PMT can be used to assess in situ stress changes.

The "overconsolidation effect" which is caused by soil compaction is of great importance for the realistic prediction of settlements. If this effect is not taken into consideration, excessive and unrealistic settlement estimates will be made.

6.2 Increase of Soil Strength and Stiffness with Time

Another important factor, which is not always appreciated, is the increase of soil strength and soil stiffness with time, e.g. Mitchell (1998). Post densification CPT results suggest that natural and man-made deposits of clean sand may gain in strength with time. The mechanism causing this phenomenon, which is not related to volume change or porewater pressure dissipation, is not yet well understood and may not occur to the same extent at all compaction projects. However, the increase is often significant and should be taken into consideration. In the absence of reliable prediction methods, it is recommended to verify post densification improvement of soil properties by field trials. The CPT and DMT are effective and efficient for assessing the initial conditions and for evaluating changes after ground improvement.

7 SUMMARY AND CONCLUSIONS

Deep compaction of granular soils is used increasingly and extensive research is presently under way. In the present paper, an attempt has been made to discuss the most important aspects of deep soil compaction. In comparison to other foundation methods, such as pile driving, deep compaction requires careful planning and supervision during all phases of the project. The active design concept can provide valuable information and serves as a basis for optimisation of the compaction process.

The soil placement method can be of significance for the geotechnical properties of reclaimed land, to be compacted. Therefore, if possible the placement process should be supervised in order to avoid contamination with fine-grained particles. Soil investigation methods, which are frequently used in connection with compaction process, are described and their advantages and limitations discussed. The cone penetrometer is the most suitable investigation and monitoring tool. An important aspect of data interpretation is the correction of the cone resistance with respect to the mean effective stress. This procedure eliminates an important uncertainty associated with the CPT. Sleeve friction measurements are usually used for the determination of the friction ratio. However, the increase of the sleeve friction after compaction can provide valuable information regarding changes of lateral stress. Recent developments of the cone penetrometer, such as pore pressure sounding (piezo cone) or the seismic down-hole cone offer new possibilities for compaction control.

The most common design aspects for compaction projects are determination of relative density, settlements due to static loading and soil strength, which governs stability and bearing capacity. Another important application is the mitigation of liquefaction hazards in seismic areas.

The most common deep compaction methods have been discussed, and recent developments with respect to equipment and/or test execution were addressed. Probably the most important recent developments have taken place with respect to the use of hydraulic vibrators. These can be used for surface compaction (vibratory compaction plate) or deep vibratory compaction (MRC method). Important improvements have also taken place regarding the use of electronic monitoring systems. These can be used to supervise and verify the execution of compaction projects, but can also assist the machine operator during the execution of the work.

An attempt was made to outline the basic principles, which govern the energy transfer process during compaction. It was shown that sand exhibits strain-softening behavior. At a strain level of about 10^{-2}%, permanent deformations occur, which are mainly responsible for the compaction effect. It is possible to estimate the required particle velocity for achieving plastic deformations in the zone surrounding the compaction probe. The proposed hypothesis appears to capture some of the most important aspects of soil compaction.

Finally, some important changes of geotechnical properties, which can occur during compaction, are discussed. Probably the most important, and not yet generally appreciated effect is the permanent increase of the lateral effective stress, which results in a overconsolidation effect, which is important for a realistic settlement analysis. Another factor of practical importance is the increase of soil stiffness and strength with time. However, no rational explanation has been given for this effect. It is therefore difficult to incorporate this aspect at the design stage. Instead, the time-effect must be investigated on site by trial compaction.

REFERENCES

- (1985). Canadian Foundation Engineering Manual, 2nd Edition. Canadian Geotechnical Society. BiTech Publishers Ltd. 456 pp.
Baguelin, F. J., Bustamante, M. G. and Frank R. A., (1986). "The pressuremeter for foundations: French experience". Proceedings of In Situ 86, a Speciality Conference sponsored by the Geotechnical Engineering Division, ASCE, Geotechnical Special Publication No. 6, pp. 31-46.
Belotti, R., Ghionna, V., Jamiolkowski, M., Lancellotta, R. and Manfredini, G., (1986). "Deformation characteristics of cohesionless soils from in situ tests". Proceedings of In Situ 86, a Speciality Conference sponsored by the Geotechnical Engineering Division,

ASCE, Geotechnical Special Publication No. 6, pp. 47-73.

Broms, B. B., (1986). "Penetration tests". Fourth International Geotechnical Seminar, Field Measurements and In-situ Measurements, Nanyang Technological Institute, pp. 1-29.

Choa, V., (1996)."Soil Improvement in Large Reclamation Projects". Proceedings of the Twelfth Southeast Asian Geotechnical Conference, Kuala Lumpur, 6-10 May, 1996, pp. 23-42.

Jamiolkowski, M., Ghionna, V. N., Lancelotta R, and Pasqualini, E. (1988). "New correlations of penetration tests for design practice". Proceedings Penetration Testing, ISOPT-1, DeRuiter (ed.) Balkema, Rotterdam, ISBN 90 6191 801 4. pp. 263-296.

Janbu, N. (1963). "Soil compressibility as determined by oedometer and triaxial tests Proceedings, European Conference on Soil Mechanics and Foundation Engineering, Wiesbaden, Vol. 1, pp. 19-25 and Vol. 2, pp. 17-2 1.

Leonards, G. A. and Frost, J. D. (1988). "Settlement of shallow foundations on granular soils. ASCE Journal of Geotechnical Engineering, Vol. 114, No. 7, pp. 791-809.

Lunne, T., Robertson, P. K. and Powell, J. J. M., (1997). "Cone penetration test in geotechnical practice". Blacker Academic & Professional, 312 pp.

Massarsch, K. R., (1983). "Dynamic and Static Shear Modulus", Discussion Session 10. Soil Dynamics," Proceedings, 10th Internat. Conference on Soil Mechanics and Foundation Engineering, Stockholm, 15-19 June, 1981, Proceedings, Vol. 4, pp. 880-881.

Massarsch, K. R., (1991). "Deep Vibratory Compaction of Land Fill Using Soil Resonance", Proceedings, "Infrastructure'91", Intern. Workshop on Technolgy for Hong Kong's Infrastructure Development, December 1991, pp. 677-697.

Massarsch, K. R., (1994). "Settlement Analysis of Compacted Fill". Proceedings, XIII. International Conference on Soil Mechanics and Foundation Engineering, New Delhi, India, Vol. 1, pp. 325-328.

Massarsch, K. R. and Westerberg, E., (1995). "The active design concept applied to soil compaction". Bengt B. Broms Symposium in Geotechnical Engineering, Singapore, 13-15. December 1995. Proceedings, pp. 262-276.

Massarsch, K. R., Westerberg, E. and Broms, B. B., (1997). "Footings supported on settlement-reducing vibrated soil nails". Paper submitted for publication, XIV, International Conference on Soil Mechanics and Foundation Engineering, Hamburg 97, Vol. 3, pp. 1533-1539.

Mayne, P. W., Jones, J. S., and Dumas, J. C. (1984). "Ground Response to Dynamic Compaction". Journal of the Geotechnical Engineering Division, ASCE, Vol. 110, No. 6, June, 1984, pp. 757-774.

Mitchell, J. K., (1981). "Soil improvement — State-of-the-Art", Proceedings, 10th ICSMFE, Stockholm, Vol. 4, pp. 509-565.

Mitchell, J. K., (1986). "Ground improvement evaluation by in-situ tests". Proceedings of In Situ 86, a Speciality Conference sponsored by the Geotechnical Engineering Division, ASCE, Geotechnical Special Publication No. 6, pp. 221-236.

Mitchell, J. K., Cooke, H. G. and Schaeffer, J., (1998). "Design considerations in ground improvement for seismic risk mitigation". Proceedings, Geotechnical Earthquake Engineering and Soil Dynamics III, Seattle, Wa., ASCE Geotechnical Publication No. 75, Vol. 1, pp. 580-613.

Nelissen, H. A. M., (1983). "Underwater compaction of sand-gravel layers by vibration plates". Proceedings ECSMFE Helsinki, Vol. 2, pp. 861-863,

Peck, R. B., Hanson, W. E. and Thornburn, T. H., (1974), "Foundation Engineering". John Wiley & Sons, p. 312.

Robertson, P. K. and Campanella, R. G. (1983). "SPT-CPT correlations", Journal of the Geotechnical Engineering Division, ASCE Vol. 109, No. 10, pp. 1449-1459

Schmertman, J. H., (1970). "Static cone to compute static settlement over sand". American Society of Civil Engineers, ASCE Journal of Soil Mechanics and Foundation Engineering, Vol. 96, SM3, pp. 1011-1043

Schmertman, J. H., (1986). "Dilatometer to compute foundation settlements". Proceedings of In Situ 86, a Speciality Conference sponsored by the Geotechnical Engineering Division, ASCE, Geotechnical Special Publication No. 6, pp. 303-321.

Seed, H, B., (1976). "Evaluation of soil liquefaction effects on level ground during earthquakes". Liquefaction problems in Geotechnical Engineering, Proceedings of ASCE Annual Convention and Exposition, Philadelphia, pp. 1-104.

Seed, H. B. and De Alba, P., (1986). "Use of SPT and CPT Test for evaluating the liquefaction resistance of sands". Proceedings of In Situ 86, a Speciality Conference sponsored by the Geotechnical Engineering Division, ASCE, Geotechnical Special Publication No. 6, pp. 281-302.

Woods, R. D., (1986). "In situ tests for foundation vibrations". Proceedings of In Situ 86, a Speciality Conference sponsored by the Geotechnical Engineering Division, ASCE, Geotechnical Special Publication No. 6, pp. 336-375.

6

The Fourth Tidal Wave of Geotechnical Engineering

Qian Qi-Hu

Science and Technology Committee of General Staff, P. R. China

1. SEVERAL WAVES IN THE DEVELOPMENT OF GEOTECHNICAL ENGINEERING

Human history has witnessed several tidal waves of geotechnical engineering exploitation. The first one can be traced back to the early days of human history, when human beings lived in caves so as to be sheltered from wind, snow and storm, and the teeth and claws of beasts. During the Industrial Revolution, the advent of the second tidal wave was announced by large-scaled constructions of mine works, urbanization of capitalist countries symbolized by the underground embedment of such municipal engineering facilities as drainpipes and flow pipes, and the building of urban subways. Meanwhile, underground fortifications and hydropower stations were greatly developed. The third tidal wave of geotechnical engineering is characterized by the extension of human activities to underground, which result in constructions of underground commercial streets, underground parking lots, and various underground recreation and sports facilities such as art exhibition house, music hall, gymnasium, swimming hall and skating yard. Since the end of the twentieth century, the upgrade and development of the fourth tidal wave has been promoted by formulation of continuable development strategy by the United Nations and each country in the world and the heightening of people's sense of environmental protection and their awareness of the importance of ever lasting facilitation of resources. The characteristics of the fourth tidal wave of geotechnical engineering are the improvement of urban environment and the protection of natural ecological environment. The high waves of geotechnical engineering rise one by one. The later waves do not substitute or exclude the former ones but improve and perfect them. The construction of underground municipal works and subways during the second tidal wave have been continuing in the 3rd and 4th wave, of course, in a higher level. The construction of underground commercial streets, garages and various recreation and sports facilities in the 3rd tidal wave will go on in the 4th one and in more cities. Over the past 30 years, excavation techniques for tunneling in solid rock or in complicated geology conditions have made great progress. The solidity of rock, which used to be regarded as a disadvantage, turns into advantage, for it helps increase the span of underground cave and enable rock to stand higher speed of tunneling. Since 1990s, some advanced techniques, such as the directional and orientation technique of underground construction (DOT), the underground prospecting technique (UPT) and the automatically controlling construction technique of tunneling and lining, have been applied to practice in varying degrees. DOT is represented by laser guiding technique and GPS orientation technique, while UPT is represented by remote sensing technique and by through-ground radar technique. As a result, the 4th tidal wave is becoming wider in scope, larger in scale and higher in level.

2. THE FOURTH TIDAL WAVE: PROTECTION OF NATURAL ECOLOGICAL ENVIRONMENT AND IMPROVEMENT OF URBAN ENVIRONMENT

In the present world, population is increasing. Resources are using up. Cultivated land and forest are reducing rapidly. Soil is deteriorating. Desert is creeping. Furthermore, greenhouse effect, acid rain and environmental pollution become more serious. All these environmental problems hinder the development of world economy. They are especially severe problems facing the economic development of our country. The loss due to the environmental pollution and ecological destruction is more than 200,000 million yuan per year. Geotechnical engineering, which is treated as a friend to the environmental engineering, is playing important roles in the protection and improvement of environment. Of course, it will play more roles and become one of the principal measures in continuable development.

The prime culprit of forming greenhouse effect and acid rain is over-consumption of such resources as coal and oil. Thus, the effective way to decrease the area of acid rain and to abate greenhouse effect is to save resources. To do so, the geotechnical engineering plays an important part. Building energy consumption is defined as the total yearly energy consumed by one building after it is set up. According to statistics, building energy consumption, which takes a high proportion in General National Energy Consumption (GNEC), takes about 30 percent of GNEC in European or American countries, and the energy consumed by heating, ventilating and air-conditioning of buildings takes 19.5 percent of the NGEC. Underground space has the advantage of heat insulation, hence the avoidance the influences of surface temperature due to wind, rain and sunshine. It is found from measurements that the temperature hardly vary in days one meter down the earth surface and keeps constant at a depth of 5-meter. As a result, underground buildings consume less energy than the upground ones. Based on large amount of tests carried out in U.S.A. to compare the underground buildings to the upground ones, the excess energy-saving ratios of the underground buildings in five tested districts are: 48% in Minnesota and in Boston, 58% in Salt Lake City, 51% in Louisville and 33% in Houston. The highest energy-saving ratio is that of underground buildings set up among abandoned limestone mine in Kansas City. Compared to the upground buildings, it is 60%, 70% and 47.9% respectively for service buildings, storehouses and manufactures.

It should be specially pointed out that the exploitation and utilization of geotechnical engineering open up a wide and effective way for the use of natural clean energy resources, especially the use of recoverable resources. Solar energy is one huge clean recoverable energy resource, but it varies unstably with seasons. Solar radiation is rich only in summer and needs storing seasonally. It is usually the best even the unique choice to store heat in underground water, rock or soil. To use underground space as the storehouse of natural ice in winter for airconditioning in summer is economical as well as clean and recoverable, as being used in North Europe. Because rock and soil are thermally stable and airtight, underground storage hardly loses heat and cold, and need no thermal insulation materials. Furthermore, the advantage of self-support of rock results in simple cave structure and low maintenance cost, and brings about the effective usage of natural power resources and large amount of residual industry heat thereby. For example, a 15,000 m^3-hot-water reservoir is built in rock cave in Alaska, about 150 kilometers northwest Stockholm, Sweden. The heat source is from waste incineration. The reservoir joins up with local heat offering system by one heat transformer. The temperature range of heat storage is from 70 to 150 degree centigrade. The construction was invested 4,000,000 dollars and built up in 1982, and the corresponding tests were accomplished in 1984. It works as the heat offering system in Alaska and helps save more than 400 m^3 oil every year. In Germany and Japan, the technique for underground storehouse of compressed air is now in research. For example, a compressed air storehouse was built in salt rock stratum in Germany. The power is 290,000 kW and the stored air pressure is 8,000,000 Pa. The technique for underground magnetic storehouse of superconductivity, which is a spirally arranged annular cave, is cooperatively studied by Japanese and American researchers. Both techniques can help store residual electricity energy of low peak load for the use in high peak load, hence the power and power consumption of the power station are saved. The research of producing unexhausted and non-polluting energy—deep heat-dry-rock (HDR) electricity generation—is carried out in the United States and in Germany. The first 10 MW HDR power station was set up on Fenden Mountain in Kardeila by Los Alamos Lab of the United States in June, 1984, which mainly consisted of two drills of more than 4,000 meters deep and their connective holes. First, the HDR is pressed to make numerous cracks by watercracking method, forming fissure path and heat storage space. Then 21,300 m^3 cold surface water is poured at a time, forming circulation at drills' bottom. The cold water is then rapidly turned into high-temperature and high-pressure steam, which directly enters air-turbine for generating electricity. To build cold storage under the ground can greatly save energy resources, and need few heat insulation material. Moreover, underground cold storage employs simpler temperature regulation system than that on the ground, and consumes much less running- and maintenance-cost by 20-50%, according to statistics.

In our country, the shortage of water resource is increasingly serious. It mainly results from the unbalanced distribution of water resource in space and time. In many places, a great deal of fresh water of rainy season flows freely into seas owing to the deficient water-storing facilities. Besides the construction of reservoirs in upper- or middle-reaches of big rivers, the water shortage condition can be partially improved by making use of loose rock stratum, cracks in rock, caves in rock or dried subterranean aquifer, as in Norway and in Finland, or by making artificial underground rivers, reservoirs and snowmelt trough which can store surplus rain and snow in rainy seasons for the need of dry seasons, as in some Japanese cities such as Tokyo, Yokohama, Sapporo, Nagoya.

The two primary sources of urban air pollution are pollutants such as sulfur dioxide and suspended granular formed in coal burning and pollutants such as nitrogen-oxide and carbon monoxide resulting from tail gas of automobiles. The latter has become the major cause of air pollution with the increasing number of automobiles in large cities. In order to eliminate the former pollutants, the aboveground buildings should be taken place by the underground ones because underground building can directly decrease energy consumption and pollution. In addition, it is a primary solution to replace coal with nature gas as the change of fuel resource structure, or to change disperse heat offering system into concentrating system. Therefore, it is necessary to built up underground heat- and gas-offering pipe net. For example, 120 km long tunnel for heat offering was constructed in Stockholm in Sweden. One of the fundamental solutions of automobile tail gas is to decrease the quantity of vehicles entering downtown, and in this respect, the experience of some European or American countries can be referred to. From the experience of Canada, it is found to be effective for air-pollution control and automobile quantity diminishing to set up underground traffic system, which combines underground network of communication lines to underground passageway, to underground parking lots, and to suburb trains which enter city center by underground railway. The underground network of railway tracks in Montreal consists of two east-west subway axis, two north-south subway axis, subway rings and two underground suburb railways stretching into downtown. Almost 60 high buildings, which are synthetically used for commerce, office and residence, were connected with the platforms of underground network of communication lines by more than 150 underground entrances and the corresponding passages. In central urban district, most people move by underground traffic, whereas, people outside the central district go to work or go into business by suburb railways or by cars, which could pull up at subway stops on the rim of the central district. And then, they go to their destinations by subways or enter buildings by underground passages. Of course, the cars can be parked in underground parking lots. As a result, the automobiles moving in the central area decrease to minimum quantity, and the tail gas given off by cars is minimized at the same time. In some European countries, in order to lower the degree of traffic congestion and to reduce air pollution in cities, no-stopping electric charging system is installed at every crossing from suburbs to town, which charges the urban district traffic jam fee at traffic peak. The actual situation indicates that this traffic mode and the measure are effective, especially in northern America and in northern Europe, where the winter lasts a long time and is often with North Pole wind or storm. This traffic mode ensures various city activities carried out normally, making a show of its advantage.

In order to reduce air-pollution, trees should be planted in and around the city. Civil afforestation can effectively improve air quality and clear poisonous matters. The green blocks in cities can lower wind speed, hold up floating dust and, due to the photosynthesis, absorb carbon dioxide and release oxygen. Many plants can absorb poisonous matters such as sulfur dioxide and Benzopyrene. It is also well known that the greens have the function of killing bacteria in air. Moreover, green plants help to lower air temperature, increase relative humidity and mitigate urban Heat Island Effect. When vegetation ratio of a city reaches 50%, the pollutants in air can be basically under control. Thus, the target as pointed out in the 21st Century Agenda of China is: "To devote major efforts to developing city afforestation, and by 2000, the plant cover ratio will have reached 30% and the public green area for per capita shall have reached 7 m^2 or so." The United Nations suggests that the public green area of a city be 40 m^2 for per capita. Thus, more urban ground needs clearing out for planting. For example, in an especially big city with a population of 5,000,000 people, 20 m^2 of public green area for per capita necessitates 100 km^2 of land, which leads to a large scale of underground construction to clear out urban ground for environment improvement. In Washington, the uniform distributed green area of 40 m^2 for one person makes the city like a garden. There is a green area of 70 m^2 for every person in Warsaw City, which is entitled Green City in the world and is a real city of ecological garden. It is welcome to see the afforestation in quite a few cities in our country and to see the appearance of public green squares and civilian squares. However, it would not last long if the underground space were not exploited at the same time. Because the total cultivated land warning system has been implemented in our country, in order to guarantee the total cultivated land in dynamic equilibrium, the land for urban construction is very limited and insufficient.

It cannot do without geotechnical engineering to harness urban water condition and waste condition. In our country, among the 136 rivers which flow through cities, there are 105 rivers whose water pollution seriously exceed standard, and the groundwater is contaminated in more than 50% of the corresponding cities. In our country, the municipal waste lump is as much as over 6,000,000,000 tons, and it is common for a city to be surrounded by garbage. In some advanced cities of developed countries, the unified system of sewage collecting, conveying and disposing, and the unified system of garbage classifying, collecting, conveying and managing, are set up underground. In Sweden, the sewage disposing plants of drainage system are all set up under ground. In the city of Stockholm alone, there are totally 200 km long drainage tunnels and six large-scale sewage disposal plants in all, and the disposal ratio is 100%. Of course, there are underground sewage disposal plants in middle- and small-sized cities as well. As a result, the urban source of water has been protected and the Baltic Sea is immune from pollution. Sweden is the first country where transporting rubbish by pipes is tested, and in early 1960s, the air blowing system was in research for this purpose. For instance, in a resident district where 1700 people live, a set of rubbish pipe transporting system making use of air blowing method was set up in 1983, which was estimated to serve sixty years. Because the recovering system was set up to cooperate with the disposal system, the investment could be recovered in 3~4 years. In Helsinki, the sewage disposal plant, which is built in a 1,000,000 m^3 rock cave beneath a future resident district, works sophisticatedly and efficiently and can dispose civil daily sewage of 700,000 residents and industrial waste, and thus helps to save the valuable construction land above ground and to remove the smell sent forth by sewage disposal. In Florida, U.S.A., natural classification- and collection-systems of garbage have been installed in the basements of high buildings in recent years. Owing to the sealing property of underground space, the unified system of waste collection and disposal mentioned above has the function of decreasing the urban pollution from sewage and refuse to the lowest degree.

It is an important way of bringing urban landscape environment under control to carry out underground constructions and to make use of underground space. In many developing countries, including our country, the rapid and extensive urbanization results in crowded structural space, such as forest of high buildings, crisscross overhead roads, shafts and wires densely covering streets for lighting, electricity offering and communications, vehicles parking freely by roadside, etc. The deficient space is unpleasant and makes people uneasy. However, the exploitation of underground space can make the three-dimensional urban structure closer, enhance the intensity degree of cities and decrease people's requirement for traffics in cities and help the urban service meet the demands of the residences, hence make more comfortable urban environment. In developed countries, underground space has been successfully used to keep the urban environment pleasant and to protect the historical culture landscape. In some advanced cities, public tunnels for municipal pipelines—common channels—are widely built. For example, a 130-km-long common channel has been buried under the city of Moscow. In Japan, the total length of common channels is planned to reach 526 km. Setting up common channel can remove the spiraling state of wires, shafts and transformers over town, and increase cities' capability to resist disasters. The construction of underground commercial streets and underground installations of recreation, sports catering has improved the condition of crowded structural space and concentration of steam of people above ground. Furthermore, it can enlarge the open space in cities. For instance, in the central area of Lille in Paris, one former crowded center of food transaction and wholesale has been turned into a multi-purpose public square where plants cover most area, beneath which various activities such as commerce, recreation, traffic and sports proceed, forming a largescaled underground synthesis. The total area of the synthesis is over 200,000 m^3 and consists of four floors. The Louvre Palace, which is in the center of Paris, is world famous, and its original architectural style must be maintained. There is no extensional land for expansion projects. Under these circumstances, designers made use of the spaces beneath the Napoleon's Square, which is surrounded by palace buildings, to achieve the modernized transformation of this classical architecture. A large number of underground garages solve the problems of crowded traffics and vehicles' casual parking by roadsides. There are 383 underground garages built in Paris, which can accommodate more than 43,000 vehicles. The biggest European garage with 4 floors is under Feiyor Road and for 3000 vehicles. The urban construction of underground high-speed and large-flow network of communication lines can not only mitigate the historical problem of crowded traffic, but also solve the landscape problem of overhead roads. On the Central Avenue in Boston, U.S. was ever built overhead roads in 1930s. It has been demolished and rebuilt as underground highways in 1990s due to the issues of landscape, noises and shake. From the viewpoint of urban environment, even overhead tracked high-speed traffic is not a proper plan because of its vibration, noise and electromagnetic radiation. In cities with a population over 2,000,000, 77.5% tracks of the tracked traffic mode are for subways, and their running length is 90.5% of the total length of tracked traffic. Even the overhead light-track above ground should transform to be under the ground when entering downtown. Such installations as underground automobile ways, grade separation for vehicles, pedestrian crossing and walking street can separate the steam of people from that of vehicles and are helpful to decrease the aboveground steam of people and of vehicles. Moreover, they are important to ease the crowded traffics and to improve the ground landscape. The subways in Moscow, which have the largest number of passengers in the world, transport 2.6 billions of passengers every year, and attract a large steam of people to underground. In Japan, one hundred and forty-six underground streets have been set up in at least 26 cities, and the number of people passing in and out is about 12 millions every day, which is one-ninth of the total population of Japan. In the city of Houston, the underground walking streets, the whole length of which is 4.5 km, link up 350 large-sized structures. In the southern city of Dallas, 29 underground walking streets has been built up, which can avoid the influence of summer high temperature, to link up the main public buildings and activity centers. Finland is a northern country of few and scattered population. However, it highly prizes the urban environment of low density of buildings and the green natural environment. In Helsinki, which is surrounded by seas on three sides, only structures with less than six floors are permitted to be built so as to protect the urban environment as mentioned above. Meanwhile, underground spaces are widely used in order to keep the urban intensity and density as well. A great quantity of facilities of culture, sports, commerce, recreation and even churches are built underground.

In the areas of high density and high urbanization, the underground expressways and high-speed railways are being designed and constructed in order not to disturb the peace and the natural ecological environment between suburbs and towns by expressways or by high-speed railways. As a subway project in urbanized area of new generation, the magnetic suspension train (similar to the body of plane) which will run in underground tunnels of partial vacuum is in research in Switzerland, with designed running speed of 400 km/h and designed passenger number of 800. It is planned to accomplish the main research projects and acquire the special permission of technique application in 1989. In Japan, underground expressway system is being constructed in the area of Tokyo.

3. GEOTECHNICAL ENGINEERING AND ENGINEERING GEOLOGICAL ENVIRONMENT

Geotechnical engineering, which helps protect natural ecological environment and improve urban environment, is essentially a friend engineering of environment. On the other hand, it changes the original engineering geological environment. This change may be favorable. For example, in our country, the engineering of moving the water of Yangtse River from south to north and moving the water of Brahmaputra to the north, which are in expounding and in proving, can stop the soil deterioration and change desert into fertile farmland. Of course, the constructions should stand the tests of history. However, the change of original engineering geological environment may also be unfavorable and may result in hazards to the geological environment. For instance, geotechnical engineering may destroy underground water system and lead to dry-up of springs, collapse of the ground surface and cracks in soil; the tunneling breaks down the rock structure and thus causes landslide or mud-rock flow; the excavations disturb soil stratum and result in subsidence of urban ground surface. Because the underground constructions are irreversible, the changes mentioned above are irreversible as well and the results may be serious. Therefore, full attentions should be paid and thorough demonstration and analysis, and if necessary, emulation or stimulation should be held to the engineering sensitivity of bedrock, of soil layers and of underground water systems before constructions. Comprehensive engineering investigation with modern techniques is required as well. When making program designs, the problems of harmful pound and beneficial protection aroused by geotechnical engineering should be fully demonstrated. The disadvantageous pound must be treated specially in construction in order to remove danger, to minimize its unfavorable influence on the geological environment, and to derive the maximum benefits.

We should not refrain from doing something for fear of a slight risk. We should not echo the noises in the 4th tidal wave of underground engineering either. For example, the International Anti-Dam Committee put forward such slogan as "Let rivers flow freely". They oppose dams in rivers. They denounce the Aswandam in Egypt. They protest the SanXia Engineering in Yangtse River and even request the premier of China on trial. There are also people of civil engineering field point out that geotechnical engineering is essentially harmful to engineering geological environment. The extreme environment protectionism has completely confused the essentials and the nonessentials. They have ignored the fact that the Aswandam helps change desert into fertile farmland and increase cultivated land by times. They cannot bear to see that SanXia Engineering will improve the adverse circumstances of Yangtse River which is flood almost every year. However, the fourth tidal wave of underground engineering will develop smoothly in the direction of remarking nature for continuable development of human.

4. SEVERAL TRENDS IN THE 4TH TIDAL WAVE OF GEOTECHNICAL ENGINEERING

(1) Synthesization: The Leading Trend in Urban Underground Space Exploitation

Synthesization is first characterized by the appearance of underground syntheses. In some metropolises of Europe, North America and Japan, during the process of new city proper construction and old city proper re-exploitation, underground syntheses of various scales have been set up, standing for one of the tectonic styles of metropolis modernization. Secondly, the synthesization of city shows the combination of underground walking street system and fast-track system with expressway system, and the combination of underground syntheses with shift hub of underground communications. Thirdly, synthesization expresses not only the difference in functions of ground and underground space, but also the combined mode of their coordinated growth.

(2) Layered and Deepened Underground Space

In some advanced cities of developed countries, the shallow parts of underground space have been almost fully developed. Along with the progressive perfection of the technique and the corresponding equipments for excavating in deeper parts, the geo-exploitation is proceeding to deeper stratum to make synthetic use of the underground resource of space. For instance, in Minnesota University of the United States, the underground structure of the department hall for Art and Mining Engineering Department is as deep as seven floors. In Vancouver, Canada, the underground garage has 14 floors, and its total area is 72,324 m^2. The project of making use of underground space deeper than 30 m, is studied mainly by Japanese researchers. The exploitation of deep underground space resources has become the main task of future urban modernization. Following the deep-seated geotechnical engineering, the trend of layer division becomes stronger and stronger. In the layered space, people and their service area are taken as the center. The steam of people and vehicles are divided. The municipal lines, sewage disposal and waste treatment are located in different layers. Various underground communications are installed in different layers in order to decrease the interruption and to ensure the full and comprehensive use of underground space.

(3) Underground Urban Communication and Inter-communication between High-density and High-urbanization Cities

Underground communication may be the emphasis of future geotechnical engineering. In the 21st century, people will pay more and more attention to environment, beautification and comfort. People's sense of and requirement for urban environment will rise step by step as well. Formerly built overhead roads will be moved underground. The high-speed and tracked underground traffic will be the best choice for metropolis and for communication between high-density and high-urbanization cities. At the same time, the high-speed underground traffic will also be developed in urbanized areas. It will come true that magnetic suspended train runs in partial vacuum tunnels.

(4) Various TBM and Shield as General Techniques for Rapid Tunneling

With the extensive exploitation of underground space, the chance of constructing long- and big-tunnel and encountering unfavorable soil stratum will increase. Thus, the speed and safety of tunneling must be more strictly controlled. It can be well estimated that TBM and shield will play more and more important parts in tunneling.

(5) Digitalization in Drilling-Blasting Tunneling

Digitalized tunneling is controlled by predetermined computer program. One operator can operate three drilling rods at one time. The location and depth of the drilling hole can be controlled by designed programs, and the axis orientation of excavating could be simultaneously made certain by laser. As a result, the over-excavated section will be minimized and optimized, and the speed of tunneling will be raised.

(6) Smaller Section of Subway Tunnel and Lower Cost

Because linear motor has been applied to drawing subway trains, the dimension of walking chassis can be decreased, which results in a decrease of the section area of subway tunnel by more than a half. The cost of subway construction is lowered accordingly.

(7) Rapid Development of Micro-tunnel

With the diameter of 25~30cm, micro-tunnel cannot be entered by man. It is excavated and retained by remote control at the entrance. This method is fast, accurate, economical and safe, so it fits for pipes installed under high buildings, science spots, historical sites, expressways and express railways, and rivers. Up to now, 5000 km-long pipe has been built up by applying micro-tunnel technique. As the steady increase of underground pipes, micro-tunnel will be widely used in future.

(8) Extensive Use and Development of Public Tunnel for Municipal Pipelines in the 21st Century

With the development of the modernization of city and life, the various types, densities and lengths of pipe increase rapidly. The common channel will be inevitable.

(9) Increased Use of Three "S" Techniques in Underground Space Exploitation

Because of the modernization of underground orientation, geological information and reconnaissance, global planet system (GPS), remote sensing (RS) and geological information system (GIS) will be spread in the use of underground space.

7

Recipe to Define Near-Field Earthquake Motion for Structural Design

Tadanobu Sato

Professor, Disaster Prevention Research Institute, Kyoto University, Gokasho Uji, Kyoto 611 -0011, Japan

ABSTRACT

After the brief introduction of the strong motion characteristics of the 1995 Hyogoken Nambu earthquake and the history of destructive earthquakes occurred in Japan as well as chronological change of earthquake resistant design codes, we summarize what we learned from this earthquake to define near source motions and the earthquake disaster prevention measures proposed by the Japan Society of Civil Engineers. To respond the question how to define the design earthquake motion taking into account near source effects we develop simple methods to estimate acceleration response and phase spectra at the ground surface in the region near earthquake source. To estimate earthquake motion on the ground surface a formula of earthquake motion at base rock level was derived in the frequency domain. The amplification effect of the ground was introduced by using multiple reflection theory. We applied the spectral moment method to estimate an acceleration response spectrum. Actual acceleration response spectra are carefully compared with theoretically estimated response spectra at several observation points where the strong earthquake motions were recorded during the 1995 Hyogoken Nambu earthquake. The concept of group delay time is used to model a phase characteristic of earthquake motion. The time delay caused by rupture propagation on fault plane which is assumed to be expressed by a train of impulses. The effect of transmitting path and amplification effect due to the local soil condition on phase is assumed to be given by minimum phase function. Using the modeled phase spectrum we propose a method to simulate an input earthquake motion compatible with design acceleration response spectra.

Key Words: Response Spectra, Phase Spectra, Hyogoken Nambu Earthquake, Design Earthquake Motions

INTRODUCTION

Chronologies of Earthquake Damages and Earthquake-resistant Design Codes in Japan

The history of destructive earthquakes and change of earthquake resistant design codes is listed in Table 1 (Iemura, 1999). It can be seen that the earthquake resistant codes have been upgraded after the occurrence of several destructive earthquakes. Before 1970, seismic design of structures in Japan merely guaranteed the elastic strength of structures against load equivalent to 0.2 design seismic coefficient in the horizontal direction. During the 1968 Tokachioki earthquake shear failures of short column of structure elements were observed. In the first half of the 1970s, numerous studies had been conducted on the cause of shear failure and concluded that shear failure must be avoided at all costs and structure must be more ductile to prevent the collapse of structure from the seismic force exceeding the structural strength. New regulation on structural details to increase the number of transverse hoops were introduced to improve the ductile behavior of structural elements and structural system as a whole in the early 1980s. Damage survey of the 1995 Hyogoken Nambu earthquake revealed that the main cause of structural damage was shear failure of RC piers and frames. Further, steel piers experienced local buckling in response to seismic force exceeding the design level, with two cases of complete collapse. Before this earthquake steel piers had been believed more ductile than RC piers. Most of the structures that suffered major damage however were pre-1970 construction, and did not implement the lessons of 1968 Tokachioki earthquake.

The 1948 Fukui earthquake was a typical near field earthquake which directly struck Fukui City and caused very severe damage to buildings and infrastructures. Due to extremely high collapse rate of wooden houses (almost 100% in most area of Fukui City), JMA (Japan Meteorological Agency) established the new seismic intensity scale of VII in addition to VI which was the highest intensity before Fukui earthquake. The new project to develop the strong motion accelerograph (SMAC) was started after this earthquake. Damage survey report conducted by the general headquarter office (GHQ) of Far East command concluded that the maximum ground acceleration reached 0.6 G based on the survey of turning over rate of tombstones. The seismic design coefficients were not raised even though this knowledge was well known to research community because there were no engineering technologies to develop the earthquake resistant design method with reasonable construction costs to sustain this level of seismic force. From this reason the GHQ report of the Fukui earthquake had not been open to the public.

In order to design structures against such extreme earthquake motion, inelastic design method have been developed since 1960s. The basic concept of the inelastic design method is to give structures dynamic strength and ductility, making the structural system to be able to absorb seismic energy. If we could assign substantially large ductility to a structural system it would be possible to make a structure as resistant to earthquake damage as if its dynamic strength was increased 10 times (e.g. from a structure with the seismic coefficient 0.2 to that with 2.0). The ductility oriented seismic design method was adapted to the Building Standard Law in 1981 and Highway Bridge Specification in 1990 with maximum seismic coefficient of 1.0. However the strong motion record observed at JMA Kobe Station during the 1995 Hyogoken Nambu earthquake required even the allowable ductility factor of 10 in the short period range (0.1-0.4 seconds). In this range former elastic design level and allowable ductility factor is therefore requested to raise up to two times of former ones.

Table 1. Chronologies of earthquake damage and earthquake-resistant design codes in Japan

(Iemura, 1999)

1891	Nobi Earthquake (M8.0; hypocenter directly below)
1923	Great Kanto Earthquake (M7.9; hypocenter under sea and directly below) Great fire occurred
1925	Seismic design code for urban structures Design seismic coefficient: 0.1
1939	Road bridge specifications developed Design seismic coefficient: 0.2
1943	Tottori Earthquake (M7.2; hypocenter directly below)
1944	Tonankai Earthquake (M7.9; hypocenter under sea)
1945	Mikawa Earthquake (M6.8; hypocenter directly below)
1948	Fukui Earthquake (M7.1; hypocenter directly below) SMAC developed
1950	Building standard law Design seismic coefficient: 0.2

(cont.)

(cont.)

1956	Highway bridge specifications developed Design seismic coefficient: 0.1-0.35
1964	Niigata Earthquake (M7.5; hypocenter under sea) Liquefaction failures occurred
1968	Tokachioki Earthquake (M7.9; hypocenter undersea) Motion recorded at Hachinohe; RC columns sheared are destroyed
1971	Aseismic design of road bridge specifications developed Design seismic coefficient: 0.1-0.24 Modified seismic coefficient introduced Building Standard law: Stipulates that shear reinforcing bars be strengthened
1978	Miyagiken-oki Earthquake (M7.4; hypocenter under sea) Lifelines disrupted
1980	Highway bridge specifications developed (proposed new aseismic design act) Ductility requirement examined
1981	Building Standard law: stipulates capacity and ductility requirement Design seismic earthquake (M7.7; hypocenter under sea)
1990	Highway bridge specifications Aseismic design spectrum revised; dynamic analysis; bearing capacity; and possible occurrence of earthquakes three times as strong as previous ones taken into consideration Design seismic coefficient: 1.0 (inelastic design) Base-isolated structure studied and buildings constructed based on such structure
1992	Manual for base-isolated design of highway bridges prepared
1993	Kushiro Earthquake (M7.8; hypocenter under sea) Hokkaido Nanseioki Earthquake (7.8; hypocenter under sea)
1994	Hokkaido Toho-oki Earthquake (M8.1 hypocenter under sea) Sanriku Haruka-oki Earthquake (M7.5; hypocenter undersea)
1995	Hyogoken Nambu (Southern Hyogo) earthquake (M7.2; hypocenter directly below urban area)

Strong Motions Observed during the 1995 Hyogoken Nambu Earthquake

The 1995 Hyogoken Nambu earthquake (M=7.2) struck the densely populated corridor between the cities of Osaka and Kobe at 5:46 am local time, Tuesday, 17 January 1995. As of December 29, 1995 the earthquake resulted in 6,308 deaths and 43,177 persons injured. The respective numbers of completely and partially collapsed dwellings are 100,302 and 366,114. The estimated cost of damage is 10 trillion yen: 5.8 trillion yen to buildings, 1.4 trillion yen to port and harbors, 450 billion yen to highways, 420 billion yen to gas and power supply systems, and 340 billion yen to railways.

The earthquake generated a large number of strong motion recordings over a wide variety of geological site conditions, including rock and soil, as well as in structures of varying types. Several agencies, such as the Committee of Earthquake Observation and Research in the Kansai Area (CEORKA), Osaka Gas, Kansai Electricity, Japan Railway, and the Building Research Institute of the Ministry of Construction, maintain strong motion instrumentation networks. A plot of the peak horizontal ground acceleration vs. the shortest distance from the fault zone of the mainshock is shown in Fig. 1 (Ejiri et al., 1995). This is referred to the mean attenuation relationship for hard soil by Joyner and Boore (1981), for a moment magnitude of 6.9.

The cities of Kobe, Ashiya, and Nishinomiya are located between the Rokko Mountains and Osaka Bay. The ground motions are believed to be amplified in the plain at the base of the Rokko Mountains. Fig. 2 shows the distribution of active faults and the area assigned the seismic intensity of 7 (JMA scale), defined as the zone in which the collapse of wooden houses was more than 30%. The severely damaged area is 1-2 km to the south and parallel to the active faults

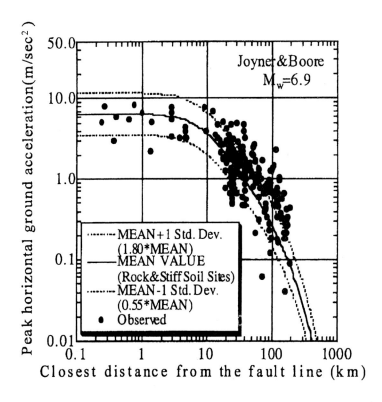

Fig. 1. Relationship between peak acceleration and shortest distance to fault (Ejiri et al., 1995). Thick line: Attenuation relationship proposed by Joyner and Boore (1981); Broken lines: The line with the standard deviation σ.

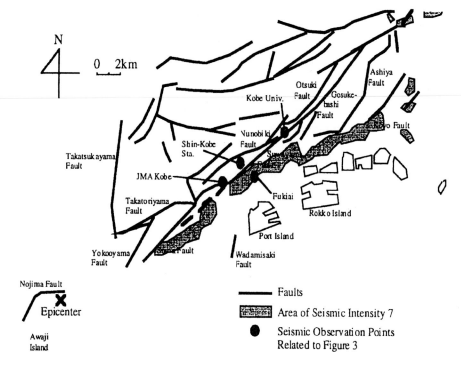

Fig. 2. Map of faults and area of seismic intensity 7 (Suetomi, 1995).

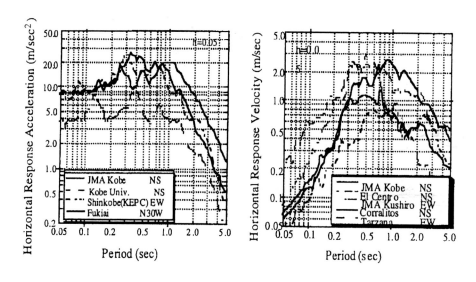

Fig. 3. Comparison of accleration and velocity response spectra (Ejiri et al., 1995).

and has a width of about 2 km. The acceleration and velocity response spectra for the Kobe Maritime Meteorological Observatory and other sites are shown in Fig. 3 (Ejiri et al., 1995). Maximum response acceleration exceeds 2G for structures with a predominant period of 0.3-0.9 sec. A comparison with the design spectra for level-2 seismic force defined in the code of earthquake resistant design of bridges (1990) with actual response spectra is shown in Fig. 4 (Ejiri et al., 1995). The actual spectra for hard ground exceeds 2G which is almost twice the level of the design spectrum. The design spectrum for soft ground is almost the envelope of the real response spectra.

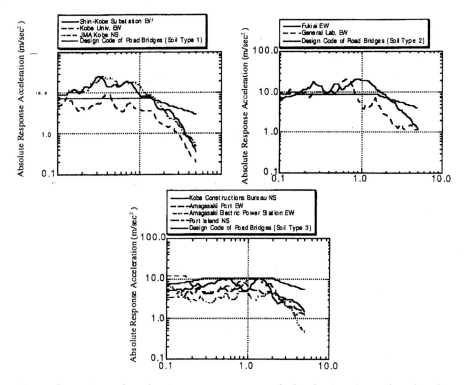

Fig. 4. Comparison of acceleration response spectra calculated using observed earthquake motions and design spectra of the level-2 seismic force (Ejiri et al., 1995)

Essential Lessons Learned from the Strong Motions of the 1995 Hyogoken Nambu Earthquake

Based on the analyses performed by many researchers using observed strong motion records after the onset of 1995 Hyogoken Nambu earthquake, following items are recognized to be essential issues to understand the strong ground motions for intra-plate earthquakes.

1) Scenario method analyses (not seismic risk analyses) taking into account the activity of faults are essential to estimate earthquake damage during a

future earthquake with same magnitude of the Hyogoken Nambu earthquake. To do this the institutions that are responsible for evaluating the activity of faults should make quick action to define active faults which seems to cause severe damage to densely populated urban areas.

2) Reliable method to estimate active fault mechanism of future earthquakes should be developed based on the statistical analyses on mechanisms of past inland earthquakes.

3) The method to define the level of design earthquake motion should be developed in the region where fault activities could not clearly define. How to take into account the fact into new design standards that the level of elastic response spectrum reached to 2G during the Hyogoken Nambu earthquake.

4) Reliable attenuation relations in the near-source region (several kilometers from the fault line) should be developed for peak acceleration, peak velocity, peak displacement, Fourier amplitude, duration and response spectra. A clear definition of duration useful for engineering studies needs to be defined. A comparable full set of attenuation relations should be developed for the vertical component.

5) The definition of distance is important issue especially for estimating ground motion in the near-source region.

6) No clear difference can be seen between the attenuation relations of PGA and PGV derived from data set in California and those of the Hyogoken Nambu earthquake. This does not mean that the attenuation characteristics of subduction earthquakes show same tendency.

7) In the new design standards we state clearly the ground condition at where the design earthquake motion is defined, e.g., on the ground surface or a certain depth at where firm ground condition to be guaranteed.

8) Not only the shallow part of the ground structure (depth less than 100 m) but the deeper structure (up to the depth of layer with shear wave velocity 3 km), especially irregular profile of the ground, is also significant to evaluate the amplification effect of propagating seismic wave.

9) Site effects are important and need to be taken into account for the attenuation relationship. The classifications of site conditions into "soft soil", "stiff soil", soft rock" and "hard rock" are qualitative. Quantitative definition of site class is needed to ensure consistent site categories. Definitions of site categories based on the average shear-velocity at the site should be developed.

10) Nonlinear response characteristics of shallow part of the ground is also important to evaluate the amplification effect of earthquake motions.

11) Not only the prediction model of Fourier amplitude and/or response spectra but that of phase spectrum should be developed to simulate strong ground motion to be used for design purpose.

Recommendation Proposed by Japan Society of Civil Engineers on Earthquake Resistance for Civil Engineering Structures

Based on the tremendous damage caused by the 1995 Hyogoken Nambu earthquake the JSCE compiled a following proposal on desirable earthquake resistance capacity for civil engineering structures from an academic point of view.

1) Earthquakes and earthquake motions for earthquake-resistant design

Earthquake motions in near field of an active fault with a JMA magnitude 7 has not been incorporated into conventional earthquake-resistant design standards. The very strong earthquake, which had a maximum acceleration of about 8 m/s², a maximum velocity of about 1 m/s, and a maximum displacement of about 30-50 cm, were widely observed near the fault.

The severity of the damage is attributed to the extremely strong earthquake motion, force beyond the design criterion, that directly struck above ground structures built before the introduction of elasto-plastic design, as well as underground structures which had been considered relatively safe.

Two types of earthquake motions should be considered in assessing the aseismic capacity of civil engineering structures. The first is likely to strike once or twice while it is in service (Level 1). The second type is very unlikely to strike a structure during the structure's life time, but when it does, it is extremely strong (Level 2). Level 1 earthquake motion is the level in which structures are not damaged when these motion strike. Level 2 earthquake motion is the level in which an ultimate capacity of earthquake resistance of a structure is assessed in plastic deformation range. In design systems of road bridges before the Hyogoken Nambu earthquake Level 2 earthquake motion was defined as design earthquake motion with an elastic response of 1 G on standard ground. Because the earthquake motions during this earthquake were very destructive, re-evaluation of Level 2 earthquake motions needs for very strong earthquake motions generated in the near field of inland faults. Level 2 earthquake motions, therefore, are determined based on identification of active faults that threaten an area and assumptions of source mechanism, through comprehensive examination of geological information on active faults, geodetic information on diastrophism, and seismological information on earthquake activity.

2) Earthquake-resistant design methods

The expected aseismic performance of civil engineering structures are defined and design methods for achieving this performance are proposed. In the proposal civil engineering structures are categorized as i) above-ground structures, ii) in-ground structures, iii) various types of foundation and soil structures. For Level 1 earthquake motions no damage should occur to any structures. Accordingly, the dynamic response during of this level should not exceed the elastic limit. For Level 2 earthquake motions important structures and structures requiring immediate restoration in the event of an earthquake should, in principle, be designed to be relatively easily repairable; even if damage is suffered in the inelastic range. Accordingly the maximum earthquake response of such structures must not exceed the allowable plastic deformation or the limit of ultimate strength. For other structures, complete collapse should not occur even if damage is beyond repair. Accordingly deformation during an earthquake of this level should not exceed the ultimate deformation.

3) Aseismic diagnosis and aseismic reinforcement

Earthquake resistance diagnosis of existing civil engineering structures is in two stages: primary diagnosis using approximate methods and secondary diagnosis using detailed methods. Primary diagnosis should be based on damage to structures caused by the Hyogoken Nambu earthquake. After ground conditions and the ages, design standards, and outlines of the structural characteristics are examined, structures requiring a detailed aseismic capacity examination by secondary diagnosis are selected. In secondary diagnosis, the bottom line in judging the aseismic capacity of a structure is that it does not collapse even if damage beyond repair. On site measurements and testing, and surveys on the ground conditions should be conducted, and the seismic capacity of the structure to withstand the earthquake motions must be evaluated through redesigning and/or numerical analysis.

4) General seismic safety plan

In Japan open spaces formed by streets, roads, and parks are limited in most urban areas and certain areas are densely packed with houses on small lots that do not satisfy the present earthquake design codes. A regional seismic hazard assessment system must be developed. The basic requirement in urban/regional plans is safety from the disaster. To increase the seismic safety of society an urgent review and revision of planning standards for urban infrastructure is needed. Delays in rescue operation and fire-fighting aggravated the Hyogoken Nambu earthquake disaster, and revealed the inadequacy of current emergency management system in Japan. Measures for pre- and post-disaster mitigation must be developed. The schedules for seismic reinforcement of existing infrastructures, and planes to post-disaster reconstruction must be evaluated based on the proper cost-benefit evaluations.

ANALYTICAL PROCEDURE FOR ESTIMATING RESPONSE SPECTRA

We estimate response spectra to investigate the dynamic response characteristics of structures located in the severely damaged area. The response of a single degree of freedom system (SDF) in the frequency domain is derived by multiplying the strong motion at the base rock level with the transfer function of the soil and the frequency transfer function for a single degree of freedom system. A simplified analytical method is introduced which predicts earthquake motion at base rock level near the source region, where the effect of the fault dimension cannot be ignored. For this purpose, a large event is assumed to be synthesized from small events on the fault plane. This method was first proposed by Hartzell in a simple formula, and others have attempted to improve his method (Hartzell, 1978; Muramatsu and Irikura, 1981). We used Irikura's revised model (Irikura, 1983), which is based on the idea that slips on the fault plane during a large event can be replaced by the spatial distribution of slips from small events. An ω^{-2} scaling is used in the high frequency range.

Site effects are important when predicting earthquake motion on the surface. Assuming a shear wave velocity of 3.4 km/s in the basement, the amplification effect at the surface is evaluated from the transfer function using multiple reflection theory. The damping effect through the propagation path is also taken into account.

Estimation of the Earthquake Motion at the Basement Level

A fault model of the Haskell type is assumed. The model is described by a rectangular fault with fault length, L; fault width, W; rise time, τ; final dislocation, D; and rupture velocity, v_r. A large event can be synthesized by superposing small events and taking into account the time delay caused by the rupture propagation.

The fault plane is divided into n elements, each of which corresponds to the area of a small event. The motion of the large event, g_{0L} (t), at the observation point is expressed by (Irikura, 1983)

$$g_{OL}(t) = \sum_{i=1}^{^nL} \sum_{i=1}^{^nW} g_{OS}\left(t - t_{ij}\right) + \sum_{i=1}^{^nL} \sum_{i=1}^{^nW} \sum_{k=1}^{(^nD - 1)^{n'}} g_{OS}\left(t - t_{ijk}\right) \qquad (1)$$

where $g_{0s}(t)$ is a small event, n_L, n_W, and n_D are the number of subdivisions corresponding to the fault length, fault width, and dislocation, and t_{ij} and t_{ijk} are time delays expressed by

$$t_{ij} = \frac{\left(R_{ij} - R_0\right)}{v_s} + \frac{\xi_{ij}}{v_r} \qquad t_{ijk} = t_{ij} + \frac{k\tau}{\left(n_D - 1\right)n'} \qquad (2)$$

in which R_0 is the distance between the observation point and the hypocenter, R_{ij} the distance between a point on the basement and the initial point of rupture at ij element, ξ_{ij} the distance between the hypocenter and the initial point of the rupture at ij element, the shear wave velocity of the medium, v_r the rupture velocity, τ the rise time, and n' is an arbitrary integer to eliminate aliasing. The following relation is assumed for the number of superposition

$$n_W = n_L = n_D = n = \left(\frac{M_{0L}}{M_{0S}}\right)^{\frac{1}{3}} \tag{3}$$

in which M_{0L} and M_{0S} are seismic moments for the large and small events, respectively. In the following we assume $n' = n = 3$.

The Fourier transform of Eq. (1) is

$$g_{0L}(\omega) = \sum_{i=1}^{n_L} \sum_{j=1}^{n_W} g_{0S}(\omega)\exp\left(-i\omega t_{ij}\right) + \sum_{i=1}^{n_L} \sum_{j=1}^{n_W} \sum_{k=1}^{(n_D-1)n'} g_{0S}(\omega)\exp\left(-i\omega t_{ijk}\right) \tag{4}$$

in which $g_{0S}(\omega)$ is the Fourier spectrum of a small event given by

$$g_{0S}(\omega) = \frac{R_{\theta\phi}}{4\pi\rho v_s^3} \frac{s(\omega)}{R_{lm}} \tag{5}$$

in which $R_{\theta\phi}$ is the radiation pattern, ρ the medium density, R_{ij} the distance from a fault element to the site, ω an angular frequency, and $s(\omega)$ the source spectrum. We adopt the formula proposed by Geller (1975) for the source spectrum of each small event with ω^{-2} rollout in the high frequency range. Because the number of superposition of the second term on the right-hand side of Eq. (4) is its slope in the high frequency range is ω^{-3}. The first term on the right-hand side of Eq. (4), however, has a logarithmic slope of -2 in the high frequency range, hence $g_{0L}(\omega)$ shows ω^{-2} characteristic.

From Irikura's revised model, the source displacement spectrum, $g_{0L}(\omega)$, has a logarithmic slope of -2 in the high frequency range. The source spectrum of acceleration is flat in that range. To prevent the integral of the second order spectral moment from becoming infinite we introduce f_{max}, the corner frequency of the acceleration spectrum in the high frequency range. According to Hanks (1982), the spectrum drops off sharply for frequencies between 20-30 Hz on a rock site. A value of f_{max} of 25 Hz was selected.

Path Effect

In Eq. (5) only geometrical spreading proportional to the reciprocal distance, $1/R$, is considered. S-wave damping may be estimated by (Aki and Richards, 1980)

$$g_s(\omega) = g_{0s}(\omega)\exp\left(-\frac{\omega R_{ij}}{2v_s Q}\right) \tag{6}$$

where $g_s(\omega)$ is the Fourier transform of ground motion. The Q-value for the Kinki region (Akamatsu, 1980) is approximated by

$$\log Q^{-1} = -0.7\log\frac{\omega}{2\pi} - 2.2 \tag{7}$$

Amplification Effect

To estimate the amplification we collected the standard penetration test data of soil profiles in the Kobe area and we drew the contours of $N{=}50$. The frequency transfer function in the near-surface soils, $H_s(\omega)$, is calculated based on the multiple reflection theory assuming horizontally layered soil. The ground motion at the surface, $g_{0L}(\omega)$, is expressed by

$$g_L(\omega) = H_d(\omega)H_s(\omega)g_{0L} \tag{8}$$

in which $H_s(\omega)$ is the frequency transfer function from the base rock with a shear wave velocity of 3.4 km/sec to the soil layer with the N-value of 50, Nakagawa et al. (1996) determined the depth to the base rock in the Kobe area by use of the Bouguer anomaly as shown in Fig. 5. The geological profile above this base rock is assumed to be similar to one of the profiles shown in Fig. 6. When the information of the deeper part of ground profile is not available the value of $H_d(\omega)$ is assumed to be 1.75 as proposed by Midorikawa and Kobayashi (1988).

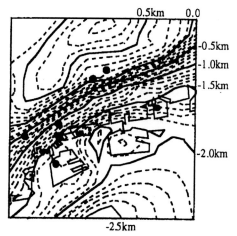

● Seismic Observation Points for Analyses

Fig. 5. Configuration of top surface of the base rock in the Kobe sedimentary basin proposed by Nakagawa et al. (1966)

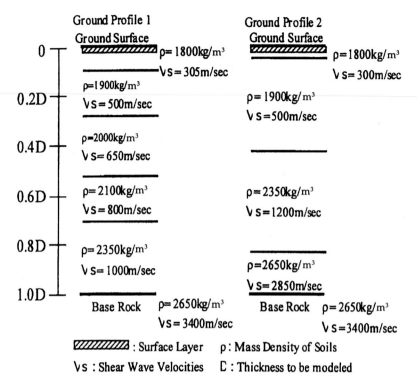

▨▨▨ : Surface Layer   ρ: Mass Density of Soils

Vs : Shear Wave Velocities   D : Thickness to be modeled

Fig. 6. Horizontally layered ground models for deeper parts of ground profiles

Response Spectrum

The response of a single degree of freedom system at the ground surface $X(\omega)$ is expressed by multiplying the frequency transfer function of SDF, H_x $(\omega, T_x h_x)$ with g_{0L} (ω) as follows

$$X(\omega) = H_x(\omega, T_x, h_x)H_s(\omega)H_d(\omega)g_{0L}(\omega) \tag{9}$$

in which T_x is the natural frequency of SDF and h_x is the damping constant. The power spectrum of the response of SDF is

$$P(\omega) = \frac{1}{T}X(\omega)X^*(\omega) \tag{10}$$

in which T is the duration of the stationary part of the ground motion and * indicates the complex conjugate.

The duration, T, is defined as

$$T = T_e - T_b + T_\alpha \tag{11}$$

in which T_b and T_e are the arrival times of the seismic waves originating from the initial and final rupture elements, and T_a is the duration of the seismic wave generated in the last rupture element (Kawashima et al., 1984).

Once the power spectrum of the large event at the rock site has been obtained, the expected absolute peak acceleration response of the SDF, $S_a(T_x, h_x)$, is calculated as follows (Der Kiureghian, 1980):

$$S_a(T_x, h_x) = p\sqrt{\lambda_0}, \quad p = f(v_e, \lambda_0, \lambda_1, \lambda_2) \tag{12}$$

in which p is the peak coefficient, v_e the reduced ratio of the zero-crossing given as a function of the three spectral moments and the duration, and $\lambda_m (m = 0, 1, 2)$ are the spectral moments of zero and the first and second orders given by

$$\lambda_m = \int_0^\infty \omega^m P(\omega) d\omega \tag{13}$$

If soil softening occurs we must consider the nonlinear response of the soil layer in the estimated frequency transfer function. We use a Hardin-Drnevich type relationship between the equivalent shear strain and the equivalent shear modulus as well as the damping constant as shown in Fig. 7. The equivalent shear strain is assumed to be 0.65 times the peak shear strain at the center of each layer. The peak shear strain is calculated through the same procedure to estimate the peak acceleration for Eq. (12). Iterative calculation of the peak strain is confirmed to reach the converging error of strain less than 1% of the peak strain at the previous iteration step.

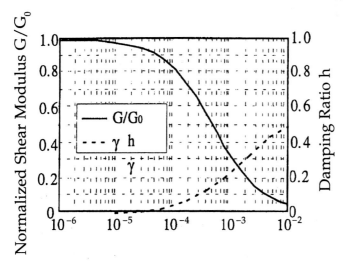

Fig. 7. An example of shear strain dependence of equivalent shear modulus and damping ratio

ESTIMATION OF RESPONSE SPECTRA FOR KOBE EARTHQUAKE

Equation (4) can be extended to multiple fault rupture mechanisms. The 1995 Hyogoken Nambu earthquake was composed of three sub-events based on Kikuchi and Irikura (Kikuchi, 1995; Pitarka et al., 1995). Kikuchi's model is composed of three point source sub-events obtained from inversion of teleseismic body waves. We assume the seismic moments of the first, second and third subfaults to be $M_0 = 1.81 \times 10^{19}$, 0.30×10^{19} and 0.59×10^{19} Nm (Kikuchi, 1995). The lengths of the first, second, and third subfaults are assumed to be 29, 16 and 20 km. The ratio of the length to the width of each subfault is assumed to be 0.5. The strike and dip angles for each subfault are given in Fig. 8. The locations of the three subfaults are shown in Fig. 8 (Kikuchi, 1995). We subdivided each subfault into nine subelements of equal area. Rupture of the first subfault is assumed to start at the bottom center element, and the ruptures of the second and third subfaults at the bottom southwest corner element. Two seconds after the start of the first rupture the second subfault ruptures and two seconds later the third rupture is assumed to occur. Pitarka et al (1995) propose a source rupture mechanism to explain the short period ground motion near the source region. We assume three fault planes on the basis of their solution. The strike and dip angles as well as seismic moment are given in Fig. 9. The location of three subfaults are also shown in Fig. 9. The length of the first, second, and third subfaults are assumed to be 15, 7.5 and 7.5 km (Pitarka et al., 1995). The ratio of the length to the width of each subfault is assumed to be 0.5. We subdivided each subfault into nine subelements of equal area. Rupture of the first subfault is assumed to start at the northeast of the bottom corner element, and the ruptures of the second and third subfaults from the bottom southwest corner element. Two seconds after the start of the first rupture the second subfault ruptures and two seconds later the third rupture is assumed to occur.

The locations of seismic observation points are shown by A to G in Fig. 10, together with the observed peak accelerations at these points.

At each site we calculate the Fourier spectra of surface ground motion by using Eq. (8) to evaluate the amplification effect. Seven records were available for the calculation of response spectra in Kobe city. The estimated acceleration response spectra are compared with actual response spectra from records. Fig. 11 (a) and (b) show the results obtained using Kikuchi's and Irikura's fault rupture models at point C. The heavy, thin and broken lines are respectively the real response spectrum and the estimated response spectra using soil profiles of type 1 and type 2 (Fig. 6). There is not much difference between real response spectra and estimated ones except in the period range from 0.1 to 0.5 sec.

The estimated response spectra are very similar even though we assume fairly different fault rupture mechanisms and ground profiles. At point B where liquefaction occurred we can see very good agreement between the real and estimated absolute response spectra (Fig. 12). Thus equivalent linear analysis provides an efficient tool for estimating the seismic response of liquefied ground. The comparison of real and estimated response spectra at point D is shown in

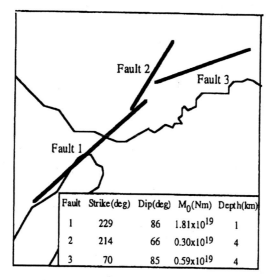

Fault	Strike(deg)	Dip(deg)	M_0(Nm)	Depth(km)
1	229	86	1.81×10^{19}	1
2	214	66	0.30×10^{19}	4
3	70	85	0.59×10^{19}	4

Fig. 8. Locations of three subfaults and source parameters determined by using Kikuchi's source mechanism

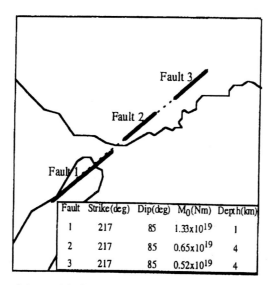

Fault	Strike(deg)	Dip(deg)	M_0(Nm)	Depth(km)
1	217	85	1.33×10^{19}	1
2	217	85	0.65×10^{19}	4
3	217	85	0.52×10^{19}	4

Fig. 9. Locations of three subfaults and source parameters determined by using Irikura's source mechanism.

Fig. 13. This point is located near the south edge of the belt of seismic intensity 7 where the irregular profile of deeper part of ground structure is said to play a predominant effect on the amplification of ground motion in the period range of 0.5 to 2 sec (Irikura, 1995). Fairly large discrepancies can be seen between real and estimated spectra especially when we use Kikuchi's fault rupture model.

For Irikura's fault rupture model the level of the estimated response spectrum is lower than the real one in the period range higher than 0.5 sec and the opposite can be seen in the period range lower than 0.5 sec.

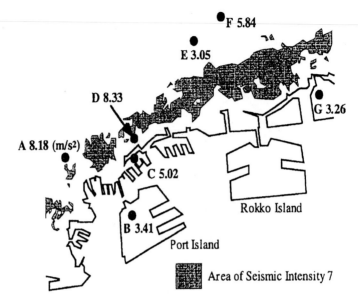

Fig. 10. Locations of observation sites and peak accelarations at observation sites

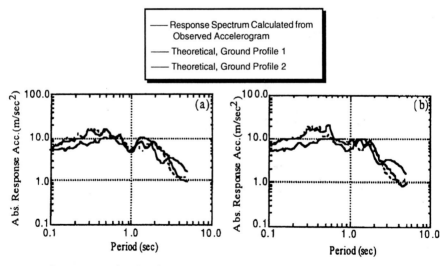

Fig. 11. Comparison of real and estimated acceleration response spectra at the observation point C for the case of using (a) Modified Kikuchi's source mechanism, and (b) Modified Irikura's source mechanism

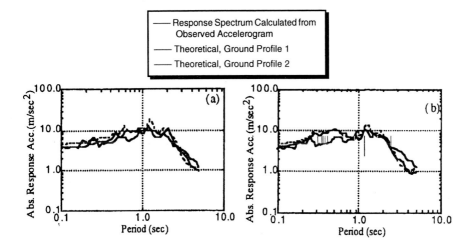

Fig. 12. Comparison of real and estimated acceleration response spectra at the observation point B for the case of using (a) Modified Kikuchi's source mechanism, and (b) Modified Irikura's source mechanism

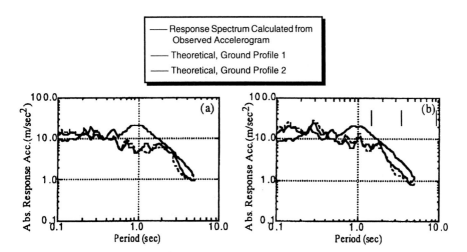

Fig. 13. Comparison of real and estimated acceleration response spectra at the observation point D for the case of using (a) Modified Kikuchi's source mechanism, and (b) Modified Irikura's source mechanism

METHOD TO ESTIMATE PHASE SPECTRUM

Based on a modeling of phase increment at each circular frequency, Osaki et al. (1978) studied the nonstationary characteristics of strong earthquake motions. The concept of group delay time is introduced to extract the duration and average arrival time of earthquake motions (Izumi and Katukura, 1983). This concept was firstly applied to simulate earthquake motions for the given initial phase and group delay time function (Kimura, 1983). The stochastic analysis was performed to obtain the average group delay time and its standard deviation using narrow banded earthquake motions and a method to simulate nonstationary earthquake motions was developed (Satoh et al., 1997).

A real earthquake motion contains effects of the rupture process, transmitting path and local site condition. These effects appear in both amplification and phase characteristics but the effect of phase on the time history of earthquake motion is stronger than that of amplification characteristic. In this section we assume that amplification characteristic is defined by the response spectra and develop a method to simulate the phase spectrum based on the concept of group delay time taking into account the fault rupture process, transmitting path and local site condition. By assuming that the fault rupture process is modeled by a train of impulse, the group delay time caused by the seismic fault rupture is calculated. The phase shift due to the transmitting path and local site condition is assumed to be expressed by the Hilbert transformation of a frequency transfer function from source to site of which information is usually provided through the microtoremore or micro-earthquake observations, such as using a ratio of amplitude spectra between horizontal and vertical components (Nakamura, 1989).

Theoretical Base

The group delay time is defined by the derivative of phase spectra $\phi(\omega)$ with respect to circular frequency ω as follows:

$$t_{gr}(\omega) = \frac{d\phi(\omega)}{d\omega} \qquad (14)$$

Because the phase spectrum possesses linear characteristic we can add group delay times from different origins as follows:

$$t_{gr}(\omega) = t_{gr}^{p}(\omega) + t_{gr}^{h}(\omega) \qquad (15)$$

in which $t_{gr}^{p}(\omega)$ is the group delay time obtained from the train of impulses and t_{gr}^{h} is from effect of the transmitting path.

Modeling of Phase Characteristics Due to the Fault Rupture Process

By deconvoluting Eq. (4) with respect to the source time function and the train of impulses $p(t)$ expressing the rupture process of earthquake fault we obtain the following expression for $p(t)$

$$p(t) = \sum_{i=1}^{n_L} \sum_{i=1}^{n_W} a_i \delta\left(t - t_{ij}\right) + \sum_{i=1}^{n_L} \sum_{i=1}^{n_W} \sum_{\cdot k=1}^{(n_D - 1)n'} \frac{1}{n'} a_{ij} \delta\left(t - t_{ijk}\right) \qquad (16)$$

Equation (16) can be simplified as

$$p(t) = \sum_{i=1}^{N} a_i \delta\left(t - t_i\right) \qquad (17)$$

in which $N = n_L + n_W + (n_D - 1)n'$. The Fourier amplitude $A_p(\omega)$ and phase spectrum $\phi_p(\omega)$ of $p(t)$ are

$$A_p(\omega) = \sqrt{\left\{\sum a_i \sin\left(\omega\, t_i\right)\right\}^2 + \left\{\sum a_i \cos\left(\omega\, t_i\right)\right\}^2} \qquad (18)$$

$$\phi_p(\omega) = \tan^{-1}\left(-\frac{\sum a_i \sin(\omega\, t_i)}{\sum a_i \cos(\omega\, t_i)}\right) \qquad (19)$$

The group delay time of the train of impulses is obtained by substituting Eq. (18) into Eq. (14)

$$\frac{d\phi_p(\omega)}{d\omega} = \frac{-\sum_{i=1}^{N} a_i^2 t_i - \sum_{i=1}^{N} \sum_{j=i+1}^{N} a_i a_j \left(t_j + t_i\right)\cos\left\{\omega\left(t_j - t_i\right)\right\}}{\sum_{i=1}^{N} a_i^2 + 2\sum_{i=1}^{N} \sum_{j=i+1}^{N} a_i a_j \cos\left\{\omega\left(t_j - t_i\right)\right\}} \qquad (20)$$

If the value of $a_i a_j / \sum a_i^2$ is assumed to be small the group delay time is simplified as

$$\frac{d\phi_p(\omega)}{d\omega} \approx -t_p - \sum_{i=1}^{N} \sum_{j=i+1}^{N} \left(a_i a_j / \sum_{l=1}^{N} a_i^2\right)\left(\tau_i + \tau_j\right)\cos\left\{\omega\left(\tau_j - \tau_i\right)\right\} \qquad (21)$$

in which

$$t_p = \sum_{i=1}^{N} a_i^2 \, t_i \Big/ \sum_{i=1}^{N} a_i^2 \, , \quad \tau_i = t_i - t_p, \quad \tau_j = t_j - t_p \tag{22}$$

The amplitude spectrum of the train of impulses at the zero frequency is proportional to the seismic moment of an earthquake fault. Therefore if we treat an earthquake caused by ruptures from multiple faults the intensity of each impulse generated from each fault can be modified to be proportional to the seismic moment of each fault.

Modeling of Phase Shift due to the Transmitting Path

The amplitude characteristics of frequency transfer function expressing the effect of transmitting path and local soil condition is assumed to be given a priority and defined by $A_h(\omega)$. Its phase spectrum $\phi_h(\omega)$ is assumed to be expressed by the minimum phase transfer function of $A_h(\omega)$. This assumption is not valid when the seismic wave has multiple paths between the observation site and seismic source. The phase spectrum is derived by using the Hilbert transformation which defines the relation between the amplitude and phase spectra

$$\phi_h(\omega) = -\frac{1}{\pi} \int_{-\infty}^{\infty} \frac{\ln(A_h(y))}{\omega - y} dy \tag{23}$$

Taking derivative with respect to ω the group delay time $t_{gr}^{\ h}(\omega)$ is obtained

$$\frac{d\phi_h(\omega)}{d\omega} = \frac{1}{\pi} \int_{-\infty}^{\infty} \frac{\ln(A_h(y))}{(\omega-y)^2} dy = \frac{1}{\omega^2} * \left\{ \ln(A_h(y)) \right\} \tag{24}$$

SIMULATION OF THE 1995 HYOGOKEN NAMBU EARTHQUAKE MOTION

The phase spectrum at Fukiai observation site is modeled by using the proposed method. Combining the observed amplitude Fourier spectrum (assumed to be given) and the modeled phase spectrum we resimulated an earthquake motion and compare with the observed motion at Fukiai site.

To express rupture process of the 1995 Hyogoken Nambu earthquake we assume three vertical sub-event faults as shown in Fig. 9 in order to calculate a train of impulses at the base rock level just beneath the observation site. To count for effect of transmitting path in the crustal rock we considered only the direct arrival of S wave and for the site condition we take only into account the vertical soil profile from base rock to the ground surface. The depth of base rock was read from the contour map of base rock (Fig. 5) and the soil profile from the base rock to engineering base layer is assumed to be given in Table 2. The surface ground profile shallower than the engineering base layer is modeled base on the soil exploration data. The amplitude of frequency transfer function at observation site is assumed to be given theoretically by using multiple reflection theory of S wave, not based on empirical or engineering estimation. Because the purpose of this section is to show applicability of our proposed method we assume that the Fourier amplitude spectrum of earthquake motion is given at Fukiai observation site as shown in Fig. 14. The case that amplitude characteristic is defined by the response spectrum is discussed later.

For the first case we assume that all amplitude of impulses generated on the sub-faults are assumed to possess unit scale, $a_l = 1.0$, in Eq. (17). The result is shown in Fig. 15. The group delay times of train of impulses $p(t)$ calculated using Eqs. (20) and (21) are shown in Fig. 16a and b. The duration of the train

Table 2. Soil profile model at Fukiai observation site

Layer	Depth (m)	Shear velocity V_s (M/S)	Unit weight $\gamma(tf/m^3)$
1	0~6.0	257.7	1.80
2	6~50.0	340.8	2.00
3	50.0~500.0	500.0	1.90
4	500.0~1000.0	1200.0	2.35
5	1000.0~1200.0	2850,0	2.65
6	> 1200.0	3400.0	2.65

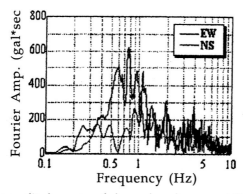

Fig. 14. Fourier amplitude spectra of observed accelerogram at Fukiai observation site

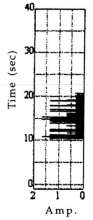

Amp.

Fig. 15. Train of impulses

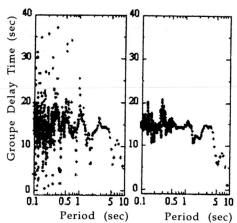

Fig. 16. Group delay time of train of impulses
(a) Obtained by Eq. (20), (b) Obtained by Eq. (21)

of impulses is about 10 sec and the group delay time is also distributed between 10 to 20 sec. The result obtained using approximate expression given by Eq. (21) shows less fluctuation than that obtained by Eq.(20).

The amplitude of frequency transfer function is calculated by multiple reflection theory as shown is Fig. 16. The phase shift due to the local site condition is defined by the minimum phase function, i.e., the Hilbert transformation of the amplitude transfer function as defined by Eq. (23), and group delay time is calculated using Eq. (24) as shown in Fig. 18. The group delay time has small peaks at peaks of the amplitude transfer functions. From these figures the phase characteristic of earthquake motion is strongly affected by the rupture process of seismic fault not by the local site condition. The phase spectrum obtained by integrating group delay time is shown in Fig. 19.

The earthquake motion at Fukiai station is simulated using both the phase spectrum shown Fig. 19 and the observed Fourier amplitude spectrum shown in Fig. 14. The case using exact and approximated group delay times are shown in Fig. 20a and b, respectively. Figure 21 is the observed earthquake motion. The duration time of simulated earthquake motion using exact group delay time becomes longer than that of observed motion. The peak acceleration, on the contrary, becomes smaller due to the earthquake motion energy spreads in long time range.

APPLICABILITY FOR NONLINEAR RESPONSE ANALYSES

The similarity of time histories between simulated and observed earthquake motions is essential requirement to confirm the efficiency of proposed method to estimate phase spectra. From the stand point of structural response the non-linear behavior of structural system is strongly affected by the input earthquake motion especially its phase characteristic. The strength demand spectra are

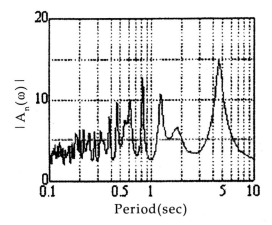

Fig. 17. Amplitude of frequency transfer function
from the base rack to the ground surface

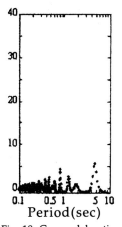

Fig. 18. Group delay time
obtained using Eq. (24)

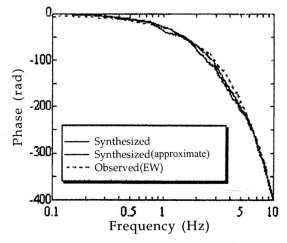

Fig. 19. Comparison of observed and calculated phase spectra

calculated for several required ductilities. These spectra are obtained through
a non-linear response analysis of single degree of freedom system. The
constitutive relation is a bi-linear model of which second stiffness is 10% of the
initial stiffness. The results are shown in Fig. 22 as the ductility being a constant
parameter. The vertical axis is defined by the yield strength divided by the total
weight of structure. The horizontal axis is the natural period calculated using
initial stiffness of the system.

Comparison of strength demand spectra obtained by using simulated and
observed ground motions shows good agreement between them. This means
that the non-linear structural response characteristics can be simulated very
well by using the simulated earthquake motion based on the proposed method.
It has been said that the non-linear response of structural system is strongly

affected by the phase characteristic of input earthquake motion. The proposed method to simulate the phase spectrum grasps the essential phase characteristic of strong earthquake motion.

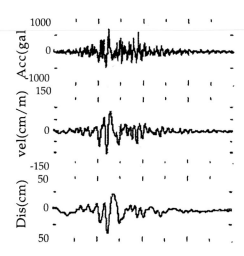

Fig. 20 (a). Simulated time hitory using exact group delay time

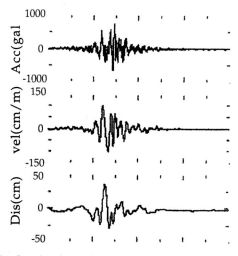

Fig. 20 (b). Simulated time history using approximate group delay time

SIMULATION OF INPUT EARTHQUAKE MOTION COMPATIBLE WITH DESIGN RESPONSE SPECTRUM

The concept of response spectrum is often used for usual seismic designs of civil structures. To check the dynamic behavior of designed structure, we need an input earthquake motion compatible with the response spectrum used for

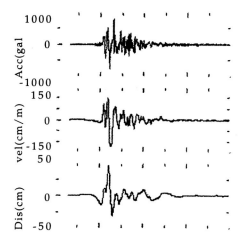

Fig. 21. Observed time history of earthquake motion

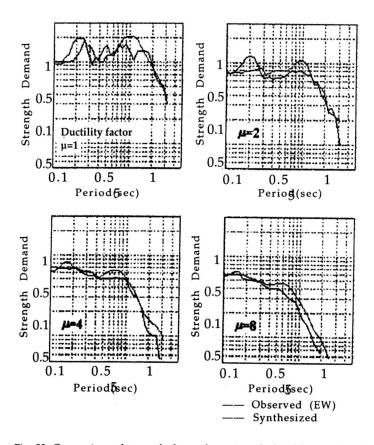

Fig. 22. Comparison of strength demand spectra calculated by simulated
and observed earthquake motions

design. In this session we simulate a design earthquake motion compatible with a spectrum used to design railway structures. The simulation follows the four steps:

1. The group delay time at Fukiai observation site shown in Fig. 16a is used.
2. It is not practical to use the detail amplified frequency transfer function as defined in Fig. 17. We simplify this amplitude transfer function as show in Fig. 23. The several peaks of amplitude transfer function are modeled by triangular shapes as that a top of triangle is coincide with a peak amplitude, Y_0, and two side lines pass the points on the amplitude transfer function with amplitudes $Y_0 / \sqrt{2}$. The base line of amplitude transfer function is assumed to be 2.0. For the first case we take into account only one peak at $T_g = 4.55$ s with the amplitude of $Y_0 = 14.65$. For the second case we take into account three peaks, ($T_g = 4.55$ s, $Y_0 = 14.65$), ($T_g = 1.24$ s, $Y_0 = 10.78$) and ($T_g = 0.83$ s, $Y_0 = 12.57$).
3. The ground condition at Fukiai site is classified into G2 site in the railway seismic design standard (Railway Code 1996). We use the G2 design spectrum (Fig. 24) to simulate a design earthquake motion.

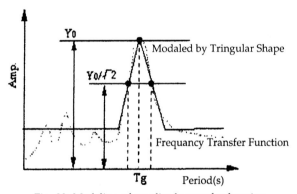

Fig. 23. Modeling of amplitude transfer function

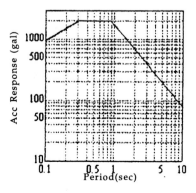

Fig. 24. Design spectrum

4. The Fourier amplitude spectrum of simulated earthquake ground motion $A(\omega)$, is modified until the response acceleration spectrum matches with the design response spectrum. The initial value of $A(\omega)$ is assumed to be given by the quasi-velocity response spectrum with zero damping constant at G2 site.

The simulated earthquake motions are shown in Fig. 25 and 26. The case of one predominant peak being modeled is in Fig. 25 and the case of three predominant peaks is in Fig. 26. Comparing both figures with Fig. 21 we conclude that the proposed simple method can simulate the design earthquake motion in which the rupture process, transmitting path and local site condition is effectively reflected in its phase characteristics.

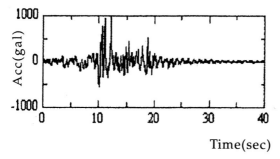

Fig. 25. Design spectrum compatible earthquake motion with singel peak modeling of amplitude transfer function

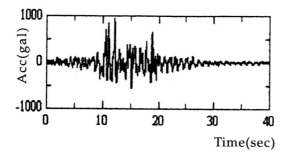

Fig. 26. Design spectrum compatible earthquake motion with three peaks modeling of amplitude transfer function

CONCLUSIONS

The chronological change of earthquake resistant design codes in Japan was briefly introduced, and the lessons learned from the 1995. Hyogoken Nambu earthquake on defining near source motion for structural design were summarized. To respond the raised questions, simple methods were developed to estimated surface ground motion and to model phase characteristics of earthquake motion.

Acceleration response spectra at ground surface were calculated theoretically for the 1995 Hyogoken Nambu earthquake, including the effect of fault extent, path effect and amplification due to the shallow and deep parts of ground profiles. The estimated acceleration response spectra at observation sites were compared with the real spectra obtained from recorded earthquake motions. At several places the level of the estimated acceleration response exceeded 2G in the period range from 0.2 to 1.5 sec.

A method to model the phase spectrum of earthquake motion was proposed based on the concept of group delay time. A rupture process of an earthquake fault was assumed to be expressed by a train of impulses. The phase shift due to the transmitting path and local site condition was defined by the minimum phase function of the amplitude spectrum of frequency transfer function. We investigate the efficiency of proposed method to estimate phase spectra by using it to simulate the time history of earthquake ground motion at Fukui observation site where the strong ground motions were recorded during the 1995 Hyogoken Nambu earthquake.

REFERENCES

Akamatsu, J. : 1980, Seismic observation at the Sumiyama seismic station (2)—On the natures of attenuation and spectra of coda parts of local earthquakes, Ann. Disas. Prev. Res. Inst., Kyoto University, 23B-1, 107-114 (in Japanese).

Aki, K. : 1984, Origin of f_{max}, Preliminary Report, (prepublication manuscript).

Aki, K. and Richards, P.G. : 1980, Quantitative Seismology, Vol. 1, Freeman, 168-169.

Der Kiureghian, A. : 1980, Structural response to stationary excitation, Jour. Am. Soc. Civil. Eng., Eng. Mech. Div., 106CEM6, 1195-1213.

Ejiri, J. Goto, Y. and Toki K. : 1995, Strong ground motion characteristics of the 1995 Hyogoken Nambu earthquake, Proc. 23rd JSCE Earthquake Engineering Symposium—1995, 237-240 (in Japanese).

Geller, R. J. : 1975, Scaling relation for earthquake source parameters and magnitudes, Bull. Seism. Soc. Am., 65, 1073-1095.

Hanks, T. C. : 1982, f_{max}, Bull. Seism. Soc. Am., 72, 1867-1879.

Hartzell, S. H. : 1978, Earthquake aftershock as green's function, Geophys. Res. Letters, 5(l),1-4.

Iemura, H. : 1998, How to all achieve acceptable performance of civil infrastructures against extreme earthquake ground motion—from ductility demand to structural control design. The Fifth Workshop on Consensus on Acceptable Risk in Urban Seismic Hazard, Los Angeles, University of Southern California, Dec. 13-16.

Irikura, K. : 1983, Semi-empirical estimation of strong ground motions during a large earthquake, Bull. Disas. Prev. Res. Inst., Kyoto University, 33, Part 2, No. 297, 63-104.

Irikura, K. : 1995, Preliminary analysis of aftershock array data at Higashinada Ward, Kobe City, DPRI News Letter, Special Issue, 9-11.

Joyner, W.B. and Boore, D.M. : 1981, Peak horizontal acceleration and velocity from strong motion records from the 1979 Imperial Valley California Earthquake, Bull. Seism. Soc. Am., 71, 757-783.

Kawashima, K., Aizawa, K. and Takahashi, K. : 1984, Attenuation of peak ground motion and absolute acceleration response spectra, Proc. 8th World Conf. on Earthq. Eng., II, 257-264.

Kikuchi, M. : 1995, Report of 112th meeting of the Coordinating Committee for Earthquake Prediction (in Japanese).

Midorikawa, S. and Kobayashi, H. : 1988, Isoseismal map in near-field with regard to fault rupture and site geological conditions, Trans. of A. I. J., 290, 83-94 (in Japanese).

Muramatsu, I. and Irikura, K. : 1981, Estimation of strong ground motion during a large earthquake using observed seismograms of small events: A study of predicting damage by a large earthquake in the Tokai district, Research Report of the Japanese Group for the Study of Natural Disaster Science, A-56-3, 33-46, (in Japanese).

Nakagawa, K., Shiono, K., Inoue, N. and Sano, M. : 1996, Geological characteristics and geotechnical problem in and around Osaka Basin for basis to assess the seismic hazards, Soils and Foundations, Special Issue for the 1995 Hyogoken Nambu earthquake, 15-28.

Pitarka, A., Irikura, K. and Kagawa, T. : 1995, Source complexity of the January 17, 1995 Hyogoken Nambu earthquake determined by near-field strong motion modeling, preliminary results, Journal of Natural Disaster Science, 16, 31-38.

Suetomi, I. : 1995, Personal Communication.

8

Environmental Geotechnical Problems caused by Urban Underground Engineering Activities

Sun Jun

*Institute of Geotechnical Engineering, Tongji University,
Shanghai, P. R. China*

With the increasing geotechnical engineering activities, principal environmental problems caused by geotechnical engineering are the deformation of soil layers, too large displacement and differential settlement and so forth. Now, what we must do further is to forecast, control, prevent and eliminate these problems effectively. Though we have done a lot about these problems and got some theoretical basis as well as experience, we still have encountered many difficulties when we have to solve some complex problems. This paper is based on the author's several years' research on theoretical and practical work and is to serve as a reference of the readers. All advice and criticism are welcome.

PART I: BASIC COGNITION

1. Environmental Geotechnical Problems Caused by Urban Underground Engineering Activities

1) It is well- known that we should ensure not only the security of construction, but also the security of surroundings. Environmental geotechnical problems that we are talking about are the influence caused by engineering activities on shallow foundations, road beds, roadways and municipal facilities such as

underground pipeline systems.
2) It is the principal task to forecast, control, prevent and eliminate the harm caused by geotechnical engineering, e.g., too large settlement and too large differential settlement of foundations in soft soil.
3) The prime environmental geotechnical problems induced by urban underground engineering activities are as follows:
- Constructing process of subway shields;
- Circuitously constructing process of shallow large-diameter jacking pipes along old urban areas;
- Excavation of deep and large foundation pits
- Soil-extrusion effects from driving precast driven piles and static penetrated piles;
- Influence caused by later construction on completed subways and piles of crossroads;
- Others, such as well sinking, abstraction of water, the influence caused by new buildings on old buildings, and so on.
4) The present severe conditions:
- The passing of shields under pile foundations or around pile foundations
- The upper and lower shields in the same interzone advancing in the same direction or athwart direction
- The oblique or right crossing of new shields and old/completed shields that are very close
- The excavation and construction of piles above completed subways
- Simultaneous or consequent construction of two foundation-pits adjacent to each other. Excavating one pit can cause the deformation, even jam of pile holes of the other pit.
- The construction of deep or complex excavations, which are close to dangerous buildings, the buildings with shallow foundations or the buildings needing protection.
- Circuitously constructing process of shallow large-diameter jacking pipes through old urban areas, etc.

There are some other problems. Here we take some problems encountered in the shield construction as the examples:
- When two shields are very close, the latter shield has to be constructed in disturbed soil and the former shield will be disturbed again.
- When a new shield is approaching an old completed shield, because the soil around the old shield has been strengthened by grout, the new shield must be constructed in disturbed soil. Consequently the old shield has to endure another disturbance.
- The secondary consolidation settlement of a new building's pile foundation in the soft soil will beget the inclination of completed vicinal subways toward the building. And this will quicken the abrasion of rails and even endanger the security of transportation.
- Because there is a shield built under a pile foundation, the load capacity

of the piles will decline and the piles will settle again.
- Because of the construction of an excavation above a completed subway, the subway will upheave and endure a second disturbance of deformation, etc.

All the aforementioned points can be called the mutation of soil mechanical behavior induced by the engineering activities and its influence on completed underground works, which are brand-new subjects.

2. The Outline of Solutions and the Requirement of Research Work

1) Because of the complexity and comprehensiveness of these problems, it is impossible to solve these problems only by theoretical calculation. We need to abide by the following steps:

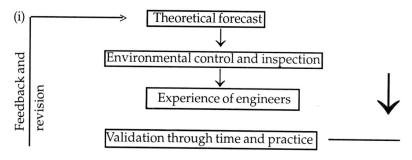

Here, artificial or virtual reality simulation and forecast techniques are very important in computer management.

    (ii) Recently the development of calculation methods, classical theories and methods about stability and displacement of foundations and the development of the methods about engineering mechanics are very useful to this kind of engineering practice, which will be discussed briefly in Part II.

2) In order to prevent and eliminate these problems by forecast and control, we need following requirements.
- In order to prevent accidents from occurring, we must adopt prevention methods as soon as possible.
- In order to plot the influenced area exactly, we must establish relevant technical standards and norms of judgement.
- In order to reduce the harm caused by construction, several suitable deformation control methods should be worked out, unlike the working control methods such as "Three-step Grouting Control Method". In our opinion, the most important are control of design, which includes the adjustment and revision of the design parameters and the rule and scope of their application.
- In order to offer accurate technical design parameters of various buildings and foundations, pipelines, roadways, and roadbeds

enduring various deformation and settlement, firm theoretical basis should be provided.

PART II: SOME PROBLEMS IN THEORETICAL FORECAST

1. Aspects of Research on Soil Mechanical Behavior Influenced by Working Disturbance

The research being carried out concerns:
- Change of soil's stress and strain, such as
  - ✧ Unloading process induced by excavation which cause the deviation of stress path
  - ✧ Strengthening by pounding
  - ✧ The extrusion of soil induced by the pile's driving process, soil's remoulding, the increment and drainage of pore pressure and soil's secondary consolidation
- Change of water capacity and void rate, such as
  - ✧ Abstraction of water
  - ✧ Rainfall, especially, consecutive rainfall which will beget sharp declination of soil's effective stress and shear strength
  - ✧ Main consolidation and secondary consolidation
- The visco-elastic-plastic deformation of soil, i.e., soil's theological behavior, such as
  - ✧ Creep — the viscous aging effect of shearing deformation and compression and the lagging of deformation
  - ✧ Stress relaxation
  - ✧ Declination of long-term strength
- The failure of soil's structure, such as
  - ✧ Cutting and excavation of soil in front of working planes
  - ✧ The replenishment of soil after excavation
  - ✧ Pile's driving process
  - ✧ Explosion
  - ✧ Improper pounding, and so on
- Change of soil's chemical components, such as
  - ✧ Various grout
  - ✧ The improvement of soil quality, various ground treatments and strengthening of foundation
- Mixing and separation of soil components which will be encountered in the above processes

2. Major Factors which Cause Settlement of Surface and Movement of Shallow Soil Layers

1) The factors of soil materials

(i) The initial stress behavior is disturbed by working process. Especially, the state of limit equilibrium is broken.

(ii) Soft clay is disturbed by construction. Because of soil's thixotropy, the disturbance will reduce the shear strength and make the load capacity decline sharply.

(iii) In the mechanical excavation process, the pore pressure will increase immediately. Subsequently, with the draining of pore pressure, the soil will remold and consolidate again, which will cause the displacement of surface and soil layers with the lapse of time and permanent settlement caused by visco-plastic effects.

(iv) Unloading caused by excavating the front and lateral soil will induce soil's stress relaxation in large areas. With the development of calculation methods, these factors can all be taken into consideration.

2) The adverse factors of construction

(i) Because the inside diameter of shields surpasses the outside diameter of pipes or because the outside diameter of jacking pipes surpasses the outside diameter of pipes, there will be circular crevice around the pipe, which will beget the loss of soil layers if grouting is delayed.

(ii) When the over-excavation, or pushing/propping work is stopped, there may be collapse of excavation surface. Or when the pressure of soil and grouting pressure have not reached the balance, there may be blow of soil or shifting sand.

(iii) Because the covering soil layers are thin and the pressure of soil and water at shield's working surface have not reached the balance, or because the designed pressure at the edge of cutting wheel of tool pipes is not as high as natural soil's pressure on excavating surface, there will be surface settlement. Otherwise, there will be surface upheaval.

(iv) Rectification of the shield's structure or tool pipes will cause the interstice, which will induce the loss of soil layers. Also, this will cause the additive deformation by pressing nearby soil.

(v) The edges of cutting wheel of selected tool pipes are not suitable for the soil in working area.

(vi) During circuitously pressing/propping process of shallow large-diameter shield or jacking pipes, there will be some problems if their curvature radius is too small.

The above (i), (iii), (iv), (vi) factors have been considered in our calculation.

3. Study of the Influence of Construction of Subway Shield on Nearby Soil

1) Introduction

When the shield is excavated, the major influence on nearby soil is the change of stress conditions and strain conditions. Change of stress conditions refers to the change of total stress and pore pressure. Change of total stress is caused by unloading and the upheaval of soil. And the change of pore pressure is caused by pressing soil and changing water level. Renato and Karel have studied some problems in the design and construction of tunnels in soft soil. Among these problems is the loss of the stability of excavating surface induced by the stress relaxation. Yi has studied the distribution of pore pressure during the excavation of shields. The working disturbance which causes the change of total stress and pore pressure can be called stress disturbance.

Stress disturbance is one of the prime factors causing the change of strain conditions. The change of strain conditions is due to soil's elastic-plastic deformation, soil's compression by stress disturbance and the aging strain induced by soil's creepage. Romo has summed up the factors causing the change of soil's strain conditions: (1) The stress change near the working surface; (2) The shear stress between the shield and soil during excavation; (3) Soil's radical displacement caused by propping tunnel and grouting; (4) The consolidation deformation induced by soil's disturbance from excavation; (5) The contraction of lining layers; (6) The steady rheologic behavior of shallow layers, and so on. Romo has worked out the all directional displacement induced by above factors and compared the results with measured values in a Mexican tunneling case. He concluded that all factors including the behavior of soil, the thickness of soil layers, the diameter of tunnel, variety of shield structure, the working conditions, interstice in the ends of shield, replenishment and grouting, etc. will influence the displacement of surface. This paper has focused on the characteristics of soil's disturbance by the excavating work of shield and the rule of the change of soil's mechanical behavior after disturbance.

2) The characteristics of soil's disturbance by shield's excavating work

(i) The characteristics of unloading during shield's excavation
With the development of soil's lateral displacement, the degree of unloading is increasing (see the right part of the curve in Fig. 2.1). When the propping load is equal to the discharged load, the unloading process in nearby soil stops. But if not propped, the soil will not stop the off-loading process until the breakage of soil mass and the collapse of the tunnel. So it is necessary to brace the tunnel. Otherwise, the effect of arching can not be employed and the bracing cannot support the weight of upper soil and subsequent creep stress.

Yi has measured the nearby soil's excess pore pressure and its distribution rule during the shield's excavation (shown as Fig. 2.2). When the pace of excavation is quicker than that of gouging out soil, the excess pore pressure increases. Otherwise, the excess pore pressure decreases. The change of excess pore pressure is shown as a vibrating process. Figure 2.3 shows the distribution of horizontal and vertical excess pore pressure during excavation. As shown in Fig. 2.3a, there is no horizontal excess pore pressure in the measured section when the measured section is 13 m away from the working surface. With the decreasing of the distance between two planes, when the distance is below 4.0 m (Plane ②), the pore pressure increases. As the shield reached Plane ③, because the pace of excavation is slower than the pace of gouging out the soil, the excess pore pressure is negative. If the pace of excavation is quicker than the

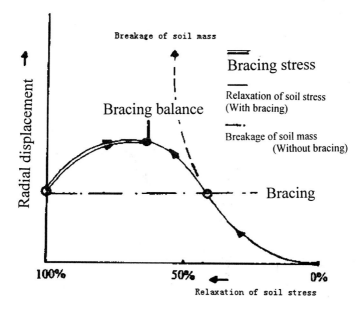

Fig. 2.1. The bracing and the relaxation of soil's stress

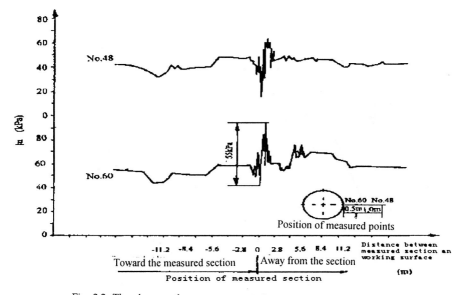

Fig. 2.2. The change of pore pressure induced by excavation

pace of gouging out soil, the excess pore pressure reaches its maximal values (Position④). When the measured position is far from the working surface, because of the water draining from between lining materials and the shield, the excess pore pressure will be negative (Position⑤). Grouting in shield's ends will increase the excess pore pressure (position⑥). The vertical distribution of

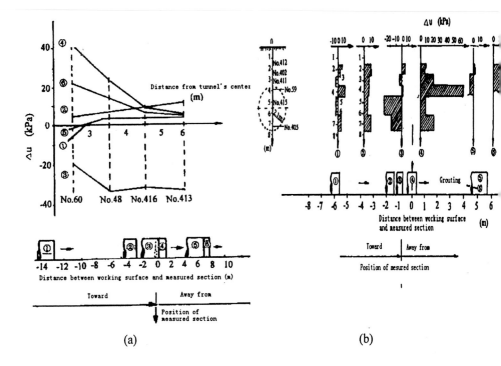

Fig. 2.3. The distribution of excess pore pressure during excavation

pore pressure is shown as Fig. 2.3b. In this figure, the vertical distribution of excess pore pressure is similar to the horizontal distribution. From the measured results, we can see that magnitude of pore pressure is influenced by the pace of excavating and the pace of gouging out soil. Moreover, the grouting outside the pipe will increase the pore pressure greatly.

3) The characteristics of soil's strain disturbance during excavation

The surface displacement caused by excavation along the shield's longitudinal direction can be divided into five different stages. I is the initial displacement; II is the displacement of surface in front of working plane (upheaval); III is the displacement during the passing of shield; IV is the displacement at shield's ends; V is the displacement caused by secondary consolidation. The initial displacement is due to the compression of disturbed soil in front of working plane and that compression induces the decrease of the void rate. The displacement in front of working plane is caused by the upheaval of frontal soil and the increase of pore pressure. The displacement during the passing of the

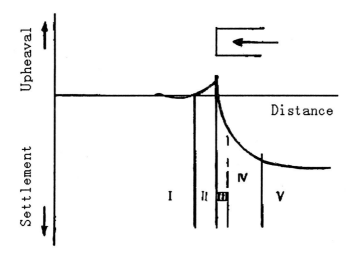

Fig. 2.4. The division of surface displacement during excavation

shield is induced by the disturbance of soil and the shear slide between soil and shield. The displacement at shield's ends is caused by the stress relaxation after soil's separating from the bracing. The displacement caused by secondary consolidation is due to the disturbance of soil and its deformation increasing with time. The causation of the above displacement is listed as Table 2.1. From this table, we can see that it is the disturbance of soil that is the prime causation for the additive displacement, because the disturbance of soil begets the change of soil's stress conditions which will induce the displacement. It should be pointed out that the displacement of soil consists of consolidation compression, the elastic-plastic shear slide and viscous-aging creepage.

By using numerical simulation methods, Romo has got the relationship of Δr (the depth of disturbed soil layers caused by excavation or lining) and δ_w (the radial displacement of tunnel wall):

$$\Delta r = \frac{E_i (1 - R_f)}{0.6 \, \sigma_f} \delta_\omega \tag{1}$$

Here, Δr is the radial distance from the wall of tunnel; E_i is the mean initial tangent modulus of disturbed soil; σ_f is the failure stress of soil; R_f is the ratio of failure, $R_f = \sigma_u / \sigma_f$, σ_u is the stress calculated from hyperbolic model; δ_w is the radial displacement of tunnel wall, whose minimum is zero and whose maximum is the void rate at shield's ends. Dazhongjianfu has got the relationship of central displacement of shield's surface and the scope of soil's disturbance from measured data (Fig. 2.5). From the figure, the larger the scope of soil's disturbance, the larger the central displacement of surface. Their relation is almost linear.

Table 2.1 The causation of soil's displacement by excavation

Displacement type	Causation	The disturbance of stress	The mechanism of deformation
Initial displacement I	Compressed soil is compacted	Pore pressure is increasing and effective stress is declining	Void rate is declining and the soil is consolidating
The displacement of front working plane (upheaval) II	The front compressed soil upheaves	Pore pressure is increasing and total stress also is increasing	Soil's compression causes elastic-plastic deformation
The displacement during the passing of the shield III	Working disturbance causes shear slide between shield and soil	The relaxation of stress	The elastic-plastic deformation
The displacement at shield's ends IV	Soil loses the propping of shield	The relaxation of stress	The elastic-plastic deformation
The displacement caused by secondary consolidation V	The subsequent aging deformation	The relaxation of stress	The creeping compression

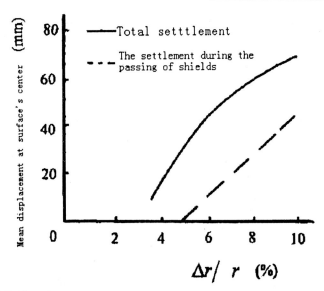

Fig. 2.5. The relationship of surface displacement and the scope of disturbance

The scope of disturbance is correlative to the propping force during the excavation, the grouting time and the ratio of the depth of covering layers and outside diameter of tunnel (H/D). From Fig. 2.6, we can get the above conclusion. The heavier the propping and the heavier the compression of front soil, the

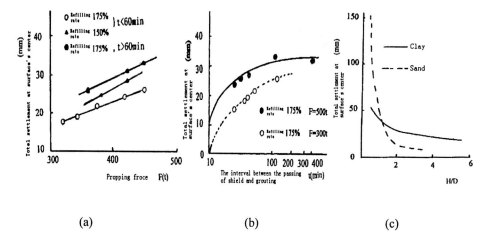

Fig. 2.6. Relationship of working conditions and the scope of disturbance

larger the central surface displacement due to the larger soil's disturbance (Fig. 2.6a). The longer the interval between the passing of shield and the grouting, the larger the central displacement due to the larger stress relaxation (Fig. 2.6b). The larger the H/D, the smaller the central total surface displacement due to the smaller surface disturbance (Fig. 2.6c).

4) The mutation of disturbed soil's mechanical qualities

Stress relaxation will cause the decline of deformation modulus (shown as Fig. 2.7). During the excavation of shield, this relaxation will make the modulus decline to 30-70% of original value. The relationship of disturbed soil's mechanical parameters and soil's strain is shown as Fig. 2.8. The greater the soil's strain, the smaller the parameters. The decline of parameters is caused by the breakage of soil's mass and the loss of sorption capacity of original soil.

5) Conclusion

(i) Soil's disturbance can be divided into stress disturbance and strain disturbance. The stress disturbance is that of soil's stress conditions by construction and mainly refers to the change of effective stress, including the change of total stress and pore pressure. The strain disturbance of soil is the disturbance of producing additive strain in soil by construction. The relationship of the scope of disturbance and the surface central displacement during the passing of shield is in direct proportion.

(ii) There will be obvious mutation in soil's mechanical qualities after disturbance. With the increase of disturbance, the strength parameters and

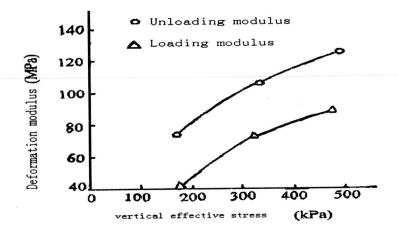

Fig. 2.7. The relationship of stress relaxation and deformation modulus

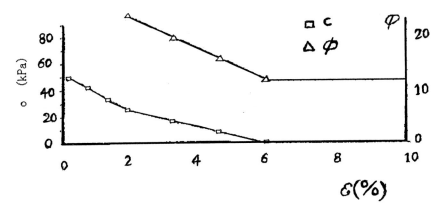

Fig. 2.8. The relationship of soil's strain and strength parameters

the deformation modulus decline. Further research will be done in quantitative description of the mutation of disturbed soil's mechanical qualities during the excavation.

4. The Six Parts of the Surface Settlement and Soil Layers' Displacement Caused by the Construction of Underground Continuous Walls

1) Elastic displacement of walls;
2) Resilience of soil for unloading in pits, plastic upheaval, the boiling of sand caused by unsuitable abstraction of water;
3) Soil's consolidation settlement outside the walls
4) The loss of soil and sand for abstraction of water (one kind of the loss of soil layers);

5) The leakage of soil and sand between walls;
6) The soil layers' inclination to grooves because of the excavation of groove walls

Among the above displacement, the surface settlement and soil layers' displacement consists mostly of the first three parts, which can be calculated. The last three parts can be reduced to their minimums through controlling the technique of construction, experience and management.

5. Five Key Problems in Theoretical Forecast

1) Non-linear coupling of clay's consolidation and rheological property

The interaction of soil's initial creepage and consolidation has been proved through indoor experiments.

Non-linear rheological property: "The ratio of consolidation" and "the flexibility of creepage" are not constant, but are:

$$J(\text{or } \eta) = f(t, \sigma)$$

2) Engineering construction mechanics and time-dependent mechanics

- The most disadvantageous conditions frequently occur at different working conditions and working stages
- Piles are driven one by one
- The excavation of subway shield and instalment of pipe pieces
- Jacking pipes are constructed straightly or deviously
- Pits are excavated step by step

"Temporal and spatial effects", i.e. deformation and stress both increase with time and spatial area during construction, can be studied by engineering construction mechanics and time-dependent mechanics.

3) Three-dimensional temporal spatial effect analysis

The protected objects: subway interzones, underground pipelines, roads and shallow foundations of buildings are arranged lengthways along the construction objects; the crossroads and viaducts are arranged crosswise.

The irregularity of the urban plots makes the pits of the buildings irregular.

Single shield shuttles and two shields advance in the adverse or the same direction.

In the design of shallow large-diameter jacking pipes, it is necessary to revise their advancing direction.

The piles are spatially spread and jumping driven.

The above are all three-dimensional conditions. Also, the deformation development during construction is non-linear temporal function, which includes:

- Soil's viscous aging deformation which refers to the relaxation of stress and strain during the excavation
- Soil rheological behavior that refers to the creepage, the stress relaxation and the decline of long-term strength which are all time-dependent functions

4) Use of half-analytic element method

Finite element method is too complex and unsuitable when it comes to the capacity of computer, the time of calculation, the pretreatment and after-treatment. The half-analytic element method is the combination of finite element method and analytic method.

Its characteristic is that only y-direction is discretizated while x, z-direction can be solved by analytic functions concerted with the deformation, which can convert three-dimensional problems into one-dimensional problems. As far as shields and jacking pipes are concerned, we must use analytic functions to fit the radial and circular displacement, which can solve the shallow surface constraint boundaries in half-infinite space problems.

Circular tunnels of double-line shields are required to be fit for these multi-continuous area problems.

Pile group, which is simplified as axisymmetric problems.

Shallow jacking pipes, which are simplified as the half-space infinite mediums with the restriction of surface.

5) Application of "The Settlement Inductive Figure", which refers to the settlement influence on each other of old and new buildings

This figure can be applied to the conditions of many irregular and arbitrary loading.

Using the "Symmetric Rule" and "Iterative Rule", from the three-dimensional analytic settlement solution in visco-elastic mediums, we can get:

(i)   The "Settlement-Displacement-Time" variation process at any location at any time in foundations;

(ii) Drawing "Chart of equal settlement" and "Spatial curved surface chart of differential settlement".

6) The allowable limit values of deformation and displacement of various protected objectives:

Constructions, completed subways, underground pipelines, urban crossroads and viaducts, road beds and road foundations, simultaneous construction of two close buildings.

The following Parts III, IV, V are the analyses of underground engineering practice, with the pipe jacking engineering cases as the focus.

PART III: PIPE JACKING CONSTRUCTION

1. Working Conditions

1) We take the construction of the jacking pipes (ϕ1650) in the project dealing with confluent waste water in Shanghai as the example. The project starts from Zhongshan Southern Road, crossing Shanghai-Hangzhou Railway, through Putuo Park and Suzhou training wall and ends at Yejiazai pumping station.

2) Soil-water balancing tool pipes are used from Well. #12 to Well. #12.2 (397.3 m long). The covering soil layer is not as thick as 12 m (minH=2.5 m). The pipes are constructed in three stages. The soil is gray and yellow clayey silt. The subjacent bed consists of saturated sand.

3) The important protected objects: Shanghai-Hangzhou Railway, two gas main pipes (ϕ500), Zhongshan Road, successfully passing through the Suzhou River bed, training walls along the river, six high line posts.

2. Soil's Vertical Settlement

Stage 1: The surface settlement before excavation plane reaches measured points. We assume:

P_0—the designed balancing soil pressure in front of tool pipes in excavation plane;

P_N—natural soil pressure;

When: $P_0 = P_N$, the displacement is very small;

$P_0 < P_N$, the displacement increases;

$P_0 > P_N$, the surface upheaves.

We can see these in Fig. 3.1.

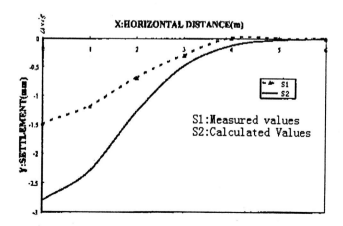

Fig. 3.1. The surface settlement before excavation surface's arrival
(The calculated values are too large)

Stage 2: The surface settlement directly above tool pipes

In this stage, because soil is propped by the tool pipes, soil has to endure the friction and the pressure from jacking pipes' cutting edge (shown in Fig. 3.2).

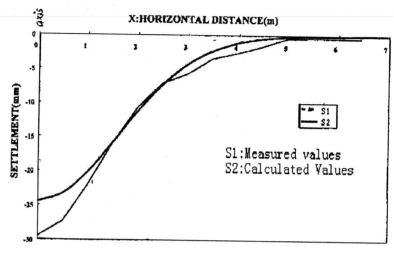

Fig. 3.2. The surface settlement directly above tool pipes

Stage 3: The surface settlement above latter pipe nodes

From the ends of tool pipes to transfixion of pipe jacking line, the settlement values also include the visco-plastic deformation during tool pipes' construction (shown in Fig. 3.3).

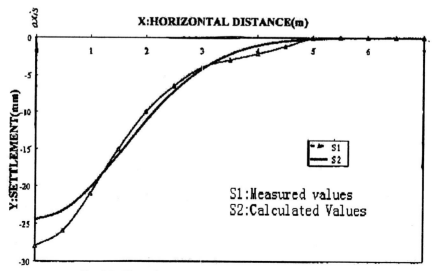

Fig. 3.3. The suface settlement above letter pipe nodes

Stage 4: The surface settlement from the ending of construction to the steadiness of deformation

As shown in Fig. 3.4, the maximal surface displacement is about 40 mm, taking place at the locations 5 m from the settlement groove.

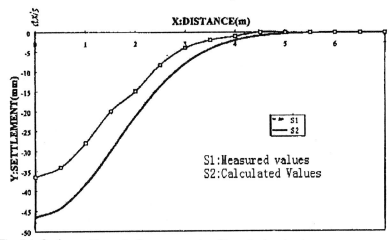

Fig. 3.4. Surface settlement after construction (the calculated values are too large)

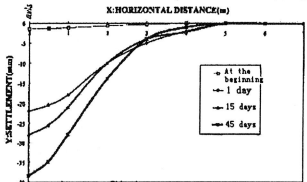

Fig. 3.5. The time-dependent variation of measured settlement in section plane

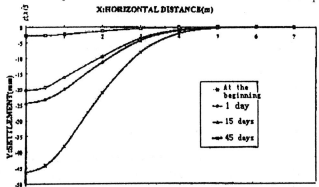

Fig. 3.6. The time-dependent variation of calculated settlement in section plane

3. The Time-Dependent Variation of Settlement

1) For a fixed section along jacking pipes, the time-dependent variation of settlement during construction is shown as Fig. 3.5 (measured) and Fig. 3.6 (calculated).
2) For particular points, the time-dependent variation of surface settlement is shown as Fig. 3.7 and Fig. 3.8.

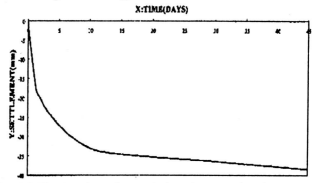

Fig. 3.7. The time-dependent variation of measured settlement

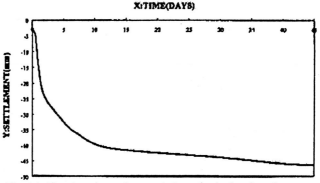

Fig. 3.8. The time-dependent variation of calculated settlement

4. The Radial Displacement of Soil Above Jacking Pipes

1) The Radial displacement of soil outside tool pipes during pipe jacking

The maximal horizontal displacement is 53 mm, as shown in Fig. 3.9 (the tool pipe's cutting edge is assumed as the origin and the advancing direction is positive).

2) Three-dimensional curved display of soil's movement

Figure 3.10 (a) and Fig. 3.10 (b) are the spatial curved display of soil in front of and behind the working plane respectively. In the figures, the depletion of radial displacement with distance in soil in front of working plane is very

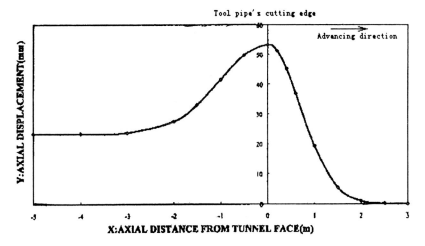

Fig. 3.9. The radial displacement of soil

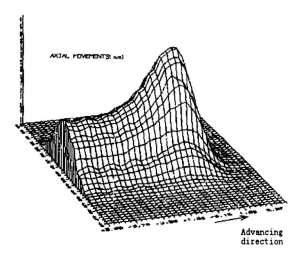

Fig. 3.10(a). The radial displacement of soil (right view)

fast—the values become zero at the distance of 2.6 m. But the depletion of back soil is very slow—the values become convergent (about 20 mm) at the distance of 23.4 m. This is because during the advancement of tool pipes, back soil endures the friction from the wall of pipes and gets to have slow-depleting radial displacement.

3) Soil's radial displacement variation with depth

In Fig. 3.11 and Fig. 3.12 (sections at the excavating plane's cutting edge), the soil with maximal radial displacement is 4.0 m deep from the excavating plane

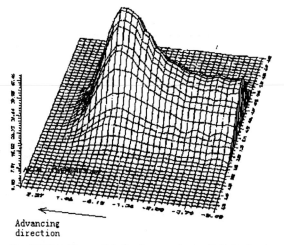

Advancing
direction

Fig. 3.10(b). The radial displacement of soil (left view)

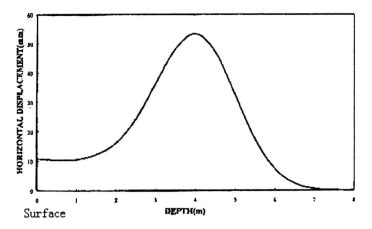

Surface   DEPTH(m)

Fig. 3.11. The radial displacement variation with depth

edge. But the horizontal displacement at surface is not the largest (only 9 mm).

5. In Order to Do Some Parametric Analyses about the Factors Influencing Soil Settlement, the Following Points have Also been Studied

1)   If the balancing soil pressure p_0 designed at the excavation plane during the jack's passing the cutting wheel surpasses the natural soil pressure in front of the excavation plane, there will be upheaval in the front surface.

2)   Soil's horizontal displacement variation with depth in the vertical plane along the jacking pipe's axis as the tool pipes arrive.

3)   The variation of pore pressure during the construction.

(The maximal radial displacement is at depth of 4.0 m from the excavating plane's cutting edge)

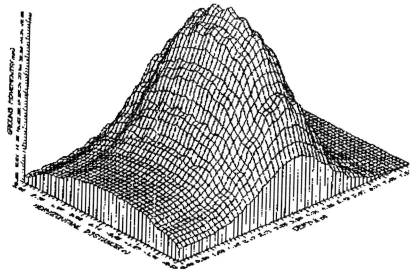

Fig. 3.12. The spatial curved figure of radial displacement variation with depth

4) The surface settlement and its variation with depth induced by rectifying shallow large-diameter jacking pipes.
5) The influence on settlement of different covering layers.
6) The influence on settlement of different soil.
7) The influence on settlement of different pressure in thixotropic mud.
8) The influence on settlement of different jacking pipes' diameter.
9) The comparison of the surface settlement with steel-jacking pipes and with concrete jacking pipes.
10) Curvilinear jacking pipes with small radius (at the sharp swerve R < 300 m).
Detailed discussion is omitted for the sake of the space of paper.

PART IV: THE CONSTRUCTION OF SHIELDS

1. The Stochastic-Fuzzy Analyses of Settlement Induced by Single-hole Shield Tunnel's Construction

The contrast between calculated values and measured values is shown in Fig. 4.1 and Fig. 4.2.

Table 4.1 The measured settlement values in section planes of D517~D590

Measured points	D517	D520	0525	D530	D535	D540	D545	D550
Measured values	−12.35	−11.98	−12.31	−12.01	−11.75	−11.40	−11.08	−10.30
Measured points	D555	D560	D565	D570	D575	D580	D585	D590
Measured values	−8.26	−6.01	−5.52	−1.75	−1.14	−0.63	−0.23	−0.72

Date: 1997.1.21.A.M.8:30; Working position: Circle 564; The negative represents settlement (mm);
Soil's stochastic parameters: $a_1= \xi_2 \approx 1.95$, $a_2= \xi_3 \approx 2.73$, $a_3= \xi_4 \approx 2.41$, $\xi_1 \approx 0.77$

Table 4.2 The measured settlement values in the bench plane of C308

Measured points	L9	L6	L3	C308	R3	R6	R9
Measured values	−1.76	−4.11	−9.52	−10.50	−8.91	−5.46	−1.38

Date: 1997.6.21.A.M.8:17; Working position: Circle 352; The negative represents settlement (mm);
Soil's stochastic parameters: $b_1 \approx 1.00$, $b_2 \approx 3.10$, $b_3 \approx 2.55$, $\eta_1 \approx 0.77$, $\eta_2 \approx 0.50$, $\eta_3 \approx 3.10$, $\eta_4 \approx 2.55$

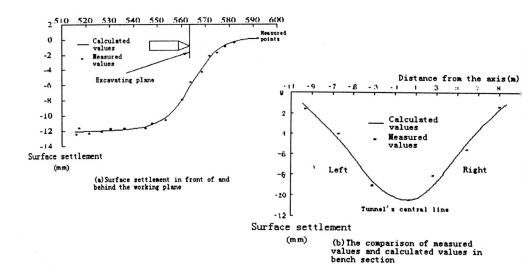

Fig. 4.1. The settlement figures in section and bench planes during constructing shields

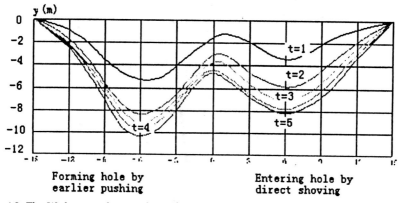

Fig. 4.2. The W-shape settlement chart of the soil above the double-hole tunnel of single shield advancing to and fro

2. Double-hole Tunnels of Single shield advancing to and fro

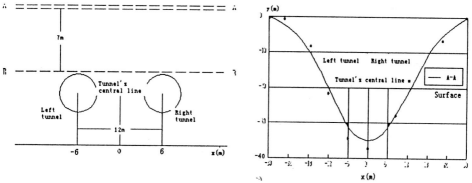

(a) The position of shield and studied section

(b) The U-shape settlement groove of sruface after excavating two-hole tunnel

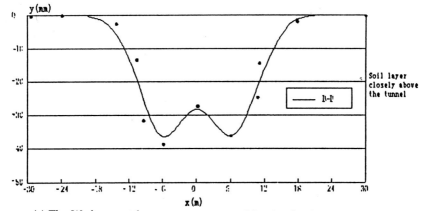

(c) The W-shape settlement groove caused by the displacement of soil closely above the tunnel

Fig. 4.3. The settlement groove of soil above double-hole tunnels of single shield advancing to and fro

3. Surface lengthways settlement during the advancing of two shields with double-hole tunnels

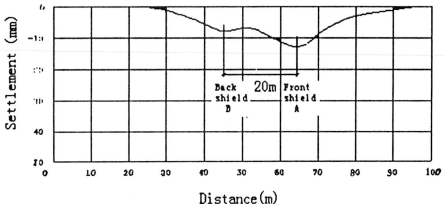

(a) I-I surface settlement during the advancing of two shields with double-hole tunnels

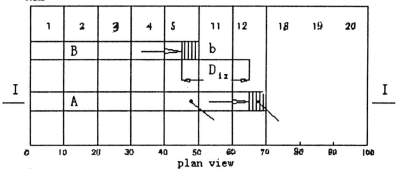

(b) I-I surface settlement when there are 20 m between two same-directional shields

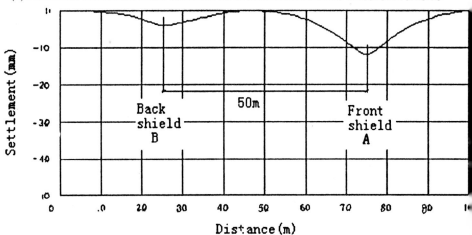

(c) I-I surface settlement when there are 50 m between two same-directional shields

Fig. 4.4 Soil's lengthways settlement during the advancing of two shields
with double-hole tunnels

4. The Surface Lengthways Settlement at Different Distance During Two Shields with Double-hole Tunnel Advancing Toward Each Other

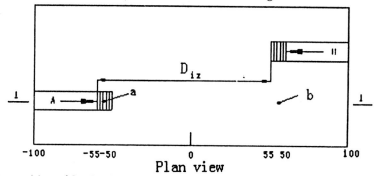

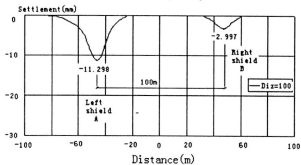

(a) The position of Section I-I and surface lengthways settlement above Shield A

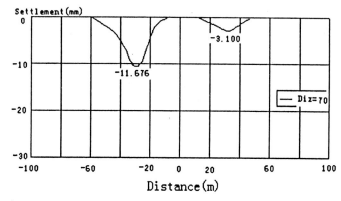

(b) The lengthways settlement in Section I-I when there are 100 m between two shields

(c) The lengthways settlement in Section I-I when there are 70 m between two shields

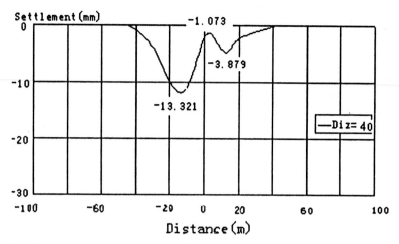

(d) The lengthways settlement in Section I-I when there are 40 m between two shields

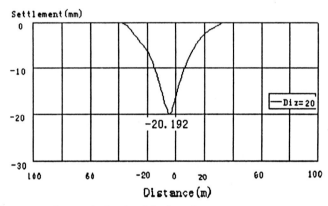

(e) The lengthways settlement in Section I-I when there are 20 m between two shields

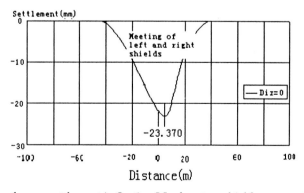

(f) The lengthways settlement in Section I-I when two shields meet each other

Fig. 4.5. Soil's lengthways settlement when two shields with
double-hole tunnel advance adversely

5. Sensitivity Analysis of Shields' Major Design Factors

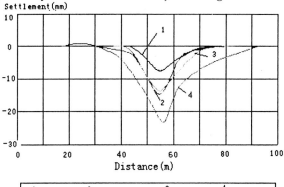

(a) The parametric comparison and analysis of measured values and surface settlement calculated from different soil's compression coefficients

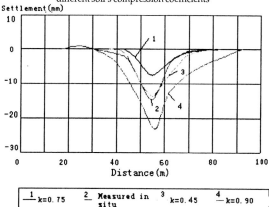

(b) The comparison of measured settlement values and surface settlement calculated from different lateral pressure coefficients

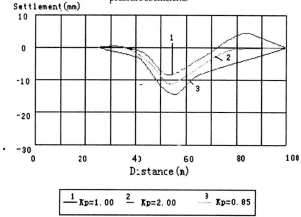

(c) The surface settlement curve under different spoiling rate
(Advising K_p =96%~104%)

Fig. 4.6. Sensitivity analysis of major design factors of shields

PART V: THE EXCAVATION AND PILE ENGINEERING

A lot of research work has been carried out about excavation engineering in China. This paper will only be confined to three-dimensional numerical analysis about construction mechanics in consideration of consider soil's nonlinear creepage.

1. The 3-D Finite Element Analysis about Excavation Construction in Consideration of Soil's Nonlinear Creepage

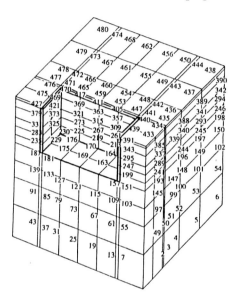

Fig. 5.1. Numbering the lattices and nodes

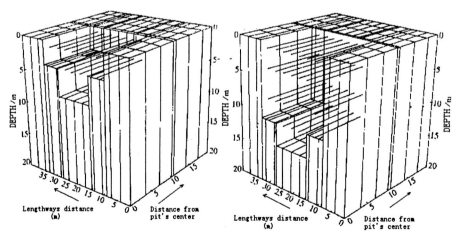

Fig. 5.2. The simulation of 3-D bracing and excavating conditions during excavation (precedence of excavation to bracing is not considered in this figure)

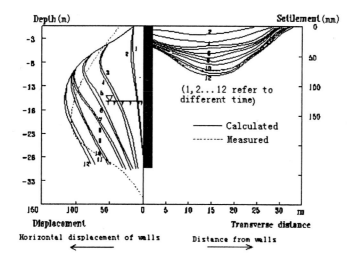

Fig. 5.3. Horizontal displacement of continuous walls and surface settlement behind walls in the pit's mid-span section (increasing with time)

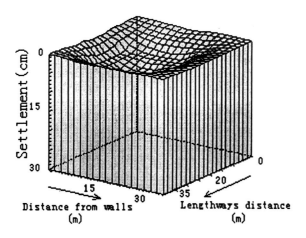

Fig. 5.4. Surface 3-D settlement basin behind underground continuous walls

2. The Construction of Piles (Influencing on the Movement of Nearby Underground Pipelines and the Subway)

The maximal settlement at the top of piles and the time list

Measured points	The minimal distance between that points and piles (m)	The vertical displacement (mm)	The maximal horizontal displacement (mm)	The date when extreme values take place	Days from the beginning
S_6	8.5	42		5.16	62
S_2	10		50	5.16	62
M_8	11.2	39		5.18	64
M_6	11.2		42	5.16	62
M_9	13		40	5.16	62
M_2	13.7	25		5.22	68
M_{11}	16	16	22	5.18	64
H_3	21	16		5.16	62
H_2	21.4	15		5.25	71
H_1	25	8		5.2	48

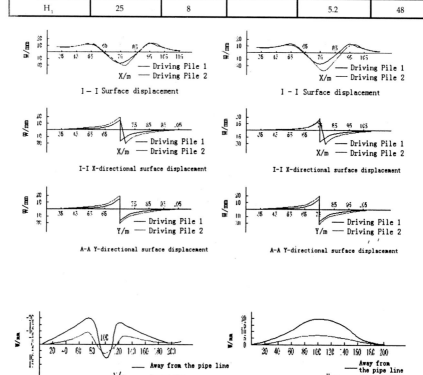

(a) Driving consecutively (away from protected objects)

(b) Jumping driving (between the protected objects)

Fig. 5.5. a) Vertical differential settlemetn of pipelines in y-z plane (8 piles/day) b) Horizontal differential settlemetn of pipelines in x-y plane (8 piles/day)

PART VI: PROSPECT OF FURTHER RESEARCH

At present, it is the prime task for geotechnical engineers to get completed and effective theories and methods which can be satisfactory to the stability and safety of constructions. Based on the 3-D artificial and factual simulation systems, they can be used for satisfying not only the theoretical connotation but also the requirement of engineering practice.

In present control systems, the task needs precise mathematical models as well as the constitutive relations. The output of systems is definite, i.e. the task needs the output to follow anticipant values. This requirement of task is so simple that it cannot satisfy the engineering problems, which are required to study from machines, to automatically respond to and deal with emergency immediately. Therefore, it is necessary to combine modem auto-control theories and artificial intelligence science and to employ suitable intelligence control methods. These methods are based on the combination of various knowledge and especially suitable for complex underground working conditions. Besides, the methods can study fuzzy objects and highly nonlinear systems very well.

When it comes to mathematical models for research, fuzzy artificial neural network technology is first adopted, with which the above-mentioned complex, non-linear, non-established systems are presented by simple relationships and is different from classic models. It is well-known that in underground engineering, there are lots of "black process", i.e., it is hard to establish classic mathematical and mechanical models. Artificial neural network technology can be used to establish a kind of "mechanism" between input and output, to do fuzzy reasoning and analysis, to forecast future prospects of the systems and to provide the basis for next decision. Moreover, it is also urgent to do some research work about the combination of artificial intelligence symbol reasoning and numerical calculation.

When it comes to 3-D artificial factual simulation research based on monitoring multimedia technology, we advise that the 3-D figure-image technology, image monitoring and imaging simulation can be used to realize the forecast and control task, which are completely different from the classic 2-D, discontinuous, static analysis. Before excavation, we first simulate excavation process, every bracing stage and the displacement in every stage, then correct and adjust the values during excavation, feedback the design and construction parameters, forecast the next conditions and the dangerous settlement's position, scope and degree, decide whether or not to adjust the parameters in next stage and the magnitude of adjusted values from the imaging analysis. So by establishing geological database, figure database and engineering characteristic parameters, we can use multimedia-monitoring 3-D factual simulation systems to demonstrate the superiority of information engineering. This is an advanced topic for civil engineers, geotechnical engineering and electrical engineers all over the world. We are trying our best to carry out this project in our field.

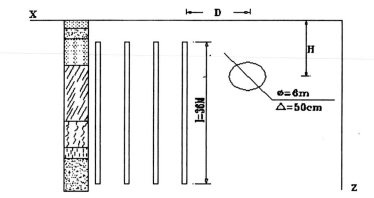

Positions of the piles and tunnels

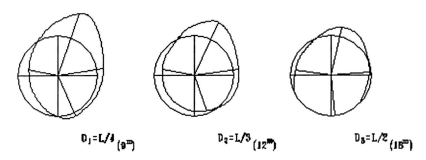

$D_1=L/4$ (9ᵐ)　　$D_2=L/3$ (12ᵐ)　　$D_3=L/2$ (18ᵐ)

Fig. 5.6. The transformation figures of the tunnel caused by pile-driving at different distance between piles and tunnel (y = 100 m) (8 piles/day; L = 36 m; Driving time: 1 hour/pile)

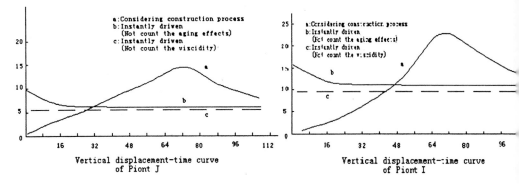

Fig. 5.7. The development of surface settlement near the piles with time

The Results and Their Applications

- The late construction of 3# and 11# subways (excavations are used for subway stations while shields are used for tunnels).
- Constructing the deep foundations of skyscrapers.
- Other constructions of geotechnical and municipal underground facilities, such as the cut excavation of Yuejiang tunnel project.
- As far as software is concerned, it can be applied to engineering projects after minor revision and correction, such as: ① The cracks caused by too large settlement and uneven settlement of building; ② The forecast, control, prevention and eliminate of geotechnical harms such as large-area settlement in highway engineering. Therefore, developing the facilities to collect data and to make imaging monitor and developing computer technological management systems in engineering will be a prosperous field in future.

ACKNOWLEDGEMENT

A great part of this paper is drawn from the "References" and the degree theses of graduate students and doctorate students supervised by author. Here I must express my acknowledgement to them.

REFERENCES

Cao Z.Y (Tongji University), Zhang Y.Q., "Half-analytic Numerical Methods", Pub. Company of National Defence Engineering, 1992.8

Fang C.Q., "Pipe Jacking Construction Induced Displacement Analysis Based on Half-analytic Element Methods", (Doctorate Thesis of Tongji University), 1998.9

Li X.Y, "The 3-D Numerical Analysis and Imaging Display of Environmental Harms Caused by Deep Pit Excavation", (Doctorate Thesis of Tongji University), 1996.6

Renato E.B., Karel R., "Tunnel Design and Construction in Extremely Difficult Ground Conditions", 1998, 8: 23-31

Romo M.P., "Elements Induced by Soft Ground Tunneling", Inter. Conf. on Case Histories in Geotech. Engrg. St. Louis Missouri, 1984, Vol. 1.

Shanghai Municipal Managing Bureau, "Construction Technological Brochure for Municipal Underground Engineering in Soft Soil", 1991

Shanghai Subway Engineering Managing Bureau, "Designing Technological Rule for Subway Engineering", 1995.4

Sun J., Huang W. (Tongji University), "The Forecast, Prevention and Cure to Environmental Problems Caused by Urban Continuous Walls Construction", Proc. of the 6th Civil Engineering Conference, pp. 641-649, China Publishing Company of Civil Engineering, 1993.5, Beijing

Wayne Clough et al., "Construction Induced Movements of In situ Walls", Engineering Structure Division, ASCE Journal, Sept. 1992 pp. 439-470

Xu Y.F., (Tongji University), "Soil's Construction Induced Disturbance Study", Proc. of the 8th National Conference on Soil Mechanics", 1999

Yi X., Rowe K., Lee M., "Observed and Calculated Pore Pressure and Deformations Induced by an Earth Balance Shield", Can. Geotch. J., 1993, 30: 476-490

Yi Z.T., "The Prevention and Cure of Environmental Harms Caused by Nearby Building's Settlement", Tongji Geoeng. Research Inst., 1995.3

Zeng X.Q. (Tongji University), "The Settlement Analysis and Calculation during the 1st Routine Tunnel's Excavation", Underground Engineering and Tunnels, Vol. 3, 1995
Zhou G.J., "The half-analytic element methods to solve influence on ground environment from extrusion effects by pile-driving", Tongji Geoeng. Research Inst., 1995. 10, Shanghai

9

Quay Walls of the Port of Rotterdam

A.F. Van Tol
Delft University of Technology and Rotterdam Public Works, Delft
Netherlands
J. de Gijt
Rotterdam Public Works, Rotterdam, Netherlands

ABSTRACT

This paper gives an overview of the development of the design of the Rotterdam quay walls. Because of the increasing dimensions of the ships during the last 40 years the retaining height of the quay walls increased as well. The larger height of the wall required heavier sheet piling, higher anchor forces and thus more intensive pile fields. This process of up-scaling resulted in the 70s in large displacements of some of the sea quay walls for coal and ores. The behavior had a decisive influence on the later designs. In the paper the present design philosophy and a new safety concept is presented. Furthermore different optimisations are shown. It appears that the constructions built according to the actual design are functioning properly and that the costs did not increase over the last ten years.

Key Words: Harbor, quay wall, sheet piling, tension pile

1. INTRODUCTION

The Netherlands is situated in the Western part of Europe at the boundary of the North Sea. This basin was filled up with sediments during the Pleistocene (diluvium), and the Holocene (alluvium). These predominantly fluvial and marine sediments were deposited by the rivers Rhine and Maas. In the Western part of the Netherlands the ground level is about or some meters below mean sea level.

The city of Rotterdam is located in the Western part of the Netherlands at the banks of the Rhine. The situation of these rivers and Rotterdam is shown in Fig. 1. The harbor of Rotterdam is the largest in the world as far as cargo throughput is concerned.

The main factor for development of any harbor and its facilities are the shipping activities within the harbor area. Therefore the Port of Rotterdam has been developing its facilities over the past 100 years in accordance to the development in shipping. These developments reached its peak in the end of the 60s resulting in larger ships and thus greater water depths. This in turn had major consequences for the construction of harbor basins, quay walls and waterways which are mainly determined by the dimensions of the ships and the cargo handling facilities. In this paper the development in the design of the quay walls will be presented.

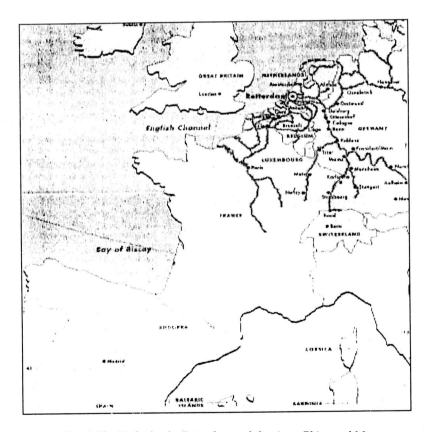

Fig. 1. The Netherlands, Rotterdam and the rivers Rhine and Mass

First an overview will be given of the history of the Port of Rotterdam with its shipping activities, important in understanding the development in quay wall design. Next the soil conditions which are always important for the design of retaining structures are described. Subsequently the design process is illustrated from the initial stages of a project up to the monitoring activities when the quay wall is in use. The construction of this kind of infrastructural facilities requires knowledge of several disciplines to optimise the design. The main disciplines involved·being geotechnical, hydraulic and structural engineering. The knowledge and experience gained during the past decades has resulted in a so-called "design philosophy" for quay walls. This also includes more recent developments in safety philosophy and computation methods. Finally several aspects of quay wall design are discussed and some examples are presented.

2. HISTORY OF THE PORT OF ROTTERDAM

In the old days landing stages or moorings for vessels were constructed as close to a port's town centre as was possible and were typically simple wooden structures of piles, jetties and quays. Also the origin of harbor of Rotterdam can be found a few centuries ago in the center of the present city of Rotterdam when it was a port of refuge for fishing boats. The 'Buizengat', one of the oldest harbors in Rotterdam, offers a reminder of that time. In the second half of the 19th century a number of harbors were dug on the southern side of the river. As a result of its privileged geographic position the port of Rotterdam has seen ever increasing volumes of incoming and outgoing goods, therefore in time the extent of these harbor areas became insufficient. Furthermore due to increasing dimensions of the ships these harbors did not have enough water depth. Expansion took place in western direction where the river had a greater water depth. Early this century new port basins like the Maashaven, Waalhaven and Merwehaven were built. The actual situation of the harbor is presented in Fig. 2.

The Waalhaven harbor constructed around 1930, was the largest dredged harbor in the world at the time. Initially used for bulk transport of coal the Waalhaven harbor at present is still used as a general cargo and container harbor. After II World War harbors and harbor basins have been constructed to create optimal conditions for merchandising and shipment. This expansion is presented in Table 1. Goods transhipment between the wars reached 40 million tons. Scale grew, and with it the importance of the investments and the thinking. Current annual port transhipment levels are 300 million tons. The quay walls played a major role in making it happen.

A feasibility study and environmental impact studies for the land reclamation for Maasvlakte II are now being carried out. It is foreseen to construct a new industrial area in the period of 2004-2006 of about 1000 ha.

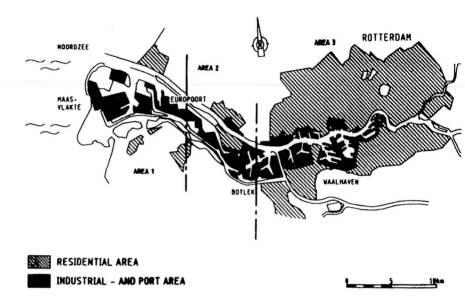

RESIDENTIAL AREA

INDUSTRIAL - AND PORT AREA

Fig. 2. Situation Rotterdam Port area

Table 1. Expansion of the Rotterdam harbor

Waalhaven	1930
Eemhaven	1950
Botlek	1955
Europoort	1958
Maasvlakte I	1968
Maavlakte II	2004-2006

In the past centuries the transportation of merchandise by ships has changed a great deal. Well into the 19th century ships were made of wood and propulsion took place with the help of sails. Around 1850 the first engines were used for propulsion and wood was replaced by iron. This development lead to the construction of specialised ships. For a long time the so-called General Cargo Ship predominated, a ship with holds containing bales, cases, crates and drums. In fact it was an expanded version of the wooden ships from former days. With the increasing goods carried the economic possibility of specialising the ships also grew. This specialisation reduced loading and unloading times considerably. Besides, it brought an improvement in the utilisation rate. The largest ships in operation at present are bulk carriers which have a length of over 400 m, a width of 65 m, a draught of about 24 m and have a tonnage of 450,000 dwt.

The dimensions of the ships largely determine the design of harbor basins. The draught of the ship determines the required depth of the quay wall; the length of the ship dictates the dimensions of the berths and of the turning circles that are necessary; and the width of the ship determines the width of the harbor basins. In Fig. 4 the increasing depth of the constructed harbors in time is presented. The earlier mentioned peak in the 60s, with increasing dimensions of the ships, is clearly visible in this graph.

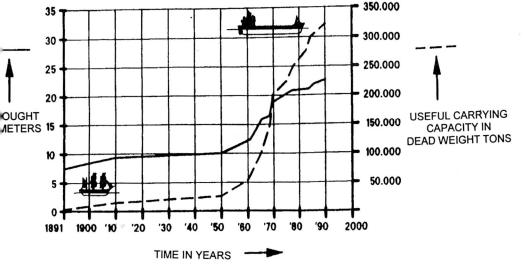

Fig. 3. The development of ship dimensions in time is shown

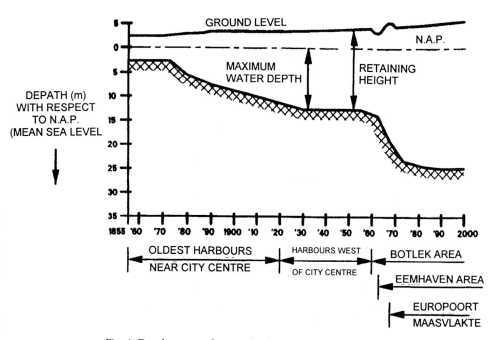

Fig. 4. Development of water depth and retaining height

It is expected that the dimensions of the bulk carriers will remain the same or even become smaller. However for the middle class ships, especially container ships, it is anticipated that their width will increase up to 50 m. The length and draught of these ships are not likely to increase that much in the future.

Quay wall constructions at the end of the 60s and 70s require therefore significantly greater dimensions due to the anticipated increase in ship dimensions. Other determining factors for the design of these type of structures is the increasing retaining height, the development in cargo handling facilities and the type and amount of surcharge on and behind the quay walls. Another development to be considered is the decreasing use of tugboats as ships will increasingly use their own engines for mooring purposes. This may consequently cause additional erosion at the bottom near the quay walls.

All these developments imply that new creative solutions and techniques had to be developed, and will have to be developed in the future, for the design and construction of new quay walls. So far the progress in the requirements for the quay walls. The other important condition is determined by the soil and (ground)water conditions. The subsoil conditions in the Rotterdam area are very important in this respect.

3. SOIL CONDITIONS

The Port of Rotterdam is located near the North Sea in the Rhine-Maas Delta. Characteristic for this area are its meandering rivers and the rise and fall of the sea level in the past. As a result soil conditions can vary significantly over short distances. From about 1700 onward geologically speaking little has changed. However since that time human activities, e.g. several land reclamations such as the Maasvlakte I, have been taking place changing the natural soil conditions.

In geotechnical respect (e.g. soil profile) the Port of Rotterdam, see also Fig. 1, can at present be subdivided in the following three areas:
1. The city area up to the river Oude Maas
2. The area between the river Oude Maas and the Maasvlakte
3. The Maasvlakte area

For each of these areas the result of a typical Cone Penetration Test (CPT), with a corresponding boring log, is shown in Fig. 5. From the boring logs it is evident that the soil profile under the city area consists mainly of very soft Holocene layers to a depth of 15 m beneath Reference Level (RL). Beneath this depth a densely packed medium to coarse Pleistocene sand layer is found. In westward direction the quality, in terms of strength and stiffness, of the Holocene layers improves due to an increasing sand content. Furthermore the top of the Pleistocene sand layer slopes downward, from RL −15 m to RL −22 m, in the direction of the North Sea.

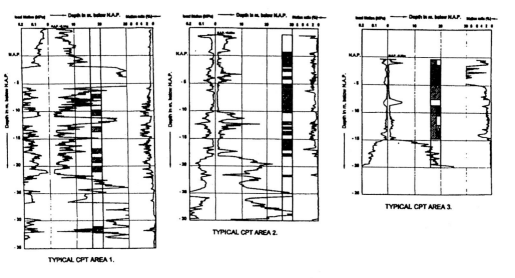

Fig. 5. Typical Cone Penetration Test (CPT) results for areas 1, 2 and 3, in Fig. 2

4. DESIGN PROCESS

The following levels can be distinguished for the design process:
- feasibility level
- preliminary design level
- final detailed design level
- preparation for tender level
- selection of contractor level
- construction level

In the feasibility level the project requirements are established, this is done in consultation with the client. In the Port of Rotterdam the client is usually the Rotterdam Port Authority. The Rotterdam Port Authority manages the harbor infrastructure, leasing the land and water (including the quay constructions) to their clients. Special requirements of these clients can be implemented in the design process.

During the preliminary design phase several alternative design options are considered. After careful consideration one of these options is selected which is then worked out in the final detailed design level. The selected option will be prepared for tender, the tender documents include a work description and drawings. For these kind of projects an open EC-tender is used. In these tender documents criteria are given which the contractor has to fulfill. After a contractor has been selected construction activities can be started. During construction the contractor is required to present detailed working plans, e.g. for sheet pile driving and concreting activities.

5. DESIGN PHILOSOPHY

5.1 General

The main principle of an optimum design is that it must fulfill the project requirements (including the environmental considerations) at a minimum of costs, not only for the design and construction but also concerning maintenance in the future. This requires that the designers are highly skilled and experienced in designing as well as in construction. The required expertise is present in the Engineering Division of Rotterdam Public Works which is a highly qualified institute as most of the quay walls, piers and jetties are designed and constructed under their supervision. Experience in the design of the deep-water quay walls has been gained since the late 60s when the depth of the waterways and retaining height of the quay walls increased considerably. The obtained experience lead to a design philosophy which is of course continuously being evaluated and improved. The design concept in Fig. 6 is the result of this ongoing evaluation. The first quay wall, the 2nd phase of the EKOM quay wall, constructed according to this design concept was constructed in 1982. Since then nearly all deep water quay walls have been constructed according to this design concept.

5.2 Development of Design Concept

The design concept applied at present has been developed through a study of different construction types of quay walls. In the comparison of these construction types factors such as reliability, multi-functionality, corrosion, maintenance, erosion, installation, construction time and costs were considered.

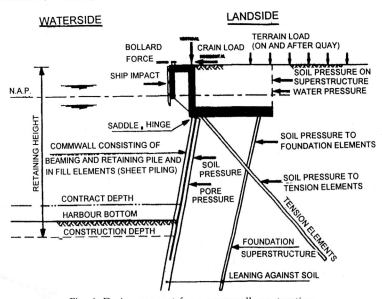

Fig. 6. Design concept for a quay-wall construction.

A quay wall, according to the present design concept, consists of a concrete superstructure supported by a sheet pile wall (e.g. combi-wall). The concrete superstructure, which also serves as a relieving floor, reduces the earth pressure on the sheet pile wall. The sheet pile wall has a supporting and retaining function. Concrete piles and steel tension piles with grout-injection (M.V.-pile; M.V. stands for Muller Verfahen) under an angle of 45° carry the main part of the horizontal load. The anchor force is considerably reduced by placing both the sheet pile wall and the concrete piles under an angle of inclination in such a fashion that the quay wall is "leaning" against the ground. The angle of inclination of the sheet pile wall also has the advantage that the soil pressures on the wall are substantially reduced. Thus with a relatively deep relieving floor and an inclined sheet pile wall a considerable reduction of the costs for the sheet pile wall and the total construction can be achieved. Furthermore the connection between the sheet pile wall and the superstructure consists of an eccentric hinge with a cast iron saddle so as to reduce the bending moments in the wall. Another advantage of an inclined sheet pile wall is that together with the M.V.-pile a kind of A-frame is formed that is very beneficial in respect to horizontal loading and creates the necessary space to place concrete piles with a large angle of inclination. The location of the M.V.-piles on the waterside of the superstructures is also beneficial because placement in the rear would require a denser field of concrete piles. The axial bearing capacity of a combi-wall is generally not critical. Other aspects to be considered are:

- the steel parts of the quay wall are always beneath the level of low tide, this in order to reduce the corrosion process.
- drainage behind the quay wall is be provided to equalise the water pressures over the quay wall as much as possible. If necessary vertical sand drains are installed behind the quay wall.
- the above presented construction is a static determined system.
- to reduce the risk of damage to the sheet pile wall during mooring of ships and in case of a collision it is necessary to have a robust concrete structure located in front (at the water side) of the sheet pile wall with a wooden guidance wall and fenders for a possible ship impact.
- with a deep lying position of the relieving floor the drivability of the combi-wall increases because the driving length is reduced.

The actual design concept as illustrated in Fig. 6 implies an important difference in the design for sea quay wall for bulk handling before 1982, as can be seen in Fig. 7 with the design of the 1st phase of the "EKOM" quay wall (a bulk terminal especially for coal and iron ores) constructed in 1974. The design concept uptill then shows a relatively high position of the relieving floor with

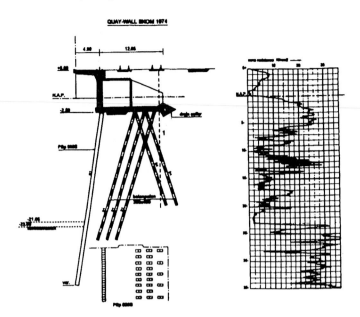

Fig. 7. Design of EKOM quay wall from 1974

a foundation consisting of a sheet pile wall and concrete compression and tension piles placed at a steep angle of inclination of 3 to 1. This type of pile frame bears the high horizontal forces which require a dense field of piles. In Fig. 7 also a top view of the piles is given.

One of the main reasons to study the different design alternatives was the fact that the design concept presented in Fig. 8, based on an up-scaling of earlier constructions to account for the increasing retaining height, showed that high surface loads of coal and iron ore in combination with the soil conditions resulted in considerable continuing displacements of the wall. This made additional anchoring for the quay wall necessary. Analysis of these displacements, much higher then expected, gave two possible explanations:

1. The toe of the concrete tension piles were placed just above the clay layer at RL-21 m, the high surface loads resulted in settlement of this clay layer and thus at the location of the tension piles (Parent, 1983, b). This lead to a situation where the pull of the tension piles on the construction was nearly equal to the force of the total shaft friction of the tension piles. This means a possible overloading of the construction which is equal to the factor of safety of the tension piles, in this case the factor of safety was equal to three. The consequence was that the compression piles were loaded above their ultimate bearing capacity. This resulted in a mainly horizontal and also vertical downward displacement of the rear of the superstructure corresponding to the horizontal and vertical upward displacement of the outer extreme of superstructure due to rotation at, and horizontal displacement of, the top of the combi-wall. This is illustrated in Fig. 8.

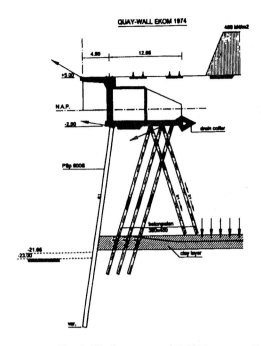

Fig. 8. Displacements of EKOM quay wall

2. The reduction of the horizontal earth pressure on the sheet pile wall, through the relieving floor, can be counteracted by earth pressure on the tension and compression piles, because the dense pile field acts as a fictive wall as illustrated in Fig. 6 (see the dotted line at the rear of the superstructure). This can lead to substantially higher anchor forces than calculated and overloading of primarily the compression piles because they have a lower safety factor than the tension piles (Parent, 1983a).

   Both explanations give a similar displacement behavior of the wall and can occur simultaneously.

   Reviewing the EKOM quay wall design it was concluded that by placing the toe of the concrete tension piles well below the compressible clay layer, e.g., at a depth of R.L.-25 m, the first situation as described above can be avoided. The disadvantage of this solution is that this probably results in higher bending moments in the concrete tension piles which may require steel tension piles to be used. Avoiding the second situation described above is more difficult and requires another design concept. These considerations and the constant increase of the retaining height was the main reason to study several construction types.

5.3 Comparative Study

The comparative study was conducted by the Engineering Division of Rotterdam Public Works (Parent, 1983a). Seven constructions types were designed and compared, as indicated in Table 2. For the comparison of the different construction types the following criteria were considered: besides the costs and the basic requirement that all types had to meet the project requirements, quality aspects played an important role, such as the vulnerability for mooring ships and collisions, but also for overloading of the construction and for the stability of the construction in case of erosion in front of the wall due to scouring through ship propellers. Also construction time, maintenance and durability were considered.

Table 2. Comparison of seven different quay wall designs

Type of quay wall	Method of execution	Comparison cost in %	Remarks on quality
Cellular Cofferdam	dry	120	Large risks in execution and use, low resilience, corrosion sensitive
Diaphragm wall	dry	150	Great resilience
Quay wall with high relief floor	dry	140	Sensitive to overload with vertical loads, sensitive to corrosion
Quay wall with high relief floor	wet	150	Sensitive to overload with vertical loads, sensitive to corrosion
L-wall at R.L.-20 m on a soil improvement	dry		Not possible in connection with raft founded objects in the neighborhood
L wall on piles floor at R.L.-10 m	dry	100	Depth yet to be optimized, great resilience, insensitive to overload with vertical loads not sensitive to corrosion
Jetty	wet	180	Vulnerable low resilience in the event of collision, sensitive to overload with vertical loads

In nearly all situations in the Rotterdam harbor, it is possible to built the superstructure of the quay wall in a dry building pit. The equipment for the installation of the sheet piles, the MV-piles and the concrete piles is working in this building pit.

The importance of the different aspects may be not equal and therefore weighing factors were introduced. The results of the study are presented in Table 2. Remarkable aspects are: the quay-wall with a deep position of the relieving floor (L-wall) founded on piles is the most cost effective; the difference between positing of the relieving floor (L-wall) at R.L.-10 m or at R.L.-6 m is minimal and has to be optimized for each specific situation. Also according to the other above-mentioned criteria a deep lying relieving floor is advantageous. The final choice of the type of construction to be designed is of course selected in consultation with the client and furthermore depends on the project requirements, composition of the subsoil and the specific situation at the building site. Thus for every new project a verification is made to confirm that the chosen design concept is also the most appropriate for that specific situation.

5.4 Safety Concept

Until recently a deterministic design method was used based on the EAU-code, with some adaptations for the situation in the area of Rotterdam and based on evaluation of monitoring data. Overall safety factors were used which take into account the uncertainties in the design method as well as the uncertainties in the loading data.

Since 1992 a semi-probabilistic safety concept has been introduced in most EC countries. This design method starts with a probability of failure. This requires a so-called reliability index β which is usually based on legislation, at least for buildings. In the Netherlands this is also the case for dikes for which the reliability index has been standardised by the authorities. For quay walls which are not a primary water barrier this index has been established in consultation with the Rotterdam Port Authorities. In the semi-probabilistic concept the probability of failure is translated into load factors, material factors, geometrical factors etc., this has been done with probabilistic calculations (Spierenburg et al., 1994). For the design of sheet pile constructions in the Netherlands different safety classes have been defined for the following construction types:
- class 1: simple retaining constructions where failure only causes limited damage, the β-value is equal to 2.5.
- class 2: retaining structures where failure results in considerable damage but a low risk for humans, the β-value is equal to 3.4.
- class 3: retaining structures where failure results in risk for human live and important economical loss, the β-value is equal to 4.2.

For these classes partial factors were calculated and published in the CUR-handbook for Sheet Pile Constructions (CUR, 1993). Nowadays these factors are commonly used in the Netherlands. Because the configuration of the Rotterdam quay walls, especially the bulk and container terminals with very high surface loads and retaining height, were not included in the CUR study an additional analysis was carried out to define partial factors for the design of these quay walls (Huijzer, 1996). The β-value for the quay walls in the Rotterdam area has been determined to be at least 3.6. The main failure made for this analysis has been the failure of the combi-wall and the anchor system. The probability of failure of exceeding the passive earth pressure was in this study much lower because of the fact that the sheet pile wall is considerably longer than the minimum length, due to optimizing the bending moments (Tol, 1995). Tables 3 and 4 give an overview of the results of this study. Table 3 presents the material factors and the factors to be applied on the action effects (bending moment an anchor force) and in Table 4 the geometrical factors are presented. In Table 3 the last column gives the "former approach" (the adapted EAU concept), column 4 the values from the CUR-handbook, column 2 the results of a pure statistical analysis based on β=3.6 and column 3 the factors to be applied for the design of the quay walls from now on. The last mentioned values were chosen after a reconsideration and between the "former approach" and the "pure" results of the analysis. An important item in this study was the requirement of a relatively low probability of failure for the connection between the anchor and the superstructure, this is because of the brittle behavior of such a connection and severe consequences of such a failure.

Table 3. Material and other factors

Parameter	Factors from prob. calculations β-3.60	Definitive factors	Factors (CUR 166, safety class II/III)	EAU factors (with some adaptations)
ϕ'	1.00	1.00	1.18	1.00
c'	1.00	1.00	1.05	1.00
δ	1.00	1.00	1.00	1.00
Section modules	0.97		–	
Yield stress	1.17		–	
Bending moment	1.15	1.30	1.00	1.50
Strength of the anchor: - connection - rot	1.46 –	1.70 1.50	1.10 –	1.8 (2.0/1.1)
Ultimate tensile capacity of the anchor	0.98	1.20	1.25	1.4 (2.0/1.25) × 0.85
$E_{pas:max}/E_{pas:mob}$	1.10	1.30	1.00	1.5

Table 4. Geometrical factors

Parameter	Definitive partial factor	Standard deviation [m]	Δ_{min} [m]	Partial factors, according to CUR 166	Δ_{min} [m]
Bottom level	1.2	0.34	0.4	2.4	0.35
Ground water table	2.0	0.33	0.7	1.2	0.25
Water Level	0.6	0.18	0.1	1.9	0.05

5.5 Computation Methods for Combi-Wall

For the design of a combi-wall the following computation methods are available:
- methods based on the limit earth pressure theory, e.g. Blum method;
- methods based on a Winkler-type of model, also called a sub-grade reaction model, with multi or bilinear springs;
- finite element methods (FEM-models), e.g. PLAXIS.

In general the Blum method is used for a quick scan of the required minimum length of the combi-wall. Next a sub-grade reaction model is used to determine the exact length and the action effects. Subsequently the FEM-model is applied to check all other failure modes.

6. ASPECTS IN QUAY-WALL DESIGN

6.1 General

In this paragraph the different aspects (sheet piles, foundation and concrete piles) which play an important role in the design of a quay wall are described. For each of these aspects examples and values will be given if possible.

6.2. Sheet Pile Wall

Since the first combi-walls were utilised as sheet pile walls for quay walls in the Rotterdam area in the early 50s, the combi-wall has been developed further mainly to improve the driving of course also to resist the retaining height increased.

In Fig. 9 the development of the combi-wall in time is presented (Gijt et al.,

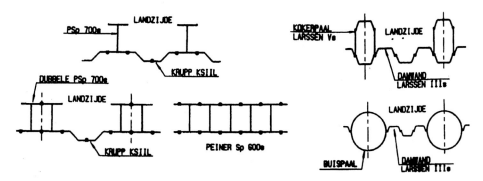

Fig. 9. Development of combi-wall systems

1996). With the extreme length of the sheet pile walls it is of eminent importance, but difficult to achieve, to prevent the occurrence of interlock openings between the sheet piles. Interlock openings have in several occasions resulted in great financial and operational loss. In sandy soil conditions under the influence of tidal waters these interlock openings produce, after a short time or sometimes after years, as liquefied sand flowing through the opening causing extreme settlement at the surface behind the quay wall, see also Fig. 10 (Weijde, 1992).

The experience gained with sheet pile walls has emphasised importance of the type of profile used when it comes to the risk of interlock openings. In Rotterdam the following systems have been applied during the last 40 years (see Fig. 8): H-piles, combi-systems of H-piles and Box-piles with three single sheet pile in between. In the latest designs however a combi-wall of the tubular piles with a triple Larssen sheet pile element in between has been chosen. In the combi-systems the primary elements have a bearing and retaining function and the sheet piles in between, which are considerably shorter, have to retain the resulting water pressure and have to be "soil-tight". To avoid piping these profiles have to penetrate far enough, in practice more then 4 m, under the bottom and take into account erosion as well as future activities.

In most quay walls in the Rotterdam harbor the tubular piles of the combi-walls have a diameter between 1.066 and 1.420 mm and a wall thickness of 16 to 22 mm, the triple piles are Larssen III profiles or equivalent. Also the type of interlock is important, for the combi-wall a Larssen type of interlock is welded onto the tubular pile.

Furthermore in the prevention of interlock openings the installation method is of vital importance. The open ended tubular piles, with a length of 30 to 35 m, are installed by vibration (when very dense layers are encountered the required energy reaches values of 1600 kN) down to the Pleistocene sand layer. The

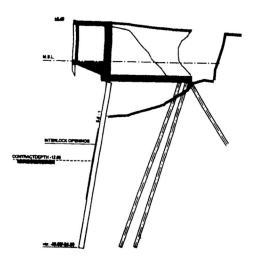

Fig. 10. Damage due to interlock opénings (Brittaniehaven)

tubular piles are then driven by impact into the Pleistocene sand layer to the required depth with at least a Diesel D62 or a S200 hydraulic hammer. This installation procedure is adopted for practical reasons as well as for accuracy, driving time and required effort, and to obtain an adequate bearing capacity. A guiding frame is used during the installation of these piles. After the installation of the tubular piles the triple sheet piles, with a length of about 20 to 25 m, are installed by the following procedure: 1) vibrating, 2) vibrating with jetting, and 3) in exceptional situations driving is permitted.

From the driving and vibration behavior of the combi-wall it is very difficult to obtain data indicating if interlock openings are present. This means that interlock openings can only be traced during dredging operations in the harbor basin in front of the quay wall. Repair of interlock openings is a very costly operation and should therefore be minimised. In order to achieve this detectors in the interlock have been developed and tested.

In practice this procedure gives satisfactory results, interlock openings are now nearly completely avoided.

6.3 Foundation

6.3.1 General

The superstructure is supported by the tubular piles of the combi-wall and by prefabricated and concrete piles and anchor piles. For anchor piles with an angle of inclination of 45° a modified M.V.-pile is used. Table 5 gives an overview of the loads and dimensions of these piles in a situation for a sea quay wall with a retaining height of 25 to 30 m and for a wall with a retaining height of 15 to 20 m.

Table 5. Overview of characteristics of the foundation

	Retaining height			
Type of pile	25-30 m		15-20 m	
Concrete pile 450 * 450 or 500 * 500 mm	$F_{s;d}$ [kN] 1800	Length [m] 20-26	$F_{s;d}$ [kN] 1800	Length [m] 20-25
Tubular pile diam. 1066-1420 mm	7000	30-35	4000	25-30
M.V.-pile	3000	35-45	2000	30-36

6.3.2 Anchor piles

The horizontal forces which act on the structure must be transferred to the subsoil. This can be realized by several types of construction elements. In the period in which the study was conducted to improve the design of large quay walls also the type of anchoring was considered. In Table 6, a summary of the

Table 6. Summary of anchoring systems

Types of anchor	Design tensile capacity ($F_{r;hor;d}$) in kN	
A frame construction (concrete)	400 to 600	per pile frame
Anchor wall	500 to 1,000	per m^1
M.V.-piles (modified)	1,500 to 2,500	per pile
Injection anchors	4,00 to 1,500	per anchor

different types of tension anchors are presented, indicating also the range in ultimate horizontal tension load capacity.

Because of the high tensile load capacity, the rigid behavior and the relative insensibility for corrosion the M.V.-pile was chosen and has since then been used for the construction of quay walls in Rotterdam. The M.V.-pile consists of a steel H-profile pile with grout injection behind an enlarged tip (Gijt et al., 1991).

The original M.V.-pile had a full tip and was difficult to drive to depth. The modified M.V.-pile comprises a reduction of the tip area, facilitating pile-driving and reducing grout consumption. The pile-drivability was also improved by selecting a larger steel cross-section than necessary for absorption of the tensile forces. The pile is equipped with two grout pipes behind the enlarged tip (see Fig. 11), after injection the cross-section of the pile is about 405 * 525 mm. In all the projects in which M.V.-piles have been used, load tests have been conducted on at least 1% of the installed piles. The ultimate shear strength derived from the load tests is equal to 1.4% of the cone resistance q_c of the CPT with a maximum of 250 kN/m^2 The M.V.-pile has a relatively high working load. This force is transferred to the concrete construction by means of numerous (up to 80) dowels with a diameter of approx. 16 mm.

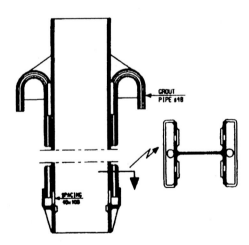

Fig. 11. Construction of the tip of the MV-pile

The M.V.-piles are driven with an S70 or S90 hydraulic hammer, because with an angle of inclination of 45°, this is more effective than using Diesel hammers.

6.3.3 Concrete piles

The prefabricated pre-stressed concrete piles have diameters of 450 and 500 mm and vary in length from 22 m to 30 m. The ultimate bearing capacity is determined according to the Dutch Code NEN 6743 for compression piles. The calculation is based on empirical relationship with the cone resistance measured with CPTs. In the Dutch Code limit values for base resistance and shaft resistance are respectively 15 MPa and 150 kPa. For the very dense Pleistocene sand at Maasvlakte it has been proved by an extensive test program that these limits are too conservative (Opstal, 1996). In the test program new limit values, as presented in Table 7, were established. It was agreed with the local authorities that these values could be used for determining the bearing capacity of prefabricated concrete pile for quay walls at the Maasvlakte location.

Table 7. Limits for shaft resistance and point resistance

Shaft resistance		Point resistance	
Dutch code	Opstal et al. (1996)	Dutch code	Opstal et al. (1996)
α_s [-] limit kPa 0.01 150	α_s [-] limit kPa 0.01 230	α_p [-] limit [MPa] 1.0 15	α_p [-] limit [MPa] 1.0 15 or[1] 0.8 20

[1] it is allowed to use the highest value of either value.

The factors in the table have the following meaning:
  α_s pile class factor for shaft resistance according to NEN 6743
  α_p pile class factor for point resistance according to NEN 6743
These factors give an empirical relationship between the ultimate resistance and the cone resistance of a CPT. The point resistance is determined by applying a factor α_p to a weighed average of the CPT cone resistance over a trajectory under and above the pile point level is used. For the shaft resistance the following relationship is used (Everts, 1997):

$$P_{r;shaft;max;z} = \alpha_s \cdot q_{cz}$$
with:

$P_{r;shaft;max;z}$ the maximum shaft resistance at depth z:
q_{cz} the cone resistance at depth z.

Experience with several projects in the Maasvlakte area has shown that utilising water jetting to install the piles down to the top of the Pleistocene sand layer is necessary. With this installation method adhesion is substantially reduced so that pile driving in the Pleistocene sand layer can be prolonged without damaging the pile head. Consequently, it is then possible to achieve the desired penetration in the Pleistocene sand layer, approximately 4 to 7 m by driving with a D55 or D62 diesel hammer. An additional advantage of this installation method is that the negative skin-friction is reduced due to the water jetting, which loosens the soil around the pile in this area, and thus improves the bearing capacity.

7. COSTS

The construction costs for a sea quay wall with a retaining height of 30 m are at present approx. 60.000 fl per m. The cost breakdown is as follows (Parent, 1990a).

For the Euroterminal, a sea quay wall of 800 m, constructed in 1989, the total construction costs were 42.4 Mfl (million dutch guilders); the steel and pile deliverance amounted to a total of 12.6 Mfl and the construction, including the driving, to 29.8 Mfl. These figures do not include dredging activities.

An interesting aspect of the improvements and optimisation of quay wall design is the development of construction costs. The costs for construction can be divided into three important parts: concrete works, steel and pile deliverance for the combi-wall and piles and all the other costs such as for pile driving, the building pit and so on. The costs for the concrete works amount to about 40% of the total costs. The important cost factor in concrete work is related to the following figures: m^3 of concrete over m^2 of formwork; the man-hours per m^2 of formwork the man hours per m^3 of formwork and the man-hours per m^3 concrete. In Fig. 12 these figures are presented for four type of quay walls. Regarding the costs of the concrete works it appears that the former EMO-wall is too complicated and that the ECT-wall has several inclined parts and therefore the superstructures of these walls are costly. It is apparent that a straight forward design, like the Euroterminal-wall and ARCO, results in is a considerable saving of labor costs (Horst et al., 1992).

Material use		Formwork/ concrete m-h/m2	Man-hour/ formwork m-h/m2	Man-hour/ formwork m-h/m3	Man-hour concrete m-h/m3
	EMO 1974	100	100	100	100
	ECT 1981	75	65	50	45
	EUROTERMINAL 1989	35	50	20	30
	ARCO 1991	25	40	10	25

Fig. 12. Development of cost of the concrete works

The costs of the combi-wall, composed of tubular piles and sheet pile elements are, including the installation, about 30-40% of the total quay wall construction.

The development of the combi-wall resulted in only a limited reduction of costs due to weight reduction.

8. CONCLUSIONS

The design of quay walls in the harbor of Rotterdam has been developed and optimized during the last 25 to 30 years. This is partly because of changing requirements, especially the increasing depth of the waterways and high surface loads, and partly by improving the design process based on the experience obtained through monitoring and evaluation. This was made possible because of the enormous growth of the harbor of Rotterdam during this period. The quay walls that were constructed since the 80s are among some of the biggest in the world.

The constructed quay walls meet the requirements and, due to a optimisation process of the constructions, are very cost effective.

REFERENCES

Everts H.J., *"Dutch national code for pile design"*, (1997), Design of axially loaded piles, European practice, edited by F. de Cock and C. Legrand, Brussels, Belgium, 1997.

Gijt, J.G. de, (1997). *"Lecture notes Port Construction Engineering"*, International Maritime Transport Academy, Rotterdam.

Gijt, J.G. de, Brassinga, H.E. (1989). *"Prediction methods for cone resistance after excavation"* Marina 1989, Southampton, pp. 309-321.

Gijt, J.G. de, Schaik, C.N. van, Roelfsema, R.E. (1991) *"Construction of a deepwater bulk terminal in the Port of Rotterdam"*. 4th International DFI Conference, Italy, Strera, 7-12 April 1991.

Gijt, J.G. de, Horst, H. van der, Heijndijk, P.J.M., Schaik, C.N. van (1993) *"Quay wall design and construction in the Port of Rotterdam"*. PIANC Bulletin. no. 80, 1993.

Horst, H. van der, Gijt, J.G. de, Heijndijk, P.J.M. (1992) *"The development of quay wall in the Port of Rotterdam"*, Ports 2000 Conference, Hong Kong.

Horst, H. van der, Gijt, J. G. de, Schaik, C.N. van (1992), *"Development in design and construction of quay walls in Rotterdam"*. Symposium Rotterdam 2010 Harbóur Expansions and Container Transport; Waterbouwdispuut; Technical University, Delft. (in Dutch)

Huijzer, G.P., *"Final report of probabilistic analyses of sheet pile constructions"* (1996), Rotterdam Public Works, Section Geotechnics, Rotterdam.

Koeman, J.W., Tuijtel, T., Gijt, J.G. de (1992) *"Planning and construction of harbour facilities in the Port of Rotterdam"*. Port Plan 2010, Presentation EAU meeting, Rostock.

Opstal, A.Th.P.J. and J. van Dalen, *"Load tests on Maasvlakte"* (1996), Cement, nr 2, feb., 1996 (in Dutch)

Parent, M.G. (1983a), *"Quay wall for the handling of ores in Europoort 1 Design"*, PT/Civiele Techniek 38, nr 2, 1983, (in Dutch)

Parent, M.G. (1983b), *"Design of Quay wall EECV"*, Otar, Technisch tijdschrift van de waterstaat kundige ambtenaren van de Rijkswaterstaat, nr 2, 1983. (in Dutch)

Parent, M.G., (1990a). *"Deep water terminal on Maasvlakte te Rotterdam 1"*, PT, Civiele Techniek, nr. 2 1990, pp. 3-6. (in Dutch)

Parent, M.G. (1990b), *"Modern Quay walls in Rotterdam"*, Cement; nr. 4, April 1990. (in Dutch)

Sijberden, H.L. (1992), *"Fendering"*. Land en Water; oktober 1992.

Spierenburg, S.E.J., L. de Quellery, E.O.F. Calle, M.Th.J.H. Smits and J.T. de Vries (1994), *"A semi-probabilistic design procedure for sheet piles"* Proc. of the XIII Int. Conf. on Soil Mech. and Found. Eng., pp 827 - 830, 5-10 Jan. 1994, New Delhi.

Tol, A.F. van, (1995) *"Design of Sheet Pile Constructions according to CUR-handbook"*, Cement; nr. 5, pp 22-27. (in Dutch)

Weijde, R.W. van der, Heijndijk, P.J.M., Noordijk, A.C., Kleijheeg, J.T. (1992) *"Scouring by ships propellers, tracing, repair, prevention"*. 10th International Harbour Congress; Antwerp.

CUR, (1993) *"Sheetpile constructions"*, Publication 166, CUR, Gouda.

E.A.U. (1996). *"Empfehlungen des Arbeidsausschusses Ufereinfassungen"*. Ernst & Sohn, Berlin,

NEN 6743 (1991). *"Geotechnics, calculation method for bearing capacity of pile foundation compression piles"*. Dutch Institute for Normalisation NNI, Delft, Netherlands.

10

Vibration Screening with Wave Barriers

Richard D. Woods

University of Michigan, Ann Arbor, MI 48109, USA

ABSTRACT

Various forms of wave barriers have been used for more than 50 years to reduce the amplitude of vibrations entering a region needing protection from excessive vibrations. The design of these barriers was at first based only on trial and error, but with a better understanding of wave propagation, design of these barriers was reduced mainly to a matter of determining the economic feasibility of the optimum barrier. Optimum barriers were shown to be void regions on the order of one Rayleigh wavelength deep. Open or void trench barriers are desirable but almost unachievable. Now, as we enter the 21st century, other choices are available from air-filled pillows to solid barriers that provide sufficient impedance mismatch to be effective barriers. Basic wave barrier principles including selection of dimensions and recent developments describing alternative barriers are presented.

Key Words: Isolation, vibrations, trenches, barriers, air bags, PC piles.

1. INTRODUCTION

There are many situations in industry, in research laboratories, and in habitats

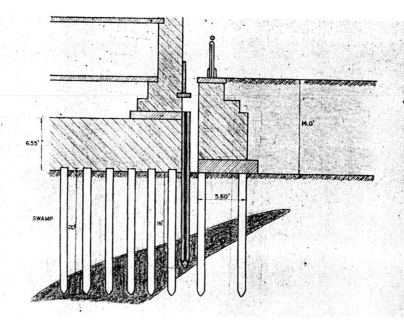

Fig. 1. Isolation by sheet-pile wall (after Barkan, 1962)

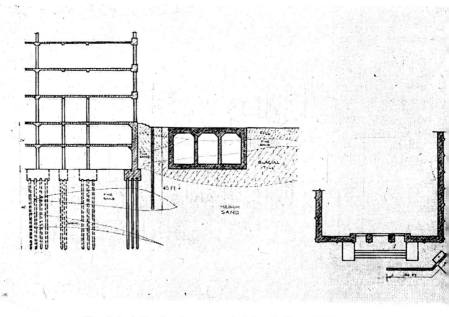

Fig. 2. Isolation by slurry trench (after Dolling, 1965)

of people where excessive vibrations are undesirable. In an early attempt by
Barkan (1962) to reduce the vibrations transmitted to an office building from a

high traffic street, he installed a wooden sheet-wall barrier, Fig. 1. Dolling (1965), Fig. 2 described another application of a wave barrier, this time a slurry-filled trench, to minimize vibrations reaching a precision printing plant from a subway. In the first case, the sheet-pile wall made some difference but the amplitude reduction was not sufficient to justify the installation. In the second case, the slurry-filled trench worked well while the slurry remained fluid, but it failed to be effective when the slurry "gelled". It was impractical to maintain the slurry, so this attempt ended in failure.

The magnitude or amplitude of "undesirable" vibrations is a topic unto itself and will not be dealt with in this work, but means of reducing vibrations with wave barriers for all of the above-cited situations are the topic of this paper. Both man-made and natural vibrations travel through the earth as elastic waves that behave in a well-known manner based on known physical principles. With an appropriate understanding of wave propagation, the selection of barriers to reduce vibrations becomes a well-defined design exercise.

1.1 Principal Methods of Vibration Isolation

Isolation from vibrations can be achieved in several ways; some of which require only siting and others require designed and constructed elements.

1.1.1 Isolation by distance

Siting a plant, laboratory, or residence far enough from a source of vibrations is a very effective means of reducing vibrations but is often not a viable choice.

1.1.2 Isolation by topography

The opportunity to isolate a sensitive operation from a source of vibration by topography is sometimes more feasible than distance. A steep canyon, Fig. 3, could easily serve as a barrier for vibrations traveling either direction simply by siting the critical facility on the canyon side opposite from the vibrations source.

1.1.3 Isolation by mechanical systems

Systems of springs and shock absorbers are often used to isolate troublesome, vibrating machines or equipment from their foundations, thereby reducing the amplitude of vibration reaching other locations by wave propagation. An isolation mass, supported on springs, can be used to isolate a vibrating machine from the floor of a sensitive manufacturing facility.

1.1.4 Isolation by design

By this, I mean that the supporting system, foundation, for a vibrating machine is designed by principles provided by knowledge of Soil Dynamics. Often the best solution is to minimize vibration at the source and selection of an appropriate foundation is part of this process.

Figure 3. Sketch of Glenwood Canyon, Colorado

1.1.5 Isolation by wave barriers

Interruptions in the wave path by barriers are also possible means of reducing vibrations traveling from one place to another through the ground. This is the only isolation method that will be covered in detail in this paper.

1.2 Wave Propagation

Elastic waves travel through the earth in basically two forms, body waves and surface waves. The schematics in Fig. 4 from Bolt (1976) show two body waves and two surface waves. In the region close to the surface of the earth where most human habitation and industrial operations occur, the surface waves are dominant. Furthermore, the geology of the earth is such that the conditions for generation of a Love Wave do not occur everywhere while those for generation of a Rayleigh Wave exist everywhere. It is the Rayleigh Wave, therefore, that holds the key to successful reduction of vibrations by wave barriers. This wave, which transmits about 67% of the total energy away from any source at the surface of the earth (Miller and Pursey, 1955), must be well understood to successfully screen surface waves.

The critical characteristic of the Rayleigh Wave, for purposes of screening of vibrations, is the distribution of the energy carried by this wave as a function of depth. Figure 5 shows the distribution of the Rayleigh Wave energy, in terms of the amplitudes of both vertical and horizontal components of motion, as a function of depth into the ground. The amplitudes have been normalized with respect to the amplitude at ground surface and normalized with respect to depth by the wavelength of the Rayleigh Wave.

This figure shows that the major portion of the energy (highest amplitudes) propagating as a Rayleigh Wave travels within a zone of about one wavelength

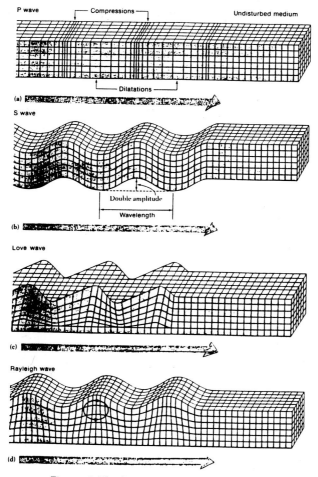

P wave

Compressions

Undisturbed medium

Dilatations

(a)

S wave

Double amplitude

Wavelength

(b)

Love wave

(c)

Rayleigh wave

(d)

Figure. 4. Elastic waves in Earth (from Bolt, 1976)

deep from ground surface. If this Rayleigh Wave, traveling near the ground surface and transmitting more than two-thirds of any energy introduced at the ground surface, is properly interrupted, vibration amplitude can be greatly reduced. It is the goal of this paper to show how the Rayleigh Wave must be screened.

2. METHODS OF EVALUATING THE SCREENING EFFECTS OF THE RAYLEIGH WAVE

There are several methods available to the geotechnical engineer to study the screening effects of wave barriers. These range from closed-form mathematical formulations to physical model studies.

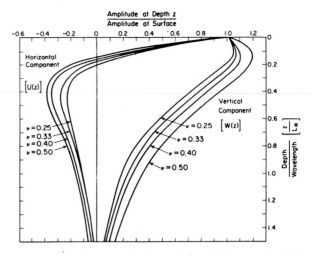

Fig. 5. Rayleigh Wave Amplitude vs. Depth (Richart et al., 1970)

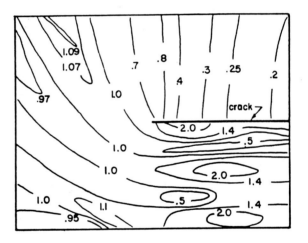

Fig. 6. Shear wave at semi-infinite crack in elastic full space
(after Thau and Pao, 1966)

2.1 Closed-form Solutions

Some problems in mechanics, including wave propagation, can be solved in closed-form while most problems need to be attacked by numerical methods. The problem of a shear wave encountering a crack in an elastic space has been solved by Thau and Pao (1966) and the resulting diagram showing dispersion of the traveling wave is shown in Fig. 6. In this figure, a plane wave with particle motion perpendicular to the plane of the page is traveling from bottom toward the top of the figure.

The plane wave encounters a "void barrier" in the form of a crack. Vertical motion contours on this figure show that some energy is reflected back from the crack and interferes constructively with the incoming waves causing higher amplitudes (contours with numbers greater than 1 in the figure). The initially plane waves bend around the end of the crack and end up traveling perpendicular to the initial wave propagation direction and diminish in amplitude (contours less than 1) in the field behind the crack (wave screening barrier).

2.2 Numerical Methods

The Finite Element Method (FEM) has been applied to this problem with success as demonstrated by the work of Haupt (1978). Haupt has shown that the screening effects of wave barriers can be effectively modeled by FEM, for example, as shown in Fig. 7 where a Rayleigh Wave is shown propagating to the right past a barrier in an elastic half-space. Haupt (1995) also showed the value of solid barriers for wave screening as demonstrated in Figs. 8 and 9.

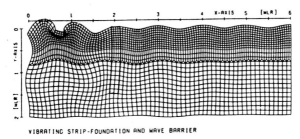

VIBRATING STRIP-FOUNDATION AND WAVE BARRIER

Fig. 7. FEM model of Rayleigh Wave propagating by trench barrier (from Haupt, 1978)

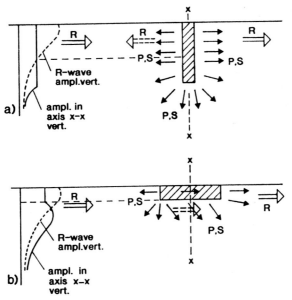

Fig. 8. Wave propagation process at solid barriers (from Haupt, 1978)

The FEM technique requires the modeling of the entire system, either as a homogeneous half-space or layered space, while the Boundary Integral Element Method (BIEM) can be used to reduce computation time by appropriate modeling of each layer with a single element, Fig. 10 (Beskos et al., 1986).

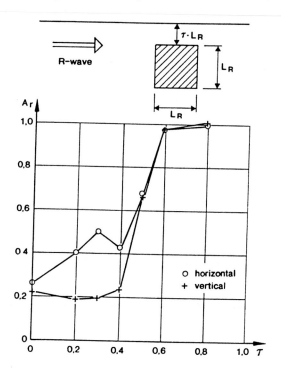

Fig. 9. Isolation effect of barrier below the surface, plane FE calculation (from Haupt, 1978)

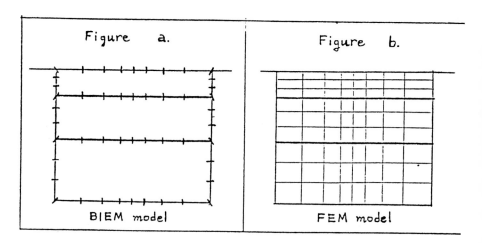

Fig. 10. FEM and BIEM (from Beskos, 1986)

2.3   Analogs

At least two simple analogs have been used to model screening of Rayleigh waves: gravity waves and acoustic waves in water. Figures 11 and 12 show gravity water wave analogs for a single row and double rows of solid barriers respectively. Plane waves coming from the top of the figure are interrupted by cylindrical barriers composed of sand-filled, 10 oz beer cans. This type of analog is not exact because water cannot sustain shear waves while the Rayleigh Wave in the ground has both shear and compression energy, but the analogous behavior of the barrier is evident.

Fig. 11. Gravity water wave analogy

Figure 13 shows an apparatus for acoustic waves in water as an analog to Rayleigh Wave barriers.

These analogs are helpful in seeing some of the basic elements of the phenomena, but models that accommodate all characteristics of the Rayleigh Wave are necessary.

2.4   Physical Models

Models of the screening phenomenon in the field and the laboratory are another very productive approach to studying the screening effectiveness of wave barriers. The remainder of this paper presents the results of several physical model studies of screening of surface waves.

Fig. 12. Gravity water wave analogy

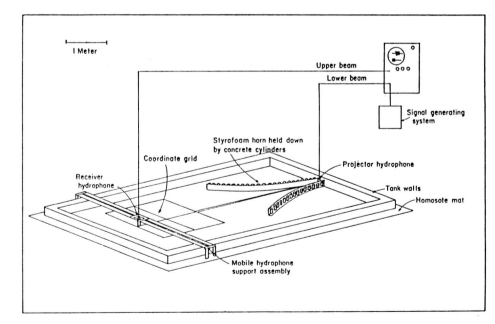

Fig. 13. Acoustic water wave analogy

3. MODEL STUDIES

Physical model studies have revealed the critical parameters that affect screening of surface waves by barriers. These studies started with the assumption that totally void barriers in the form of open trenches were the optimum barriers, then examined the physical size and shape of barriers for screening effectiveness (Woods, 1968). The conclusions formed in the research by Woods on open trenches have been confirmed by numerical modeling and by additional physical models, some full scale, but have not been contradicted. Some of the more recent research has shown, however, that solid barriers may provide some benefit in terms of screening but are never as effective as open trenches or open holes.

3.1 Field Studies of Open Trenches

A simple representation of an isolation problem is shown in Fig. 14 where a vibration source is in close vicinity to a building that is sensitive to vibration. Two approaches might be used in an attempt to reduce the vibrations reaching the building. In one, the wave barrier is placed close to the source, Active Isolation (Fig. 15a), and in the other, the barrier is placed close to the receiver building, Passive Isolation (Fig. 15b).

3.1.1 Active isolation

In the field model studies by Woods (1968) the experiments were performed at a site where the soil profile was as shown in Fig. 16. The experimental setup is shown in Fig. 17 for active isolation. With a vertically oriented vibrator positioned at the center of the experimental array, vertical vibration measurements were made at each of the locations identified with small squares, first with no barrier present and then repeated with barriers of various sizes and shapes excavated into the ground. Measurements were made at several frequencies (wavelengths) so that wavelength scaling could be applied. With data from before barrier and after barrier, plan view diagrams of Amplitude Reduction Factor (ARF) contours were

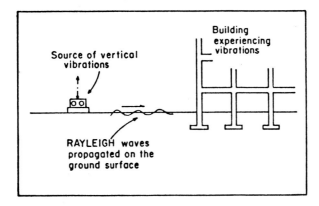

Fig. 14. Schematic representation of vibration source near building (from Liao, 1976)

developed. ARFs were calculated simply taking the ratio of after amplitude to before amplitude. Figure 18 shows an ARF contour diagram for the test setup in Fig. 17.

Note that there are some areas with amplitudes higher after the barrier than before the barrier was installed. This represents constructive interference between incoming and reflected waves. Figure 18 shows the regions near the semi-circular trench that were effectively screened and some regions where amplitudes were increased. Figure 19 shows a similar plot for a trench with an arc of 270 degrees.

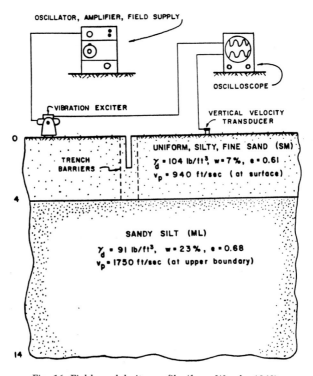

(a) Active Isolation

(b) Passive Isolation

Fig. 15. a) Active isolation barrier, b) passive isolation barrier (from Liao, 1976)

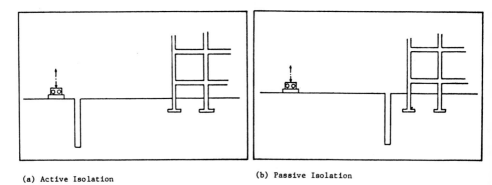

Fig. 16. Field model site profile (from Woods, 1968)

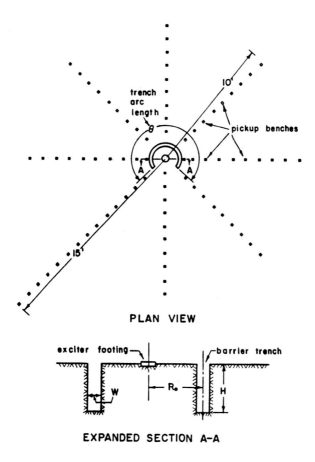

PLAN VIEW

EXPANDED SECTION A-A

Fig. 17. Experimental setup for active isolation (from Woods, 1968)

3.1.2 Passive Isolation

At that same site, passive isolation experiments were performed in which the barrier was close to the area to be screened instead of near the source. A sensitive instrument or machine that needs to be protected from incoming vibrations is shown schematically in Fig. 20.

Based partially on the active screening tests and partially on logical perception of the screening process, five basic parameters were identified as having an effect on the screening value of a barrier: depth, width, length, shape, and filling. All of the initial tests described here were on trenches with no filling and which were straight. The other three variables are shown in the perspective drawing in Fig. 21. Again the field site was laid out as in Fig. 22 and vibration measurements were made at all the locations identified with small squares both before and after excavation of a trench.

The amplitude of vibrations along a line from the source through the center

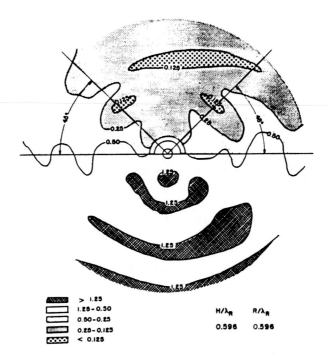

Fig. 18. ARF contour diagram for active isolation, 180° barrier (from Woods, 1968)

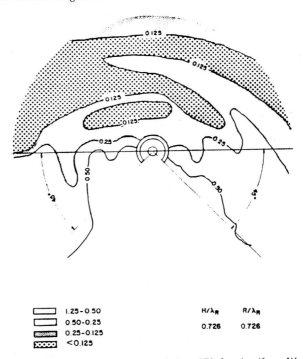

Fig. 19. ARF contour diagram for active isolation, 270° barrier (from Woods, 1968)

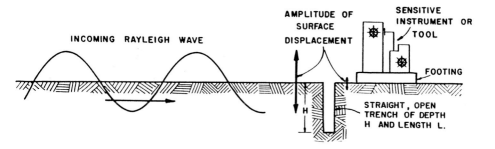

Fig. 20. Schematic representation of sensitive instrument for passive isolation (from Woods, 1968)

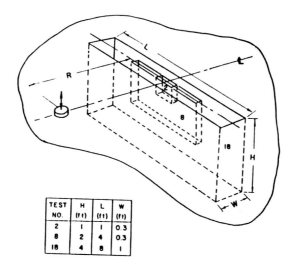

TEST NO.	H (ft)	L (ft)	W (ft)
2	1	1	0.3
8	2	4	0.3
18	4	8	1

Fig. 21. Perspective of trench with dimensions identified (from Woods, 1968)

of the barrier is one typical way of showing the effectiveness of the barrier. This is shown in Fig. 23 for three barrier sizes as well as the no-barrier situation. ARF contour diagrams for trenches of three lengths are shown in Fig. 24.

3.1.3 Preliminary conclusions

Based on the field model studies, important and basic conclusions were drawn. For *active isolation* with trenches fully surrounding the source of vibrations, a trench is effective if the amplitude of motion is reduced to 25% of the no-barrier amplitude within an annular zone extending from the outer edge of the trench for a distance of at least 10 wavelengths at the frequency to be screened. To accomplish this degree of isolation, the trench needs to be at least 2/3 of a wavelength deep. For trenches that do not fully surround the source, the effective zone is defined by removing a 45-degree slice in plan view from both ends of the trench as demonstrated in

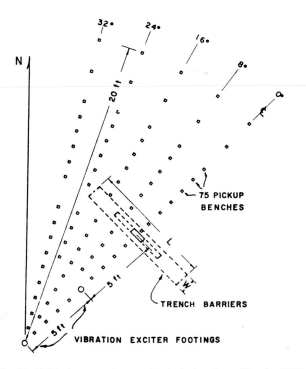

Fig. 22. Field site setup for passive isolation (from Woods, 1968)

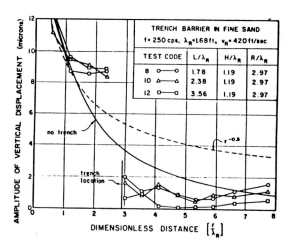

Fig.23. Centerline amplitude decay for three barriers (from Woods, 1968)

Figs. 18 and 19. For *passive isolation* a trench is considered effective if the amplitude of vertical vibration is reduced to 25% of the no-trench condition within a semicircular area with radius of one-half the trench length ($L/2$) and

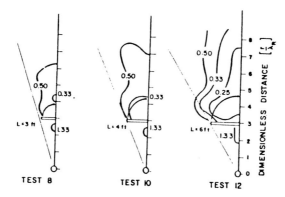

Fig. 24. ARF contour diagrams for passive isolation, (from Woods, 1968)

with center at the center of the trench. Figure 24 shows these semi-circular areas. To achieve this degree of isolation, the trench must be at least 1 and 1/3 wavelengths deep.

3.2   Laboratory Study of Alternate Barriers

The research on effectiveness of open trench barriers showed that the trenches must be relatively deep, i.e. 2/3 to 1.33 wavelengths deep. A typical wavelength to be screened is on the order of 40 to 100 feet. This means that the trenches must be quite deep. Often this depth is sufficient to encounter ground water and leads to an impossible situation with regard to maintaining an open trench. In any case, even without water, permanently keeping a trench open poses a difficult and expensive problem.

Based on the analogous screening behavior demonstrated by the cylindrical barriers in Figs. 11 and 12, it was postulated that a row of thin-wall-lined, open cylindrical holes would be an effective barrier, Fig. 25. To reduce the enormous amount of data collection required in the field model studies, a new laboratory modeling approach was employed, holographic interferometry (Woods et al., 1974), see Fig. 26 for a schematic representation of this method. Using this technique, rows of cylindrical holes and cement mortar-filled trenches were evaluated.

3.2.1 Cylindrical holes

Right circular cylinders were excavated in a moist, fine sand model in the holography laboratory to study the effectiveness of these barriers. From the holographic interferometry approach, the entire traveling wave field could be captured in a stopped-motion hologram in which each interference fringe represents a vertical amplitude difference of about one-half the wavelength of the laser light required to produce the hologram, in this case a wavelength for green light of about 5000 Å.

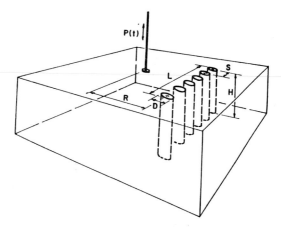

Fig. 25. Schematic redundant of rows of holes isolation (from Woods, et al., 1974)

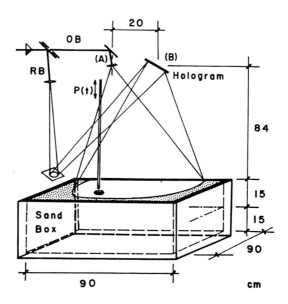

Fig. 26. Schematic representation of holographic interferometry (from Woods et al., 1974)

Therefore, the interferograms shown in Fig. 27 represent crests and troughs of traveling waves from which exact amplitudes can be determined simply by counting fringes. The difference in wave field behind the rows of cylindrical holes in Fig. 27 is obvious. The effectiveness of these barriers is dependent on the diameter of the holes (D), the depth of the holes and the spacing between the holes (S-D); all normalized by wavelength of the Rayleigh Wave of the wavelength of the propagating wave. Figure 28 is a summary of the effectiveness of the cylindrical holes, taking a void trench as a standard equal to "one" (on vertical axis) and comparing rows of cylindrical holes. Therefore, there is a family of lines in Fig. 28 showing effectiveness relative to a void trench based on cylinder diameter (D) and hole spacing (S-D), both normalized on wavelength.

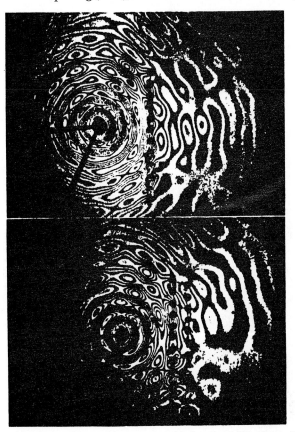

Fig. 27. Interferograms of one and two rows of holes (from Woods et al., 1974)

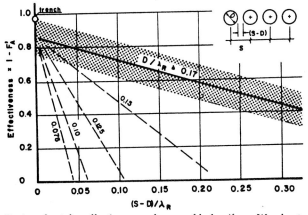

Fig. 28. Design chart for effectiveness of rows of holes (from Woods et al., 1974)

3.2.2 *Solid barriers*

Preliminary field model tests with sheet-pile type walls showed that flexible solid filled barriers were not effective, but more rigid barriers would likely be effective. The holographic interferometry approach was used again to study this possibility. An interferogram showing the wave field distortion provided by a solid barrier is presented in Fig. 29a and 29b where the actual cement/sand mortar barrier dug out of the model is shown in Fig. 29b. These tests showed that solid barriers could be effective if properly dimensioned.

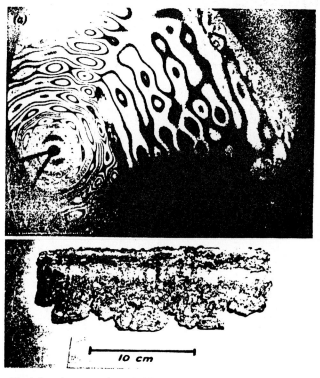

Fig. 29. a) Interferogram for solid barrier, b) sand/cement mortar barrier
(from Woods et al., 1974)

Haupt (1995) has shown in more detail the effectiveness of solid barriers which have an adequate size and impedance mismatch with the surrounding ground, see Figs. 8 and 9.

4. RECENT DEVELOPMENTS

There have been developments over the past decade that may bring the isolation-by-barrier approach into a more plausible position economically. Those developments are the air-cushion or air-bag system and a hollow pile system.

4.1 Air Cushion

Massarsch (1991) has described the development and application of an air-cushion system to, essentially, hold a trench open over an extended period of time. The air-cushion system is composed of rows of air bags mounted in a frame constructed of reinforcing bars, shown schematically in Fig. 30. The system is constructed so it may be lowered into a slurry-filled trench and each row (or layer) of air bags pressurized independently to appropriate pressure depending on active earth pressure at depth.

Figure 31 shows a small section of the overall system, and Fig. 32 shows a frame ready to be lowered into a trench. While the cost of these bags is now quite

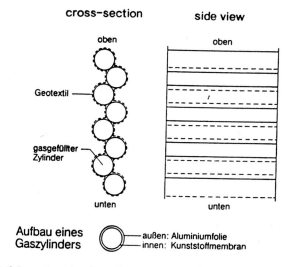

Fig. 30. Schematic redundant of action of air-cushion barrier (Schiffer, 1992)

Fig. 31. Small section of air-cushion wave barrier (from Massarsch, 1991)

expensive, the cost could be significantly reduced with a larger demand for application. An estimate of this type of installation in Phoenix, Arizona, USA in 1992 was estimated to be about US $40 per square foot of barrier wall.

4.2 PC Pile

Hayakawa et al. (1997) describe the application of hollow concrete piles as a screening barrier for application between high speed railroad lines or major highways and apartment buildings in the Osaka region of Japan. A cross section of the hollow pile is shown in Fig. 33. A pile-supported expressway viaduct in Kobe, Japan, as rebuilt after the 1995 earthquake, was screened with PC-wall piles shown during installation in Fig. 34. The effectiveness of PC-wall barriers was found to be adequate.

Fig. 32. Frame with air cushion going into slurry-filled trench to form barrier
(from Massarsch, 1991)

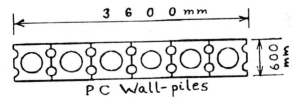

Fig. 33. Cross-section of PC wall-pile barrier (Hayakawa et al., 1997)

Fig. 34. Installation of PC wall barrier, Kobe (Hayakawa et al., 1997)

5. CONCLUSIONS

The general conclusions about barrier size and shape developed by Woods (1968) from field and laboratory physical models (Sec. 3.1.3) have been confirmed in all subsequent studies of screening of vibrations by wave barriers. Those conclusions clearly showed that effective wave barriers are difficult to achieve economically. Simple empirical equations to estimate the passive screening effectiveness of barriers based on model studies were published by Al Hussaini and Ahmad (1991). These equations, along with the general recommendations by Woods, can be used to select effective barriers if cost is not an obstacle. Recent developments, however, indicated that new methods/systems might make wave screening with barriers economical in the near future. As we enter the 21st Century there are reasons to hope that this will come about.

REFERENCES

Al-Hussaini, T.M. and Ahmad, S. (1991) "Simple Design Methods for Vibration Isolation by Wave Barriers," *Proc.: Second Inter. Conf. on Recent Advances in Geotechnical Earthquake Engineering and Soil Dynamics,* March 11-15, St. Louis, Mo, pp. 1493-1499.

Beskos, D.E., Dasgupta, B., Vardoulakis, I.G. (1986) "Vibration Isolation Using Open or Filled Trenches; Part 1: 2-D Homogeneous Soil," *Computational Mechanics,* No. 1 Springer.

Bolt, B.A. (1976) *Nuclear Explosions and Earthquakes: The Parted Veil,* W.H. Freeman and Co., San Francisco.

Dolling, H.J. (1965) "Schwingungsisolierung von Bauwerken durch tiefe auf geeignete Weise staabilisierte Schlitzze:, VDI-Bericht Nr. 88.

Haupt, W.A. (1978) "Surface-waves in Non-homogeneous Half-space," *Proc. Dyn. Meth. In: Soil and Rock Mechanics (DSMR 77),* Karlsruhe, A.A. Balkema, Vol. 1, Rotterdam.

Haupt, W.A. (1995) "Wave Propagation in the Ground and Isolation Measures," *Proc.: Third Inter. Conf. on Recent Advances in Geotechnical Earthquake Engineering and Soil Dynamics,* April 2-7, St. Louis, Mo., Vol. 2, pp. 985-1016.

Hayakawa, K., Kani, Y., Matsubara, N., Matsui, T. and Woods, R. (1997) "Ground Vibrations Isolation by PC Wall-Piles," *Proceedings of the Fourth International Conference on Case Histories in Geotechnical Engineering,* St. Louis, Mo, March.

Liao, S.S.C. (1976) "The Scattering of Elastic Waves with Application to Foundation Isolation," *Geotechnical Engineering Report 76-3,* June, Center for Urban Development Research, Cornell University, Ithaca, New York, 155pp.

Massarsch, K.R. (1991) "Ground Vibration Isolation Using Gas Cushions", *Proc. Second Int. Conf. on Rec. Advances in Geotechnical Earthquake Engineering and Soil Dynamics,* St. Louis, Missouri, March 11-15, pp. 1461-1470.

Miller and Pursey (1955) "On the Partition of Energy between Elastic Waves in a Semi-Infinite Solid," *Proc. of the Royal Society, London,* A, v.233, pp. 55-69.

Richart, F.E., Jr., Hall, J.R., Jr. and Woods, R.D. (1970), *Vibrations of Soils and Foundations,* Prentice-Hall Englewood Cliffs, N.J. 412pp.

Schiffer, W. (1992) "Schlizwandbauweise mit Gasmattenelementen zur Abschirmung von Erschutterungen", Proc. of the STUVA Symposium at Dusseldorf, Germany.

Thau, S.A. and Pao, Y. (1966) "Diffractions of Horizontal Shear Waves by a Parabolic Cylinder and Dynamic Stress Concentrations," *J. Appl. Mech., Trans ASME,* Dec., pp. 785-792.

Woods, R.D. (1968) "Screening of Surface Waves in Soils," *Journal of the Soil Mechanics and Foundations Division, ASCE,* Vol. 94, No. 4, July, pp. 951-979.

Woods, R.D., Barnett, N.E. and Sagesser, R. (1974), "Holography—A New Tool for Soil Dynamics," *Journal of the Geotechnical Engineering Division, ASCE,* Vol. 100, No. GT11, Nov., pp. 1231-1247.

11

Rayleigh Wave Dispersion and Its Applications

Shiming Wu
Tongji University, Shanghai 200092, P. R. China

ABSTRACT

Characteristic equations and their solutions of Rayleigh wave in elastic soils, fluid solid medium and saturated soils are presented. Discussion is stressed on wave propagation and dispersion in layered soils. Some applications of Rayleigh wave dispersion in Civil Engineering are introduced.

1. INTRODUCTION

In 1887, Rayleigh found a kind of wave induced from interference of compression wave and shear wave in an elastic half-space, so-called Rayleigh wave. Its velocity is independent of vibration frequency in a homogeneous and isotropic elastic half-space. The velocity attenuates more quickly in vertical than horizontal direction, that means Rayleigh wave propagates mainly along surface of half-space, that's why Rayleigh wave is called surface wave.

For layered medium as usually encountered, Rayleigh wave velocity varies with its frequency resulted from soil shear modulus varies with depth. Since 50s, many researchers have being worked in this field. Tomoson (1950) and Haskell (1953) established Rayleigh wave characteristic equation by transitive matrix, its results always lose accuracy caused from high frequency effective numeral loss (Knopoff, 1964). Dunkin (1965), Thrower (1965) and Watson (1972) made a substitution of the matrix in characteristic equation by δ matrix, thereby, voided high frequency effective numeral loss for eliminating square terms of exponential function in equation. Chen and Wu (1991) and Abo-zena (1979) presented a new method with high efficiency, as well as eliminating effective numeral loss.

All mentioned above refers to traditional analytics, based on continuity of stress and displacement in soils, most likely for soft-to-stiff type soil (modulus increasing with depth). While for stiff-to-soft type soil and intercalation (including soft in between and stiff in between) soils, there are some difficulties because of complex function. Furthermore, analytic method is usually not for non-homogeneous soils. Vrettos (1990a, 1990b) tried to extend analytic method to soils in which shear modulus increases exponentially with depth.

Finite difference method and finite element method are effective numerical approach to overcome difficulties mentioned above (Lysmer et al., 1972). Chen et al. (1991) and Xia et al. (1991) established Rayleigh wave characteristic equation in soils by means of finite element method, in which there still exists undetermined parameter, δ. Xia et al. (1992, 1993) established characteristic equation combining finite element method with analytic approach, which is quite effective for stiff-to-soft type, intercalation type and fluid-solid media.

Rayleigh wave is induced by interference of compression wave and shear wave, which exist in saturated soils. Rayleigh wave in saturated soils is of dispersion, as well as in saturated half-space (Chen, 1987 and Meng, 1981). Jone (1961) and Chiang et al. (1981) studied Rayleigh wave dispersion in saturated soils, but only one type of compression wave is taken into account, i.e., its potential function is not common solution of the problem. Xia et al. (1998) presented a numerical method for a half-space of saturated soil and studied its nature of dispersion.

Shear wave, V_s, is important both for soil dynamics and static soil mechanics. Shear wave velocity measured in small amplitude reflects shear modulus, G, in static condition relating to shear strength, τ_f (Wang, 1987). Generally, body waves (compression wave and shear wave) are treated as testing signal in field seismic test, while Rayleigh wave as a noise. Miller et al. (1995) and Xia and Wu (1994) revealed that Rayleigh wave is dominative in energy in surface vibration, i.e., the displacement caused by Rayleigh wave is much greater than that by body waves at shallow depth in a medium, resulting in low accuracy at shallow depth of a seismic test with body wave measurement. Because of domination of surface wave at shallow depth in a deposit, its dispersion curve (wave velocity vs. frequency) can be obtained, which relates thickness of soil layer and velocity in each layer. Consequently, shear wave velocity in each layer can be obtained from dispersion curve measured in the field by back analysis.

Stoneley (1924) found a type of wave in two solid half-space, so-called Stoneley wave, and studied its characteristic equation which is similar to Rayleigh wave. Koppe (1948) studied Stoneley wave in two solid media. Stoneley wave may exist in fluid-soil (saturated soil)-rock system, in the bottom of ocean, river or lake. There is a potential application of Stoneley wave in exploration of marine soil and rock, for which further studies are necessary.

2. RAYLEIGH WAVE IN ELASTIC SOIL

2.1 Characteristic Equation of Rayleigh Wave

For a layered system as shown in Fig. 1, the characteristic equation is established by means of combination of finite element method and analytics (Lysmer, 1972 and Xia et al., 1993). The system is discreted into finite layers as shown in Fig. 1(b). Stiffness matrix and mass matrix of an element as shown in Fig. 1 (c) are obtained by concentration (l→0). The stiffness matrix for solid medium.

$$[K]_e = \begin{bmatrix} \frac{2}{3}k^2h^2V_P^2 + 2V_S^2 & (V_P^2 - 3V_S^2)kh & \frac{1}{3}k^2h^2V_P^2 - 2V_S^2 & (V_S^2 - V_P^2)kh \\ & \frac{2}{3}k^2h^2V_S^2 + 2V_P^2 & (V_P^2 - V_S^2)kh & \frac{1}{3}k^2h^2V_S^2 - 2V_P^2 \\ & \text{Symmetry} & \frac{2}{3}k^2h^2V_P^2 + 2V_S^2 & (3V_S^2 - V_P^2)kh \\ & & & \frac{2}{3}k^2h^2V_S^2 + 2V_P^2 \end{bmatrix} \tag{1a}$$

and the mass matrix:

$$[M]_e = [I]_{4\times4} \cdot \frac{\rho hl}{2} \tag{1b}$$

where V_P and V_S are compression wave and shear wave velocity, respectively; $[I]$ an unit matrix with 4×4; ρ mass density; h, thickness of the element. The stiffness matrix for fluid medium can be obtained by letting V_S equal to zero in Eq. 1(a), and the mass matrix remains the same as Eq. 1(b).

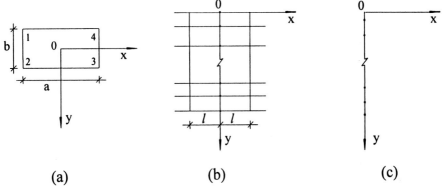

(a)   (b)   (c)

Fig. 1. A layered system

The boundary forces at the face of solid half-space can be expressed (Xia et al., 1993):

$$\begin{Bmatrix} F_{NX} \\ F_{NY} \end{Bmatrix} = \begin{bmatrix} -c^2a & 2V_S^2(1-ab)-c^2 \\ 2V_S^2(1-ab)-c^2 & -c^2b \end{bmatrix} \times \frac{\rho kl}{1-ab} \cdot \begin{Bmatrix} u_{NX} \\ -iu_{NY} \end{Bmatrix} \tag{2}$$

The characteristic equation can be established from Eq. (1) and Eq. (2) with $2N \times 2N$, if there are total N nodal points.

$$\left|[K]-\omega^2[M]\right|=0 \tag{3}$$

Equation (3) can be decomposed as a quadratic equation of wave number:

$$\left|k^2[A]+k[B]+[C]\right|=0 \tag{4}$$

where matrices $[A]$, $[B]$ and $[C]$ are function of phase velocity, c. If phase velocity, c, is known, wave number, k, can be calculated, consequently, Rayleigh wave dispersion curve can be obtained (wave length $\overline{\lambda} = 2\boxtimes/k$).

2.2 Rayleigh Wave in Soft-to-Stiff Type Soils

For soft-to-stiff type soil, shear wave velocity increases with increasing depth and the Rayleigh wave characteristic equation is a real variable function. Therefore, wave number, k, is a real and the wave amplitude does not attenuate in horizontal direction. The matrices in characteristic equation mostly depend on shear wave velocity in soil, thickness, Poisson ratio, damping ratio and mass density, therefore, Rayleigh wave dispersion is influenced by these parameters.

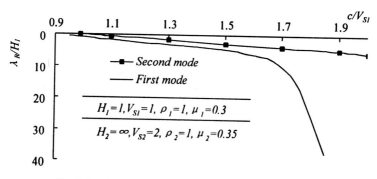

Fig. 2. Rayleigh wave dispersion in soft-to-stiff type soils

Rayleigh wave dispersion in soft-to-stiff type soil is shown in Fig. 2. The wave velocity of its first mode approaches the Rayleigh wave velocity in upmost layer at short wavelength range, while at long wavelength range it approaches the Rayleigh wave velocity in downmost layer (Chen et al., 1991; Xia et al., 1992, and Nazarian, 1984). The effective propagation depth of the first mode wave in vertical direction is about one and half wavelength. The first mode wave dominates the Rayleigh wave propagation in soft-to-stiff type soils, i.e., energy carried by the first mode wave dominates over the others (Xia et al., 1994). The first mode wave dominates over the others in steady state vibration. The signal picked up in-situ test is the first mode wave. Because of limited variation of mass density, ρ, it has little effect on Rayleigh wave dispersion, so that it can be assumed as a constant (Xia et al., 1992 and Nazarian, 1984). For homogeneous soil deposit, Poisson ratio, μ varies between 0 and 0.5, the corresponding varying range of Rayleigh wave velocity, $V_R = (0.87 \sim 0.95) V_S$, from which it can be seen that the influence of Poisson ratio on Rayleigh wave velocity is less than 9%. For a layered system, one can ignore the discrepancy affected by the difference of Poisson ratio for each layer. Damping has little effect on Rayleigh wave dispersion, if damping ratio is less than 0.1 (Xia et al., 1992).

2.3 Rayleigh Wave in Soft-Intercalation Soils

Rayleigh wave behaves differently from that mentioned above because of the presence of intercalation in deposit. Xia et al. (1993) and Gucunski and Woods (1991) discussed homogeneous three-layer system ($V_{S2} < V_{S1} < V_{S3}$). In high frequency range (first mode wavelength approximately equal to the depth of intercalation), the phase velocity of soft-intercalation layer. With contrast to soft-to-stiff type soil, energy carried by low mode waves decreases, while energy carried by high mode waves increases. The wave velocity measured in-situ will be the result superposed of multi-mode waves. The velocity measured on the ground surface will approach Rayleigh wave at upper layer, as frequency goes high. In low frequency range (wavelength of first mode greater than depth of soft-intercalation) the first mode wave dominates over the others, and the phase velocity of the first mode approaches the Rayleigh wave velocity at downmost layer. For multi-layer system (for example: $V_{S3} < V_{S1} V_{S2} V_{S4}$) or for non-homogeneous intercalation, Rayleigh wave behaves similarly, as shown in Fig. 3. The dash line in Fig. 3 represents the data measured in-situ. Propagation of Love wave is similar to Rayleigh wave in soft-intercalation soils (Xia and Wu, 1995).

2.4 Rayleigh Wave in Stiff-to-Soft Type Soils

There are two types of pavement, first, the idealized, shear wave velocity decreases with depth, i.e., stiff-to-soft type, and second, the realistic, in the upper layers, such as concrete-subgrade-soil system, shear wave velocity decreases with depth and then, in the lower layers, shear wave velocity increases with depth.

354   *Shiming Wu*

For the first type, the phase velocity of the mode approaches Rayleigh wave velocity at upmost layer, as Rayleigh wavelength decreases, while it approaches Rayleigh wave velocity at downmost layer, as Rayleigh wavelength increases (Xia et al., 1991) as shown in Fig. 4. The propagation depth of the first mode wave is greater than one and half wavelength, depending on damping ratio. The displacement in horizontal direction attenuates because of energy loss in downward propagation.

For the second type, realistic pavement system, Rayleigh wave dispersion is rather complicated. The phase velocity of the first mode approaches Rayleigh wave velocity at downmost layer, but it does not approaches the Rayleigh wave velocity at upmost layer. The first mode wave dominates over the others at low frequency range, therefore, there are multi-mode waves at high frequency range. The wave number measured in-situ is superposed of wave number of several high modes, which fluctuates with the distance between two transducers (Xia et al., 1998). The dispersion curve measured in-situ is close to the Rayleigh wave velocity at upmost layer (see dash curve in Fig. 5).

Rayleigh Wave in Transversely Isotropic Layered System

Natural deposits usually are of non-isotropic, but they can be regarded as transversely isotropic. The vertical modulus is different from horizontal one,

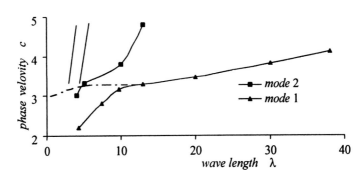

Fig. 3. Rayleigh wave dispersion in soft-intercalation type soils

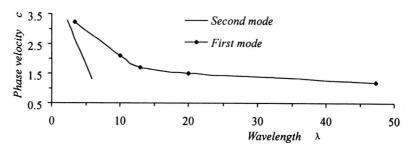

Fig. 4. Rayleigh wave dispersion in stiff-to-soft type soils

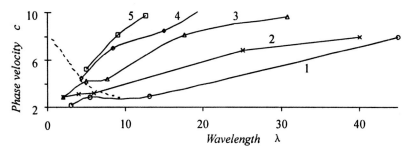

Fig. 5. Rayleigh wave dispersion in a realistic pavement system

which depends on type of soil, consolidation pressure, over-consolidation ratio and others. The Rayleigh wave velocity of an isotropic soil mainly depends on shear modulus, G, while the Rayleigh wave velocity of a transversely isotropic soil depends on two shear moduli: shear modulus in vertical direction, G_V, and in horizontal direction, G_H. The Rayleigh wave velocity in transversely half-space is independent of frequecy, while Rayleigh wave in layered transversely isotropic soil is of dispersion. Similar to the isotropic layered system, the phase velocity approaches the Rayleigh wave velocity at upmost layer in short wavelength range, while in long wavelength range, it approaches the Rayleigh wave velocity at downmost layer. Moduli E_V, and G_H, and Poisson ratio, μ, in each layer have little effect on Rayleigh wave velocity, while shear modulus in vertical direction, G_V, affects the Rayleigh wave velocity dramatically. Shear wave velocity obtained from measurement mainly reflects the shear wave velocity V_{sV} (Xia et al., 1996), and $V_{sV} = (G_V/\rho)^{1/2}$.

3. RAYLEIGH WAVE IN SATURATED SOILS

3.1 Half-Space of Saturated Soils

The theoretical model of Rayleigh wave in saturated soil involves more parameters, Chen (1987) and Meng (1981) made assumption to simplify the model for practical use. Xia et al. (1998) obtained a general solution of potential functions of Rayleigh wave in saturated soil of half-space. The potential functions of soil skeleton can be expressed as:

$$\varphi_2 = \left[A_1 \cdot \exp(-ka_1 z) + A_2 \cdot \exp(-ka_2 z)\right] \cdot \exp\left[-ik(x - ct)\right] \qquad (5a)$$

$$\psi_1 = A_3 \ \exp(-kb_1 z) \ \exp\left[-ik(x - ct)\right] \qquad (5b)$$

The potential function in water phase will be:

$$\varphi_2 = \left[A_1 \cdot B_1 \cdot \exp(-ka_1 z) + A_1 \cdot B_1 \cdot \exp(-ka_2 z)\right] \cdot \exp\left[-ik(x - ct)\right] \quad (6a)$$

$$\psi_2 = \frac{ib}{ib - \rho_2\omega} A_3 \cdot \exp(-kb_1 z) \cdot \exp\left[-ik(x - ct)\right] \qquad (6b)$$

in which A_1, A_2 and A_3 are arbitrary constants, $B_j = (\lambda + 2G) \cdot (\omega/ib)(1/V_j^2 - 1/V_{po}^2) + 1$, $(j = 1,2)$; $a_1^2 = 1 - c^2/V_1^2$, $a_2^2 = 1 - c^2/V_2^2$, V_1 and V_2 are two compression wave velocities, $b_1^2 = 1 - c^2/V_s^2$. It can be seen from Eqs. (5) and (6) that there are two terms of compression wave in potential function, which means Rayleigh wave is induced by interference of two compression waves and shear wave.

There are two kinds of boundaries of Rayleigh wave in saturated soils, permeable and impermeable, corresponding to two kinds of Rayleigh wave characteristic equations. The phase velocity in characteristic equation is affected by frequency and permeability coefficient, which can be expressed by f/f_c, in which f_c is characteristic frequency. From characteristic equation, velocity ratio, c/V_s corresponding to f/f_c can be solved, then phase velocity, c, can be obtained. By substituting ratio $|A_1|/|A_2|$ and $|A_1|/|A_3|$ into displacement and stress equations (Chen, 1987 and Meng, 1981), finally, it gives out the displacements and stresses.

Rayleigh wave front is a vertical plane, perpendicular to x-axis. Energy distribution for three body waves at x=0, can be derived. Similar to the analysis of energy distribution for modes (Lysmer et al., 1972 and Gucunski and Woods, 1991), energy distribution of Rayleigh wave for different modes in saturated soils can be obtained.

Xia et al. (1998) revealed that the Rayleigh wave velocities obtained form two different models (Chen, 1987 and Meng, 1981) show no big difference. Fig. 6 illustrates the relationship of V_R/V_s vs. μ and f/f_c (at permeable condition), where shear modulus of saturated soil skeleton, $G=85$ MPa, porosity, $n=0.6$, $\rho_s=27000$ kg/m^3, $\boxtimes = 1000$ kg/m^3, $E_w=2100$ MPa. It can be seen from Fig. 6 that μ and f/f_c have little effect on V_R/V_s, as $f/f_c < 0.1$; μ and f/f_c have a little effect on V_R/V_s, as $0.1 < f/f_c < 10$; as $f/f_c > 10$, f/f_c does not affect V_R/V_s a lot, but μ does. The relationship of V_R/V_s vs. μ and f/f_c at impermeable condition is similar to that at permeable condition, the displacement distribution of soil skeleton and water phase of Rayleigh wave is similar to that in elastic soil, its effective depth on wave propagation is about one and half wavelength.

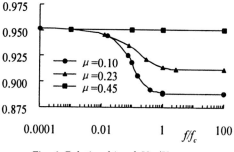

Fig. 6. Relationship of V_R/V_s
vs. μ and f/f_c

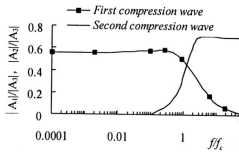

Fig. 7. Relationship of $|A_1|/|A_2|$
and $|A_1|/|A_3|$ vs. f/f_c ($\mu = 0.23$)

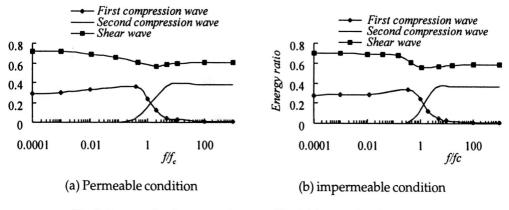

Fig. 8. Energy ratio of component waves of Rayleigh wave ($\mu = 0.23$)

Figure 7 illustrates the relationship of component amplitudes of Rayleigh wave vs. frequency (for permeable condition). Figure 8 illustrates energy variation component of Rayleigh wave for permeable and impermeable condition. From Fig. 7 and Fig. 8, the first compression wave dominates over the second compression wave which can be ignored, as $f/f_c < 0.1$; while contribution of second compression wave is greater than first compression wave (Xia et al., 1996), as $f/f_c > 0.1$.

3.2 Layered Saturated Soils

Characteristic equation of Rayleigh wave dispersion in layered saturated soil system can be established and solved by finite element method, which can be employed in studies of Rayleigh wave dispersion of first mode in layered saturated soil. The dynamic motion equations of saturated soil can be expressed as:

$$G\nabla^2 \vec{U} + (\lambda + G)\,\text{grad}\,(\text{div}\,\vec{U}) + \bar{b}\,(\vec{W} - \vec{U}) - \rho_1 \ddot{\vec{U}} = 0 \qquad (7a)$$

$$E_W\,\text{grad}(\text{div}\vec{W}) + \frac{E_W(1-n)}{n}\,\text{grad}\,(\text{div}\vec{U}) + \bar{b}\,(\vec{U} - \vec{W}) - \rho_2 \ddot{\vec{W}} = 0 \qquad (7b)$$

where $\vec{U}$, displacement vector of soil skeleton; $\vec{W}$, displacement vector of pore water; E_W, bulk modulus of water; G and λ, Lamb constants of soil skeleton; n, porosity; $\rho_1 = (1-n)\,\rho_s$; $\rho_2 = n\rho_w$; ρ_s, soil mass density; ρ_w, water mass density; $\rho = \rho_1 + \rho_2$, mass density of saturated soil; $\bar{b} = n\rho_w g/\bar{k}$

4. FLUID-SOLID MEDIUM SYSTEM

Tan (1990) studied dynamic response of a dam in frequency domain by means of Rayleigh wave propagation in fluid-solid medium system, for which finite element method was employed. For avoiding singularity of motion equation, an undetermined parameter, α, is employed in derivation of finite element stiffness matrix of fluid, which restricts its application and accuracy, and is only suitable for rigid structure with semi-infinite layers.

Idealized fluid is incompressible and no shear wave exits, i.e. shear modulus, G, is equal zero and Poisson ratio, μ =0.5. Similar to derivation of finite element in elastic medium, condensed stiffness matrix, $[K]_e$ and mass matrix, $[M]_e$, can be obtained. The stiffness matrix has the same expression as in Eq. (1), as shear wave velocity, V_s, equal zero. Then the Rayleigh wave characteristic equation can be established, which is different from that in solid medium, since discontinuity of horizontal displacement at fluid-solid interface.

The followings are Stoneley wave characteristic equations in fluid-solid two half-space system by combined finite element and analytic approach.

4.1 Fluid-Solid Two Half-Space System

For a fluid-solid two half-space system as shown in Fig. 9(a), the upper layer is fluid, for example, water, with compression modulus, λ_1, mass density, ρ_1; and the lower layer is solid with shear modulus, G, mass density, ρ and Poisson ratio, μ. For plane wave in two-medium system, its displacement approaches zero, as $y \to \pm\infty$. Therefore, displacement potential function can be expressed as:

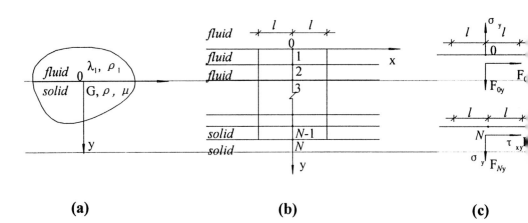

(a)   (b)   (c)

Fig. 9. Fluid-solid system

while $y \leq 0$, $\qquad\qquad\qquad \varphi_1 = A_1 \cdot \exp[ka_1 y + ik(x - ct)]$ $\qquad$ (8a)

while $y \geq 0$,
$$\begin{cases} \varphi = A \cdot \exp[-aky + ik(x - ct)] \\ \psi = B \cdot \exp[-bky + ik(x - ct)] \end{cases}$$
$\qquad$ (8b)

in which, $i = \sqrt{-1}; \varphi_1$, compression wave potential function in fluid; φ and ψ, compression and shear wave potential functions in soil, respectively; A_1, A and B, unknown constants; k, wave number (wavelength $\overline{\lambda} = 2\pi/k$); c, phase velocity; ω = kc, vibration circular frequency; $q = \sqrt{1 - c^2/V_{P1}^{*2}}$, $a = \sqrt{1 - c^2/V_P^2}$ and $b = \sqrt{1 - c^2/V_s^2}$; $V_{P1} = \sqrt{\lambda/\rho_1}$, compression wave velocity in fluid; V_p and V_s, compression and shear wave velocities in solid, respectively.

The following boundary conditions are to be satisfied as plane wave propagates at interface of two media as $y = 0$:

$$\begin{cases} u_{y1} = u_y \\ \sigma_{y1} = \sigma_y \\ \tau_{xy1} = \tau_{xy} = 0 \end{cases}$$
$\qquad$ (9)

As mentioned above, the horizontal displacement at interface can be discontinuity. The stress and displacement can be expressed in terms of potential functions in Eq. (1) (Eringen et al., 1984; Wu, 1997). Let $y = 0$ and eliminate common term exp $[ik(x-ct)]$, Eq. (9) becomes:

$$\begin{cases} aiA - B + a_1 iA_1 = 0 \\ G\left[(1 + b^2)iA - 2bB\right] - \lambda_1 iA_1(a_1^2 - 1) = 0 \\ 2GaiA - G(1 + b^2)B = 0 \end{cases}$$
$\qquad$ (10)

For existence of nonzero solution of iA_1 , iA and B, the coefficient determinant of above equation must be zero, then Stoneley wave characteristic equation will be:

$$\begin{vmatrix} a & -1 & a_1 \\ G(1 + b^2) & -2Gb & (1 - a_1^2)\lambda_1 \\ 2a & -(1 + b^2) & 0 \end{vmatrix} = 0$$
$\qquad$ (11)

The coefficients, a_1, a and b have a sense physically, only as $c < V_s < V_p$ and $c < V_{p1}$. If there exists one positive root for c in Eq. (11), and, $c < V_s < V_p$ and $c < V_{p1}$, then there exists one kind of Stoneley wave around fluid-solid inteface, which does not attenuate along x-axis, while it does exponentially in vertical direction. If no positive roots exist, it means no Stoneley wave. From Eq. (11), one can see that velocity, c, is independence of frequency, ω, which indicates the Stoneley wave in the interface of fluid and solid is of no dispersion. The velocity is a constant for any frequency.

Let $r = c/V_s$ and normalize $p = V_s^2/V_p^2$, $p_1 = V_s^2/V_{p1}^2$, and $a_1 = \sqrt{1 - r^2 p_1}$

$a = \sqrt{1 - r^2 p}$ and $b = \sqrt{1 - r^2}$, then solving for value of c can be transferred to solving for normalized r value. The eigenvalue, r, can be obtained from Eq. (11), furthermore, the coefficient vector matrix, $\bar{V} = [iA \quad B \quad iA_1]^T$. The displacement distribution of particles of fluid and solid media can be obtained (without term, $\exp[ik(x - ct)]$).

As $y < 0$, it yields :

$$\begin{cases} u_{x1} = (iA_1)k \cdot \exp(ka_1 y) \\ -iu_{y1} = -ka_1(iA_1) \cdot \exp(ka_1 y) \end{cases} \tag{12a}$$

As $y > 0$, it yields :

$$\begin{cases} u_x = k(iA) \cdot \exp(-aky) - bkB \cdot \exp(-bky) \\ -iu_y = ak(iA) \cdot \exp(-aky) - kB \cdot \exp(-bky) \end{cases} \tag{12b}$$

4.2 Two Fluid Half-Space System

The potential functions in two fluid half-space system will be:

as $y < 0$, $\qquad \varphi_1 = A_1 \cdot \exp[ka_1 y + ik(x - ct)]$ (13a)

as $y > 0$, $\qquad \varphi_2 = A_2 \cdot \exp[ka_2 y + ik(x - ct)]$ (13b)

As a plane wave propagates in the interface of two media, the boundary conditions are as follows: with $y=0$,

$$\begin{cases} u_{y1} = u_y \\ \sigma_{y1} = \sigma_y \\ u_{x1} = u_x \end{cases} \tag{14}$$

From the expression of Eq. (7), it yields $A_1 = A_2$, and from the second expression of Eq. (7), there is $k = 0$, which means no Stoneley wave exists. In other words, there are two unknowns, but three equations in Eq. (7), no nonzero solution or no Stoneley wave exist.

4.3 Fluid-Solid Layered System

As a system shown in Fig. 9(b), the upper portion is a fluid half-space and the lower portion is a solid layered half-space. Instead of analytic approach for system as shown in Fig. 9(a), combined finite element-analytic approach is employed referring to establishment of Rayleigh wave characteristic equation (Xia et al., 1993 and Xia and Wu, 1994), and Stoneley wave characteristic equation can be established. The boundary forces of fluid half-space acting on the interface are shown in Fig. 9(c). The wave potential functions of fluid are expressed as in Eq. (13a), the particle displacement and stress can be written:

$$
\left\{
\begin{aligned}
u_x &= \frac{\partial \varphi_1}{\partial x} = ik\varphi_1 \\
u_y &= \frac{\partial \varphi_1}{\partial y} = ka_1\varphi_1 \\
\sigma_y &= \lambda_1 \nabla^2 \varphi_1 = \lambda_1(a_1^2 - 1)k^2\varphi_1
\end{aligned}
\right.
\tag{15}
$$

From Eq. (15), it yields:

$$
u_x = -\frac{1}{a_1}(-iu_y)
$$

$$
\sigma_y = \frac{\lambda_1}{a_1}(a_1^2 - 1)ku_y
\tag{16}
$$

From Fig. 9(c), the boundary force of fluid acting on the solid medium, F_{oy} can be expressed, as $y = 0$:

$$
F_{oy} = -\frac{1}{2}\int_{-l}^{l} \sigma_y dx = -\frac{1}{2}\int_{-l}^{l} \frac{\lambda_1}{a_1}(a_1^2 - 1)k \cdot \exp(ikx) \cdot u_{oy} \cdot dx
\tag{17}
$$

In which, u_{oy}, displacement at the origin in Fig. 9(c). Integrate Eq. (17), and let $l \to 0$, it yields:

$$-iF_{oy} = \rho \frac{c^2 k}{a_1} \cdot l \cdot (-iu_{oy}) \tag{18}$$

Therefore, adding a term of force, as Eq. (18) shown, in the second equation of Rayleigh wave characteristic equation yields Stoneley wave characteristic equations.

4.4 Nature of Stoneley Wave in Fluid-Solid Two Half-Space System

For simplicity, normalization is employed and the ratio of shear wave velocity to compression wave velocity in solid is expressed in terms of Poisson ratio. Then it yields $p = (1-2\mu)/[2(1-\mu)]$. Let $q = \lambda_1/G$, and $n = \rho_1/\rho$, then $p = n/q$, the eigen value, r, depends on Poisson ratio of solid, μ, modulus ratio, q, and mass ratio, n $(n = \rho_1/p)$.

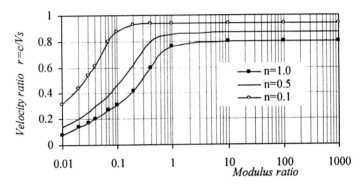

Fig. 10. Influence of q and n on r ($\mu = 0.35$)

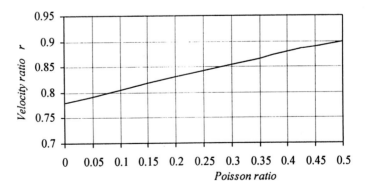

Fig. 11. The relationship of μ and r for Stoneley wave

Therefore, μ, q and n are variables of Stoneley wave.

The influence of q and n on velocity ratio r ($r = c/v_s$) as $\mu = 0.35$, is shown in Fig. 10, in which mass ratio, $n = 0.35 \sim 0.55$, and $q > 1$. It can been seen from Fig. 10 that velocity ratio, r, keeps constant about 0.8, as modulus ratio, $q = \lambda_1/G$, is lower than that for Rayleigh wave in solid half-space (see Fig. 12). Figure 11 shows the relationship of Poisson ratio, μ, V_s velocity ratio, r, as mass ratio, $n = 0.5$ and modulus ratio $q = 50$. Similar to the influence of Poisson ratio to Rayleigh wave velocity (see Fig. 12), Poisson ratio has less influence on Stoneley wave velocity.

Figure 13 shows the displacement distribution of Stoneley wave as $n = 0.5$, q = 1.0 and $q = 50.0$, in which the abscissa represents normalized displacement of solid particles on interface, and the ordinate represents normalized depth of Stoneley wave length, $\lambda = 2\pi/k$. The Stoneley wave displacement reaches its maximum on interface, and horizontal displacement shows its discontinuity on the interface. Same as Rayleigh wave in solid medium, the Stoneley wave propagates within a depth of one and half wavelength (Eringen et al., 1984),

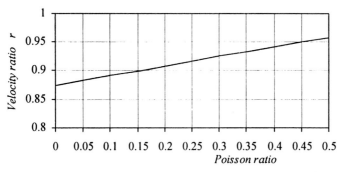

Fig. 12. Relationship of μ vs. r for Rayleigh wave

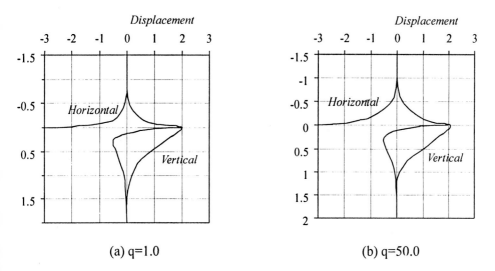

(a) q=1.0   (b) q=50.0

Fig. 13. Displacement distribution of Stoneley wave

as well as the shape of displacement distribution. In fluid medium the Stoneley wave attenuates very rapidly, and the horizontal displacement distribution is same as vertical displacement with an effective propagation depth about one wavelength.

4.5 The Nature of Stoneley Wave in a Fluid-Solid Layered System

There is a three-layer system, in which a fluid, half-space on the top, a solid medium with thickness, H_2, in middle, and a solid half-space on the bottom. Table 1 shows all parameters of the system. Figure 14 shows the dispersion curve of first mode wave of Stoneley wave and dispersion curve of first mode wave of Rayleigh wave in the same system without top fluid half-space, i.e., two-layer system. The solid dots in Fig.14 represent dispersion curve of first mode wave of Stoneley wave as fluid pressure increases to V_{P1}/V_{S2}=30. It can be seen from Fig. 14 that Stoneley wave is of dispersion, and small difference exists between two fluid pressure condition, one V_{P1}/V_{S2}=15 and the other, V_{P1}/V_{S2}=30,which draws a conclusion that modulus ratio has little influence on Stoneley wave velocity, furthermore, as mentioned in last section, the Stoneley wave velocity in a solid medium can be looked on a constant. In a layered solid system, as wavelength decreases, the wave velocity of first mode approaches to the Stoneley wave velocity of the upmost layer, while it approaches to the Stoneley wave of downmost solid layer, as wavelength increases. Usually the Stoneley wave velocity is less than Rayleigh wave velocity.

Table 1. Parameters of a three-layer system

Layer	Medium	Thickness, H_1/H_2	Poisson ratio, μ	Mass ratio, ρ_i/ρ_1	Velocity ratio
1	Fluid	∞	0.5	1.0	$V_{P1}/V_{s2} = 15.0$
2	Solid	1	0.35	1.8	$V_{s2}=1.0$
3	Solid	∞	0.35	2.0	$V_{s3}/V_{s2} = 3.0$

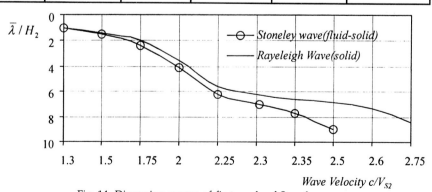

Fig. 14. Dispersion curves of first mode of Stoneley wave

Figure 15 shows displacement distribution of first mode of Stoneley wave, similar to that of Rayleigh wave of solid medium, with effective propagation depth of one and half wavelength. Stoneley wave attenuates very rapidly in fluid medium with a coefficient, $a_1 = \sqrt{1 - c^2 / V_{P1}^2}$. Figure 16 shows a dispersion curves of second mode of Stoneley wave.

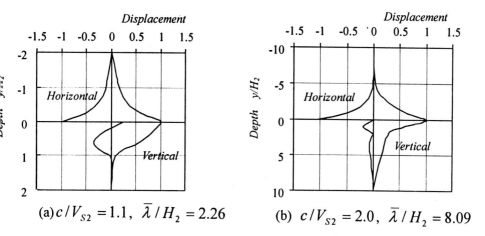

(a) $c / V_{S2} = 1.1$, $\overline{\lambda} / H_2 = 2.26$ (b) $c / V_{S2} = 2.0$, $\overline{\lambda} / H_2 = 8.09$

Fig. 15. Displacement distribution curve of first mode of Stoneley wave

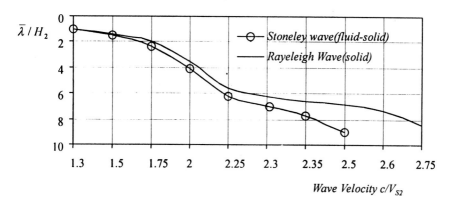

Fig. 16. Dispersion curves of second mode of stoneley wave

5. SIGNAL ANALYSIS OF RAYLEIGH WAVE

5.1 Spectral Analysis of Surface Wave with a Source

Vibration measurement with a source means a process of a forced vibration. A measurement of Rayleigh wave dispersion in shallow depth can be conducted by employing a high frequency vibration source. Figure 17(a) is a schematic diagram of Spectral Analysis of Surface Wave (SASW), and Fig. 17(b) for measurement of soil parameter on the ground with a structure on it.

Based on vibration theory, Rayleigh wave carries the most part of energy of waves caused by a pulse on the ground. The vertical components of vibration picked up by transducers 1 and 2 as shown in Fig.17(a) are dominated by Rayleigh wave, which propagates from the source outward with a wave front of cylindrical surface. If the signals picked up by transducers 1 and 2 are $x(t)$ and $y(t)$, respectively, the time for wave propagating from point 1 to point 2 is equal to the time delay between $x(t)$ and $y(t)$. Consequently, Rayleigh wave velocity can be obtained from time delay of these two signals.

Rayleigh wave velocity in a layered deposit varies with frequency or wavelength, i.e., frequency dispersion. Therefore, spectral analysis is the way to determine the time delay of signals. $G_{xx}(f)$ and $G_{yy}(f)$ are self-power spectrum of $x(t)$, and cross spectrum, $G_{yx}(f)$, from which phase difference between $y(t)$ and, $x(t)$, φ, can be obtained. Figure 18 shows a series of phase curves, each one varies form $-180°$ to $180°$ for each period. It is difficult to determine how many periods included in phase difference between two signals for a practical case. Therefore, a method that can distinguish the real phase difference automatically is needed. Unfolding the phase curves is a powerful method as shown in Fig. 18(b) to solve the problem. For a given frequency, f, one can find a phase difference φ, from Fig. 18(b), then time delay between $x(t)$ and $y(t)$ can be obtained. $\gamma(f)$ is coherence function of $x(t)$ and $y(t)$, which can be employed to judge the correlation between $x(t)$ and $y(t)$ in a certain frequency range as shown in Fig. 19. A value of $\gamma(f)$ close to 1.0 indicates good correlation between two signals. Any present of noise, non-linear of the system and multi-direction input of signal will cause reduction of coherence function value.

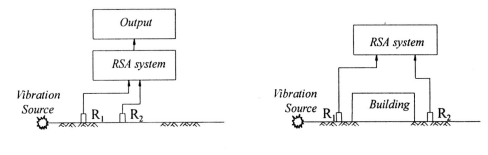

(a)   (b)

Fig. 17. Surface wave measurement

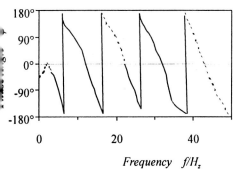

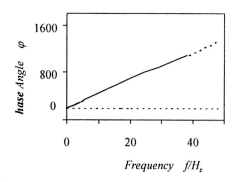

(a) Folded Phase Curves   (b) Unfolded Phase Curves

Fig. 18. Phase angle of spectrum

5.2 Frequency-Wave Number Spectral Analysis (F-K)

Microtremor is a method of vibration measurement without any certain vibration sources, as shown in Fig. 20 in which five vertical transducers are arranged in two dimensions. The distance between transducers depends on the wavelength measured. If m transducers are arranged, the self-power spectrum and cross-power spectrum of two signals picked up by transducers i and j are $G_i(f)$ and $G_j(f)$, $G_{ij}(f)$, respectively, then the frequency-wave number spectrum with high resolution, $p(f, k)$, can be expressed as (Tokimatsu et al., 1992)

$$P(f,\vec{k}) = \sum_{i=1}^{m}\sum_{j=1}^{m} B_i^*(f,\vec{k})B_j(f,\vec{k}) \cdot G_{ij}(f) \cdot \exp[ik(\vec{R}_i - \vec{R}_j)] \qquad (19)$$

in which $\vec{k}$ is a vector wave number, and

$$B_i(f,\vec{k}) = \sum_{j=i}^{m} Q_{ij}(f,\vec{k}) \Big/ \sum_{i=1}^{m}\sum_{j=1}^{m} Q_{ij}(f,\vec{k}) \qquad (20)$$

where, $Q_{ij}(f, k)$ is the inverse matrix of $\{\exp(ik(R_i-R_j)) \cdot G_{ij}(f)\}$. Each frequency-wave number (F-K) spectrum with high resolution is a 2-D wave number (k_x-k_y) space can be derived from $p(f, \vec{k})$. Taking maximum value of the spectrum yields wave number, $\vec{k}$, and propagation direction, with a relationship of f vs. $\vec{k}$, which is a dispersion curve. This approach is suitable for low frequency signal analysis (such as ground, microtremor), not for high frequency case with vibration source, where wave propagation direction is known and k is known. Figure 21 is a typical F-K spectrum.

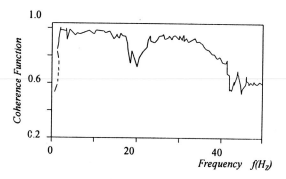

Fig. 19. Coherence function

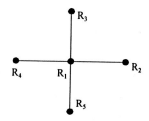

Fig. 20. Arrangement of transducer for ground microtremor

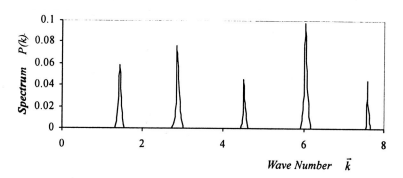

Fig. 21. A typical F-K spectrum

In the case of ground microtremor, low frequency and small wave number, $\vec{k}$, rather far frequency interval between peaks result in difficulties of judgement of accurate wave number, even though existence of multiple peaks.

6. APPLICATIONS OF RAYLEIGH WAVE IN EXPLORATION

6.1 Shear Wave Velocity Measurement of Natural Deposit

Cross-hole and down-hole methods, usually employed for shear wave measurement, need preparation of holes on the ground with high cost and long duration of time, which is specially not suitable for project with large area. Reflection and refraction wave methods are always disturbed by surface wave. Without those shortcomings, Rayleigh wave dispersion curve can be obtained from field tests operated on the ground surface (Chen et al., 1993). Shear wave velocity at different depth can be obtained by back analysis from dispersion curve (Zhou,1991 and Chui, et al., 1998). Figure 22 shows a typical Rayleigh wave dispersion curve measured in situ and a relationship of shear velocity vs. depth by back analysis form dispersion curve.

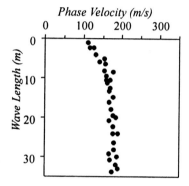 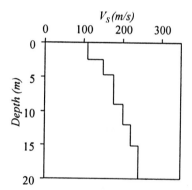

(a) Phase Velocity vs. Wavelength   (b) Shear Wave Velocity vs. Depth

Fig. 22. Shear wave measurement by means of Rayleigh wave

6.2 Quality Monitoring of Structure of Road Pavement

Usually a natural ground, especially for soft and weak soil, needs to be strengthened, such as dynamic consolidation, and Rayleigh wave method can be employed for quality monitoring, with its advantages of fast operation, non-destructive and low cost.

Yang's modulus of each layer in the road can be expressed

$$E = 2\rho V_s^2 (1 + \mu) \tag{21}$$

where ρ, mass density and μ, Poisson ratio.

There exists correlation between material strength and wave velocity, from which strength can be assessed, with a small amount of compression test in laboratory for determining coefficient of correlation.

Shear wave velocity and thickness for each layer can be obtained by back analysis from Rayleigh wave dispersion curve. Not only modulus and strength of each layer are requested for normal use, the attention should to be paid for contact condition between layers, for which coherence function, γ, can be employed. For well-contacted and homogeneous material, one can get coherence function value close to 1.0 without any back noise, while for any presence of voids, non-homogeneous and cracks within the road structure causes reduction of coherence function as shown in Fig. 23.

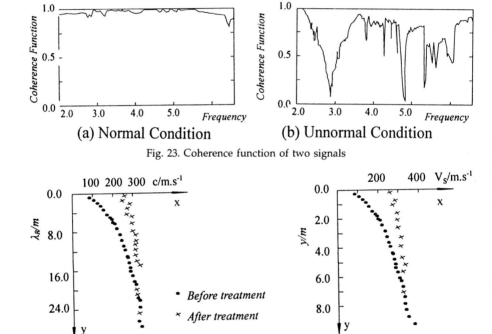

(a) Normal Condition   (b) Unnormal Condition

Fig. 23. Coherence function of two signals

(a) Measured Rayleigh Wave Velocity   (b) Measured Shear Wave Vel

Fig. 24. Comparison of shear wave velocity before and after dynamic consolidation

6.3 Assessment of Ground Wave Strengthening Treatment

A natural foundation for building a structure usually needs to be strengthened by means of different methods, such as dynamic consolidation, underpinning, vibration densification and so on, and the effectiveness of strengthening treatment is usually to be assessed. Rayleigh wave method can be applied to do so with advantages of overall assessment, fast measuring, low cost and operation during the treatment (Zhao,1994 and Kong, et al. 1996). Figure 24 demonstrates a case study of dynamic consolidation for a natural deposit.

6.4 Measurement of Shear Wave Velocity under a Structure

There exists one-to-one relationship between shear modulus and shear wave velocity, for a given medium, and there also exists correlation between shear strength and shear modulus. Measurement of shear wave velocity can be used for estimation of strength condition under an existing oil tank, Stoneley wave measurement may be employed to detect the leakage (Kashi et al., 1991).

6.5 Detection of Fissures in Concrete

Detection of fissures in concrete of existing structure, its condition and corresponding parameters is helpful for assessment of an existing structure and selection of strengthening measures if strengthening is needed. At present, ultrasonic wave method is commonly used. As a supplement, surface wave method can be employed to measure the fissure depth. As shown in Fig. 25, as surface wave propagates through the fissure, two given points, A and B, transducers can pick up vertical components of surface wave and give out phase difference. One can estimate the fissure depth from the phase difference and frequency. For low frequency or long wavelength, surface wave passes through without any reflection, and φ is linear with f, while for high frequency or short wavelength, the relation between φ and f shows non-linear as shown in Fig. 25(b), from which fissure depth, D, can be estimated. Further study is needed in this field.

6.6 Seismic CT Technique for Deep Water Bed Rock

There is a potential to use the nature of Rayleigh wave and Stoneley wave in fluid-solid system and their dispersion for analysis of bed rock under deep water (ocean, river or lake), i.e., CT technique. Further investigations, both theoretically and experimentally, are encouraged.

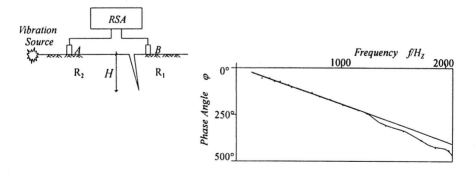

Fig. 25. Detection of fissure by surface wave measurement

REFERENCES

Abo-zena, A., Dispersion function computation for unlimited frequency value, Geophys, J.R. Astr. Soc., 1979, 58:91-105.

Chen, Y.M. and Wu, S.M., Solution of characteristic equation of Rayleigh Wave in layered soils, J. of Zhejiang Univ., 1991, 25(1): 40-52 (in Chinese).

Chen, Y.M. et al., A semi-infinite element for R-wave dispersion in a layered half-space, Proc. 7th Intl. Conf. on Numerical Method Develop. in Geot. Mech., Australia, A. ABACEMA, 1991.

Chen, L.Z., Propagation velocity of elastic stress wave in saturated soils and its applications, Ph.D. thesis, Zhejiang Univ., 1987 (in Chinese).

Chen, Y. M. et al., Quality monitoring of road structure by spectral analysis of surface wave, J. Zhejiang Univ., 1993, 27(3) (in Chinese).

Chiang, C.M. and Monstafa, A.F., Wave-induced response in a fluid-filled poro-elastic solid with a free surface—a boundary layer theory. Geophy. J.R. Atr. Soc., 1981, 66: 597-631.

Chui, J.W. et al., Application of transient surface wave in engineering geology, Chinese J. Geot. Engrg., 1998, 18(3): 35-40 (in Chinese).

Dunkin, L.W., Computation of model solution in layered elastic media at high frequencies. Bull. Seism. Soc. Am. 1965.

Eringen, A.C., Shu, S.A., Elastic Dynamics., Vol. 2, 1984, Oil Industry Press (in Chinese).

Feng, G.Z., Soil Mechanics, Press of Water Conservancy and Electric Power, Beijing, 1986 (in Chinese).

Gucunski, L.D., Woods, R.D., Selection of Rayleigh modes in interpretation of SASW test. Second International Conference on Recent Advances in Geotechnical Earthquake Engineering and Soil Dynamics, 1991.

Haskell, N.A., The dispersion of surface wave on multi-layered media. Bull. Seism. Soc. Am. 1953.

Ishihara, K., Approximate forms of wave equations for water-saturated porous materials and related dynamic moduli, Soils and Foundations, 1970, 10(4): 10-38.

Jone, J.P., Rayleigh wave in a porous elastic saturated solid, J. Acoust Soc. Am., 1961, 33: 959-962.

Jones, R., A vibration method for measuring the thickness of concrete road slabs in situ, Magazine of Concrete Research, July, 1995, 97-102 (in Chinese).

Jones, R., Surface wave technique for measuring the elastic properties and thickness of roads: theoretical development, BRIT. J. Appl. Phys., 1962, Vol. 13, 21-29.

Knopoff, L., A matrix method for elastic wave problems. Bull. Seism. Soc. Am. 1964.

Koppe, H., Uber Rayleigh-wellen an den oberflache Zweier mediea, Angew. Math. Mech., 1948, 28: 355-360.

Tokimatsue, K., Shinzawa, K and Kuwayama, S., Use of Short-period Microtremors for Vs Profiling, Journal of Geotechnical Engineering, Vol. 118, 1992.

Kong, L.W. and Yaun, H.L., Rayleigh wave motivation on strengthening soft soil by dynamic consolidation, Engrg. Expl., 1996, No. 5: 1-5 (in Chinese).

Kashi, M., Sotoodehnia, and A., Maser, K., Leak detection in large storage tanks using seimic boundary waves, Second International Conference on Recent Advances in Geotechnical Earthquake Engineering and Soil Dynamics, 1991.

Lysmer, J. and Kuhlemeyer, R. L., Finite dynamic model for infinite media. J. Engrg. Mech. Div., ASCE, 1972, 95: 859-877.

Meng, F.L., Wave propagation in fluid saturated medium, J. Geophysics, 1981, 24(1): 65-76(in Chinese).

Miller, G.F. and Pursey, H., On the partition of energy between elastic waves in a semi-infinite solid. Proc. Royal Society, London, A, 1995, 233: 55-69.

Nazarian, S., In situ determination of elastic moduli of soil deposits and pavement system by SASW method, Ph.D. thesis, University of Texas at Austin, 1984.

Stoneley, R., Elastic waves at the surface of separation of two solids, Proc. Roy. Soc., London, Ser. A, 1924, 106: 416-428.

Tan, H. H., Transmitting boundary for semi-infinite reservoir, J. Engrg. Mech. Div., ASCE, 1990.

Tan, H.H., Generalized Rayleigh wave in layered solid-fluid media, Second International Conference on Recent Advances in Geotechnical Earthquake Engineering and Soil Dynamics, 1991.

Thomson, W.T., Transmission of elastic waves through a stratified solid medium. J. Appl. Phys. 1950.

Thrower, E.N., The computation of dispersion of elastic in layered media. J. Appl. Phys. 1965.

Vrettos, C., In-plane vibration of soil deposits with variable shear modulus: I. Surface wave. Int. J. Num. Anal. Methods Geomech, 1990: 209-222.(a)

Vrettos, C., Dispersive SH-surface waves in soil deposits of variable shear modulus. Soil Dynamic and Earthquake Engineering, 1990: 255-264.(b)

Wang, J.R., Relationship between shear modulus, shear wave velocity and shear strength, consolidation degree of soft clay, ME thesis, Zhejiang Univ. 1987 (in Chinese).

Watson, T.H., A real frequency complex wave number analysis of leaking model. Bull. Seism. Soc. Am. 1972.

Wu, S.M., Wave propagation in soils, Scientific Press, Beijing, 1997 (in Chinese).

Xia, T.D. et al., Dispersion of Rayleigh wave in pavement structure, Proc. 6th Conf. on SMFE, Shanghai, Tongji Univ. Press, 1991 (in Chinese).

Xia, T.D. et al., Rayleigh wave dispersion in homogeneous soft intercalation layer, Chinese J. Vibration Engrg., 1993,6(l): 42-50 (in Chinese).

Xia, T.D. et al., Rayleigh wave dispersion in soils, Proc. 2nd East China Conf. on Geot. Mech., Zhejiang Univ. Press, 1992 (in Chinese).

Xia, T.D. et al., Nature of Rayleigh wave in saturated soils of half-space, J. Water Conservancy, 1998, 2, 47-53(in Chinese).

Xia, T.D. and Wu, S.M, Energy distribution of in-plane steady vibration wave in soils, J. Zhejiang Univ. 1994, 28 (supplement): 224-230 (in Chinese).

Xia, T.D. and Wu, S.M, Finite element method for Love wave dispersion in soils, Proc. 3rd Int.Conf. on Soil Dyn. Earthq. Engrg., USA, 1995.

Xia, T.D. et al, Nature of Rayleigh wave in road structure, Proc. 5th Conf on Soil Dyn. Dalian Univ. of Tech. Press, Dalian, 1998 (in Chinese).

Xia, T.D. et al, Rayleigh wave dispersion in non-isotropic layered soils, Chinese J. Vibration Engrg., 9(2): 191-197 (in Chinese).

Xia, T.D. et al, Dynamic response analysis by means of Rayleigh wave, Chinese J. Geot. Engrg. 18(3): 28-34 (in Chinese).

Xia, T.D. and Wu, S.M, Rayleigh wave in fluid-solid media, J. Water Conservancy, 1994, (1): 69-85 (in Chinese).

Xia, T.D. et al, Signal spectral analysis of surface wave, Proc. 5th Conf. on Soil Dyn., Dalian Univ. of Tech. Press, Dalian, 1998 (in Chinese).

Zhao, Z.Z., Assessment of effectiveness of ground treatment by spectral analysis of surface wave, Proc. 4th Conf. Soil Dyn., Zhejiang Univ. Press, Hangzhou, 1994 (in Chinese).

Zhu, Y.L., Application of Rayleigh wave in exploration, Engrg. Expl. 1991, No. 1, 67-88 (in Chinese).

12

Morphology, Ageing and Engineering Behavior of Sands

Yudhbir, J. Thomas and A. Rahim
Indian Institute of Technology, Kanpur, India

ABSTRACT

This paper reviews approaches to quantify grain morphology and microfabric and their influence on engineering behavior of sands. Grain morphology and microfabric induced by ageing are quantified in terms of simple indices which are shown to be correlated with mechanical characteristics of sands. It is stressed that correlations based on SPT/SCPT values developed for a particular type of sand cannot be extrapolated for all types of granular materials. Cyclic pre-straining and secondary compression mechanisms of ageing of sands are shown to give identical results. Strain-time response of a sand, that has experienced increased vertical strain under an effective stress level either by secondary compression or cyclic pre-straining, is shown to give an excellent demonstration of the influence of grain morphology and constancy of C_α C_c. Based on analysis of available data for uncemented aged sands, it is shown that majority of improved behavior of aged sands is primarily due to particle interlocking—planar type contacts—and that during ageing the changes in grain contact types, from tangential to planar, occur at practically constant volume. Field studies of aged sands and quantitative evaluation of their microfabric are recommended for better understanding of the mechanisms of ageing and their effect on mechanical characteristics of aged sands.

Key Words: Morphology-microfabric quantification, mechanisms of ageing, secondary compression, cyclic pre-straining, over-consolidation, aged sand, engineering characteristics, in-situ behavior.

1. INTRODUCTION

Geotechnical engineering practice, in case of sandy deposits, is generally dominated by design methods based on simple in-situ tests, predominantly Standard Penetration Test, SPT, and past experience. In many parts of the world SPT equipment and procedures do not necessarily conform to recommended International Standards and the energy transmitted to the penetrometer is generally unknown. Improved correlation charts between N_{60} (SPT values corresponding to 60% energy) and mechanical characteristics, now available as design guidelines, end up being misused or abused in situations where correct N_{60} cannot be estimated.

Furthermore, the nature of sands in terms of grain morphology and mineralogy—and their engineering geologic history—in terms of chronological age, depth of burial, past loading history, fabric, cementation etc. is now known to significantly influence their mechanical characteristics. Therefore, empirical correlations primarily developed from laboratory model tests, such as calibration chamber tests, and elemental tests for some sand or set of sands need to be cautiously applied to other natural sands especially for evaluation of in-situ mechanical characteristics. Regrettably no such caution is advised, even in standard textbooks, against indiscriminate use of such correlations.

2. A BRIEF OVERVIEW OF THE CURRENT STATUS

In recent times sophisticated in-situ and laboratory test techniques have been devised for understanding and prediction of response of sands to applied loads. These efforts have been primarily motivated by the need for more accurate prediction of liquefaction potential of saturated sand and foundation displacements. Notable attempts have also been made for quantification of the influence of morphology and ageing on the engineering behavior of sands. Even though the behavior of sands is usually studied at the phenomenological level, it is clear that the macroscopic response of sands is the result of interactions at the particulate level. The effects of particle morphology—their form or shape —are manifested at different levels. Firstly, the deformation, modification and crushing of sands are strongly influenced by the shape of the grains (in addition to grain mineralogy). Secondly, particle shape has a marked effect on the geometrical arrangement of grains (fabric) which is an important variable governing the response of sands. Lastly, morphology has a pronounced impact on the relative movements of particles and hence on compressibility and dilatancy.

Particular sand with a given set of grain characteristics can exist in different states. The response of a sand to imposed loads would be controlled by state variables characterising the geometrical arrangement of the grains and the forces acting between them. The important state variables are—the orientation of particles and contact normals, the mean and standard deviation of void ratio, the type of grain contacts, the nature and degree of cementation, the state of in-situ stress and the stress and strain history. In the past, the behavior of sands was assumed to be mainly a function of void ratio or relative density, I_D, confining stress and, if applicable mechanical over consolidation. The pioneering work of Vesic brought out the significance of compressibility on bearing capacity of foundations on/in sand and the experience gained over the last few decades has stressed the significance of grain modification/crushing and changes in fabric/structure conditioned by the phenomenon of ageing vis-à-vis shearing resistance and stiffness of sands. Grain modification, crushing, compressibility and time rate of deformation are governed by morphology of sands of a given mineralogy.

As mentioned earlier elemental tests in the laboratory and penetration tests in the field form the basis of evaluation of design parameters. Given the difficulties and costs involved in obtaining high quality undisturbed samples of sand, extensive reliance is placed on in-situ tests for the determination of mechanical characteristics of sands. However, and major problem with the in-situ tests is that they do not measure any basic constitutive parameter. Following Jamiolkowsky et al., there are three approaches towards the use of results of results of penetration tests in design: (1) in-situ tests results are directly related to the response under imposed loads: Terzaghi and Peck charts for allowable bearing pressure, and Burland and Burbidge method for predicting settlement of foundations being typical examples of this approach; (2) formulation of a mathematical model, with appropriate assumptions regarding the boundary and drainage conditions and constitutive relationship, for the response of sand during the test-pressure meter and seismic tests have been well modeled, however in case of penetration tests the results are not very satisfactory; and (3) formulation of empirical correlation between the results of in-situ tests in large calibration chamber tests and the required soil characteristics determined from laboratory tests : deformation modulus vs. cone tip resistance being a typical example.

Empirical correlations approach, using results of penetration tests, is widely used in design but, as indicated earlier their applicability is limited to the conditions under which they were developed. As is well known different aspects of behavior of sands like penetration resistance and deformation modulus are functions of grain characteristics and state variables. Any empirical correlation between two variables is only a reflection of the fact that these variables are comparable functions of the same set of factors. If the effect of grain characteristics and state variables on penetration resistance is comparable in magnitude to that on deformation modulus, then an empirical correlation between these two may be applicable to all sands. But if the effect of any factor on one variable is significant and on the other is negligible, it is clear that an empirical correlation in this case is restricted only to the sands for which it was developed. There are indications that the outcome of morphology and ageing on penetration resistance is not the same as that on other aspects of sand behavior. Hence consideration of these phenomena is of great relevance to the interpretation of penetration tests. Furthermore, correlations developed on the basis of laboratory elemental tests and calibration chamber tests on freshly deposited specimens will lead to considerable underestimation of the stiffness, shearing resistance and liquefaction potential of aged in-situ sands and such correlations cannot be used directly to predict in-situ behavior.

The preceding brief discussion of the factors governing the response of sands and the current approaches to realistic evaluation of their in-situ characteristics brings out the importance of the effects of morphology and ageing. Furthermore, design criteria are expected to become more stringent in the future and hence there is an urgent need to evolve reliable rational methods, incorporating the influence of morphology and ageing, for the prediction of in-situ behavior of sands.

In this paper some quantitative trends correlating mechanical characteristics and grain morphology/ageing are discussed. A brief and highly illustrative presentation will highlight:

3. Approaches to quantify grain morphology and microfabric
4. Morphology and engineering behavior of sands:
   - resistance to penetration
   - Shearing resistance
   - 1-D deformation under monotonic loading and cyclic loading for normally consolidated and over-consolidated states.
5. Ageing and engineering behavior of sands—some aspects only:
   - Effect of secondary compression, cyclic pre-straining and over-consolidation
   - Strain-time response of normally consolidated, cyclically pre-strained and over-consolidated specimens
   - Time-settlement response of buildings
   - Shear response of aged sands
   - Ageing, interparticle contacts and mechanical characteristics of aged sands

3. APPROACH TO QUANTIFY GRAIN MORPHOLOGY AND MICROFABRIC

In this section some of the approaches and techniques currently in use for quantification of morphology and microfabric are briefly reviewed.

3.1 Quantification of Grain Morphology

Characterization of morphology comprises two principal tasks: formulation of quantitative indices of measures and development of reliable and economic procedures for their measurement. Sedimentary petrologists have developed many concepts and methodologies for characterization of the morphology of sedimentary particles. The sedimentologists have highlighted three aspects of morphology: Sphericity, ψ—a measure of degree to which the shape of a particle approaches that of a sphere, roundness, R—a measure of the sharpness of the corners and edges of the grains, and surface texture which describes dullness or polish of the grain surface and various surface markings in the form of striations and scratches, percussion marks, surface indentations, pits etc.: these features being on a very small scale. Sphericity ψ and shape factor, *SF* are normally expressed in terms of three dimensions of the particle: longest, d_L, intermediate, d_I and shortest, d_g as

$$\psi = \sqrt[3]{\frac{d_S \cdot d_I}{d_L^2}} \qquad (1)$$

and

$$SF = \sqrt{\frac{d_S}{d_L d_I}} \qquad (2)$$

While d_L and d_I are measured on the projected image of the particle, d_s is measured by focussing the microscope alternatively on the glass plate and top of the grain-resting on its maximum projected area (using the vernier readings of the microscope) and measuring the differential deviation.

Because of the difficulties associated with the determination of roundness, R, based on measurements on projected images of the grains, the common procedure is to estimate roundness from a visual comparison of the images under microscope with standard images of grains having known values of *R*. Powers prepared such a set of images and proposed roundness criteria and values as indicated in Table 1. Rahim presents details of the procedure adopted for calculating average value of *R* for a given sand. The surface texture can be studied using optical, electron and scanning electron microscopes though it is usually described qualitatively.

Table 1. Powers roundness criteria and values

Roundness class	Description	Roundness interval	Mean roundness
Very angular	Particles with unworn fractured surfaces and multiple sharp corners and edges	0.12-0.17	0.14
Angular	Particles with sharp corners and approximately prismoidal or tetrahedral shapes	0.17-0.25	0.21
Subangular	Particles with distinct but blunted or slightly rounded corners and edges	0.25-0.35	0.30
Rounded	Irregularly shaped rounded particles with no distinct corners or edges	0.49-0.70	0.59
Well rounded	Smooth nearly spherical or ellipsoidal particles	0.70-1.00	0.84

Powers method to estimate R value is quite subjective and Rahim developed a procedure for particle shape analysis in terms form, angularity (roundness), sphericity, and shape factor using image analyser, Fig. 1, commonly employed by metallurgists for the study of surface characteristics of powers. Fig. 2 and Fig. 3 show typical distribution of grain irregularities for sands with different morphology. Results of image analysis in terms of tangent count are

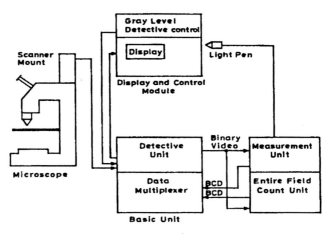

Fig. 1. Block diagram of the Omnicon Alpha image analyser with microscope

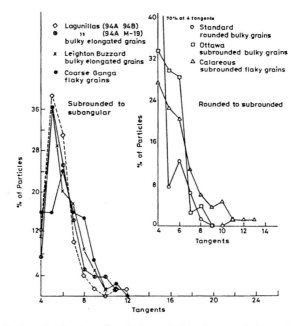

Fig. 2. Typical distribution of grain irregularities for rounded to subrounded and subrounded to subangular sands

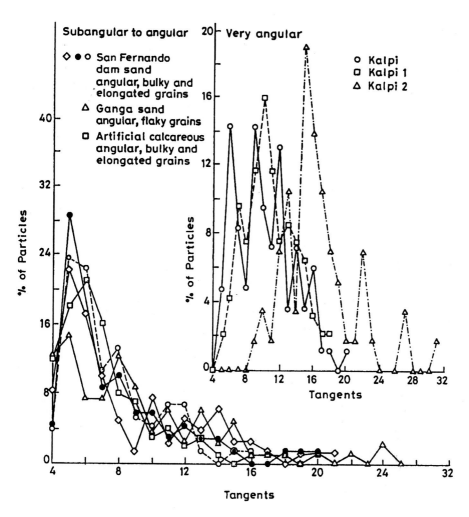

Fig. 3. Typical distribution of grain irregularities for subangular to angular and angular to very angular sands

compared with Powers roundness index R in Fig. 4; inset in Fig. 4 depicts relationship between particle shape factor and sphericity. Figure 5 depicts particle shape classification using Zingg diagram. Rahim characterized some sands using image analyser and the results are depicted in Table 2 and Table 3.

Based on critical review of earlier work by Youd, De Jaeger and others concerning the factors influencing limiting void ratios, e_{max} and e_{min}, Thomas followed De Jagear's suggestion to use e_{max} as an index of particle shape. De Jaeger argued that e_{max} is function of grain shape and coefficient of uniformity C_u, and since C_u can be easily evaluated e_{max} may be used as a shape index. Fig. 6 depicts relationship between e_{max} and R for sands with comparable C_u and highlights the influence of particle size on the value of e_{max}. A consistent method to evaluate e_{max} easily will be discussed in a subsequent section.

Results presented in this section suggest that morphology of sand grains can be readily quantified and expressed in terms of indices which may be used to correlate sand morphology with its mechanical characteristics (as discussed in later sections). For a detailed discussion see Thomas.

Table 2. Morphological characteristics of some sands

Sand	Average shape factor	Average sphericity	Average roundness
Ganga	0.539	0.678	0.15-0.25
Kalpi	0.596	0.698	0.10-0.20
Standard	0.766	0.81	0.78-0.80

Table 3. Characterization of sands

Sand	Gradation		Volume Change potential		Morphology					Mineral**	Physical description
	C_u	d_{50} mm	V_{max}	V_{min}	ψ	SF	R	E	T		
Standard	1.39	0.475	1.81	1.49	0.81	0.77	0.79	1.27	4	Q=98-100 F=0-2	Bulky, sphero-idal, rounded to well rounded grains
Calcareous	1.78	0.43	2.05	1.57	0.67	0.55	0.55	1.49	6	Q=5 C=95	Flaky, oblate, angular grains composed of shells
Ganga	2.57	0.18	2.25	1.577	0.68	0.54	0.19	2.32	11	Q=60-65 F=20-25 M=8-10 C=2-3	Falky, oblate, angular grains
Kalpi*	4.81	1.0	1.91	1.48	0.70	0.60	0.14	2.42	12	Q=40 F=40 M=1-2 C=18	Bulky, elongated, very angular grains coated with carbonate

Kalpi is not a typical river sand like Ganga since its distance of transport form source is very short and the grains are coated with carbonate deposited by the percolating ground water. These special characteristics are responsible for its typical tangent count distribution.
** Q—Quartz, F—Feldspar, M—Mica, C—Carbonate (Calcite, Aragonite)

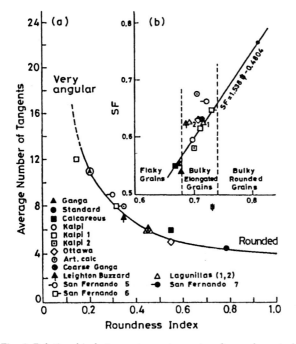

Fig. 4. Relationship between tangent count and roundness index R

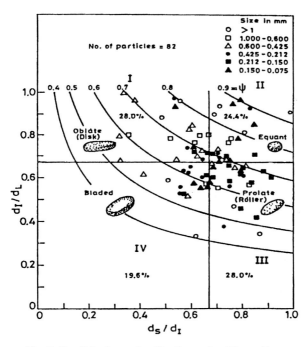

Fig. 5. Practicle shape classification using Zingg diagram

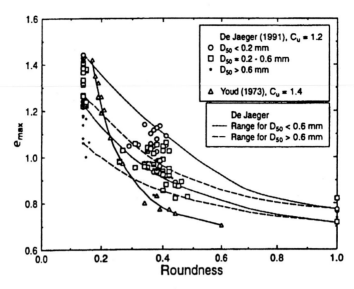

Fig. 6. e_{max} as particle morphology index

3.2 Quantification of Microfabric

The emissive and X-ray (Energy Dispersive X-ray, EDX) modes of Scanning Electron Microscope, SEM and Image Analysis, IA (to electronically process the SEM information) provides most powerful research tool to quantify microfabric indices in addition to investigations for mechanisms of ageing, and search for evidence of cementation of grain contacts in the form of crystalline or amorphous coatings. Using optical microscope and SEM images Dusseault and Morgenstern analysed the surface texture and microfabric of locked sands, and adopted a descriptive fabric classification in terms of particle contact relationship, as shown in Fig. 7 to evaluate intergranular fabric which was shown to control the behavior of these sands during shear in the laboratory and in steep slopes in the field. Dobereiner quantified microfabric of sands and weak sandstones in terms of grain contact area index, GC—defined as a ratio of summation of lengths of grain contacts and the total length (circumference) of the grain in SEM back scatter mode photographs (see Fig. 8). Palmer and Barton quantified microfabric of 8 uncemented sands of varying age in the U.K. in terms of proportion of different interparticle contacts as depicted in Fig. 9. The significance of change with time of nature of contacts vis-à-vis strength of these aged sands will be discussed in a later section on ageing and engineering behavior of sands. Yudhbir and Rahim have shown that the data in Fig. 9 can be interpreted in terms of grain contact index, I_c which is equal to 100 minus sum of percentages of planar-type contacts. They further showed that I_c and Dobereiner's grain contact area index GC are practically one and the same thing. More advanced IA models are now available, with statistical and graphic software packages, in which simple paper images and SEM micrographs can be used to easily quantify grain morphology and contact area indices by employing count and length/fragments measurement modes.

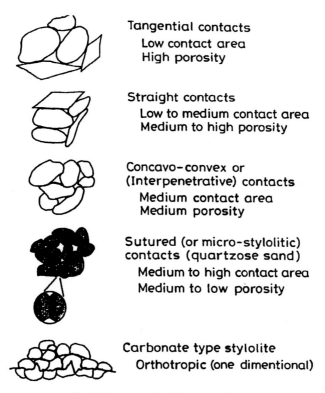

Tangential contacts
  Low contact area
  High porosity

Straight contacts
  Low to medium contact area
  Medium to high porosity

Concavo-convex or (Interpenetrative) contacts
  Medium contact area
  Medium porosity

Sutured (or micro-stylolitic) contacts (quartzose sand)
  Medium to high contact area
  Medium to low porosity

Carbonate type stylolite
  Orthotropic (one dimentional)

Fig. 7. Intergranular fabric classification

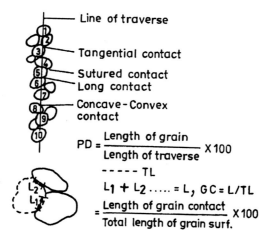

Line of traverse

Tangential contact

Sutured contact
Long contact

Concave-Convex contact

$$PD = \frac{\text{Length of grain}}{\text{Length of traverse}} \times 100$$

----- TL

$L_1 + L_2 \ldots = L, \quad GC = L/TL$

$$= \frac{\text{Length of grain contact}}{\text{Total length of grain surf.}} \times 100$$

Fig. 8. Quantification of intergranular fabric

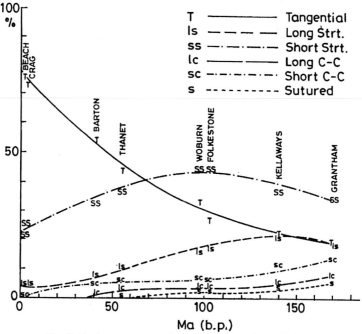

Fig. 9. Grain contact morphology vs chronological age

Oda and Oda et al. described microfabric in terms of coordination number, N, standard deviation of N and orientation of contact normals. Relationship between morphology and fabric can be examined at a fundamental level by investigating the influence of morphology on each of these elements characterising fabric of sands. Since a unique relationship between porosity n (or void ratio) and N has been indicated (see Grivas and Harr who show $n \times N$ = constant), void ratio can be used to describe fabric in place of N. Bhatia and Soliman have examined the effect of particle shape on the frequency distribution of the local void ratios, and have demonstrated that at a given I_D, rounded glass spheres had the least and the angular sand had the maximum standard deviation of the void ratio. Kuo and Frost have developed a method for quantifying the local void ratio distribution of granular materials and have presented data for a subrounded Ottawa sand. They observed that the mean of the local void ratios on horizontal and vertical planes did not show a significant variation as expected for subrounded grains. They observed pronounced deviations of void ratio on horizontal planes which could be one of the factors leading to anisotropy in fabric. It appears that these differences could be more pronounced in the case of angular sands.

4. MORPHOLOGY AND ENGINEERING BEHAVIOR OF SANDS

One of the main concerns of this presentation is to highlight the significant influence of morphology on engineering behavior of sands. We will briefly discuss influence of morphology on resistance to penetration, shearing resistance and 1-dimensional deformation under monotonic and cyclic loading for normally and over-consolidated states of sands.

4.1 Resistance to Penetration

Penetration testing in-situ and in calibration chambers in now widely recommended for characterization of mechanical characteristics of sands. Yudhbir and Thomas have highlighted the significance of grain morphology on the changes in state of sand under the trip during Static Cone Penetration Test, SCPT, in calibration chamber tests, and its implications on interpretation of these tests using state parameter approach. Shahu and Yudhbir have brought out the importance of morphology while investigating the effect of method of sample preparation at same I_D on the load-deformation response for model plate load tests.

Table 4. Mineralogical composition of some sands

Sand	Percentage of mineral content				
	Quartz	Feldspar	Mica	Carbonate	Chlorite
Ganga	60-65	20-25	8-10	2-3	2-3
Kalpi	40	40	12	18	0
Standard	100	0	0	0	0

Here results of fall cone test (see Fig. 10 for experimental setup) on sands shown in Fig. 11 and Tables 3 and 4 will be discussed. Figs. 12 and 13 show the effect of method of sample preparation on cone penetration. It will be seen that in case of angular sands method of sample preparation at same void ratio produces different microfabric which controls the penetration resistance offered by the sand. Vibrated specimens offer much stiffer resistance to penetration as compared to that by samples prepared by pluviation. Shahu and Yudhbir showed similar effects in case of plate load tests on samples prepared by vibration and pluviation for the same angular sand.

In case of rounded sands there is no significant difference in response to the cone penetration irrespective of the method of sample preparation.

As indicated earlier, e_{max} could be used as an index of grain morphology and therefore it is essential that the procedures involved in the determination of e_{max}, rather than being solely concerned with finding out absolute maximum value, should in the first place lead to a physically meaningful and consistent operational definition and also should consist of experimental procedures which are simple, repeatable and applicable to a variety of sands. Presently used procedures are not standardized and their repeatability is poor (see Thomas). Thomas has demonstrated use of fall cone test data to evaluate e_{max} as the value of void ratio corresponding to cone penetration $d = 50$ mm as shown in Figs. 14, 15, 16 for Ganga, Standard and Kalpi sands. This procedure satisfies the criteria stipulated above and because of the functional similarity between liquid limit for clays (same cone penetration of 20 mm and same shearing resistance for all clays) and e_{max} (same cone penetration of 50 mm and same resistance to penetration for all sands), the idea of defining e_{max} as value of e corresponding to cone penetration of 50 mm is aesthetically very appealing. The fact that slope of all the relationships between $1/d$ vs. e shown in Figs. 14, 15, 16 is practically same has significant implications in respect of change in penetration resistance offered by sands as a result of change in void ratio (which can be argued to be equivalent to change in the state parameter (ψ) at the extremely low operating stress levels in the fall cone tests). Thomas has argued along these lines in some detail.

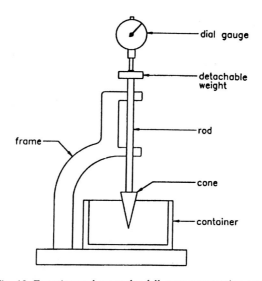

Fig. 10. Experimental setup for fall cone penetration test

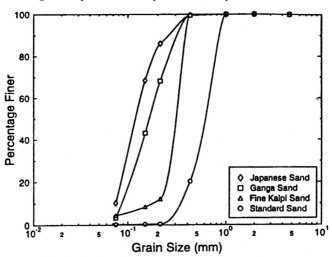

Fig. 11. Grain size distribution of sands investigated

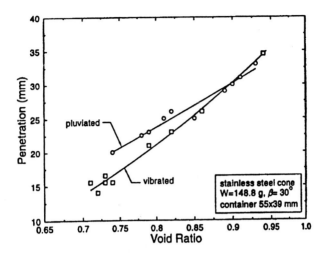

Fig. 12. Effect of method of sample preparation, angular Ganga sand

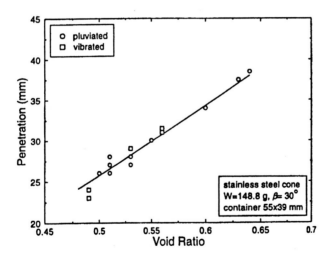

Fig. 13. Effect of method of sample preparation, rounded Standard sand

Evaluation of e_{max} based on fall cone test data thus provides an aesthetically appealing definition of e_{max} in addition to providing a simple, easy to use procedure which would ensure repeatability. A reliable estimate of e_{max} can then be used as an index of grain morphology.

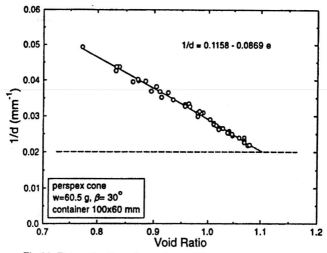

Fig.14. Determination of e_{max} from fall cone test, Ganga sand

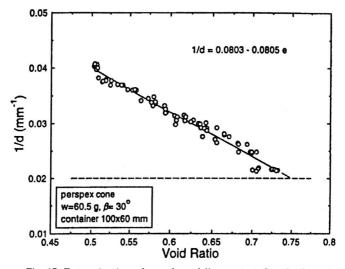

Fig. 15. Determination of e_{max} from fall cone test, Standard sand

The fall cone test data was also used to investigate the influence of morphology on bearing capacity factor N_γ versus angle of shearing resistance ϕ' relationship, by extending the dynamic analysis of the fall cone test in clays carried out by Hansbo and Houlsby to the case of sands. Fig. 17 shows this relationship for rounded standard sand and Fig. 18 depicts the similar relationship for angular Ganga sand (results of plate load test are also included). In case of rounded Standard sand the interpreted N_γ versus ϕ' relationship shows an excellent agreement with the theoretical prediction proposed by Meyerhof. For angular sands the agreement with Meyerhof's theoretical

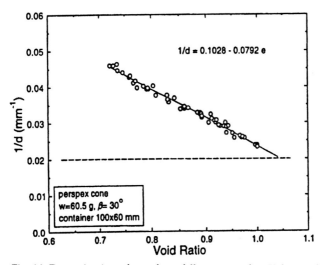

Fig. 16. Determination of e_{max} from fall cone test, fine Kalpi sand.

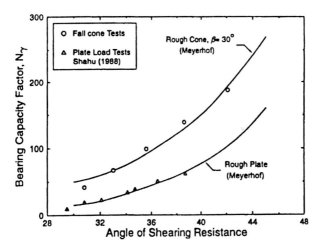

Fig. 17. N_γ vs angle of shearing resistance based on cone test and plate load test data, Standard sand.

prediction is not very satisfactory, particularly for very dense state and such a relationships may not be very accurate in case of angular sands at very low levels of confining stress.

Here again the importance of grain morphology on resistance of penetration in both fall cone and plate load tests is highlighted. The essential message is that results of dodel tests or predictive models disregarding the implications of form and angularity of sand grains cannot be extrapolated. This awareness highlights the need for caution while extrapolating correlations based on

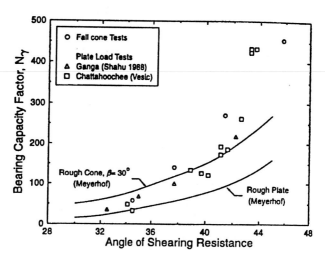

Fig. 18. N_γ vs angle of shearing resistance besed on cone test and plate load test data, Ganga sand

penetration test results, whether conducted in-situ or in calibration chambers, obtained for a particular type of sand (in terms of grain morphology, mineralogy and gradation), for all types of granular deposits. Furthermore, it is important to note that the implications of morphology on the resistance to penetration as revealed by the results of fall cone test at very low confining stress may have relevance to the bahavior of saturated sands at very low effective confining stress under conditions obtaining during liquefaction and flow failure triggered by earthquakes.

4.2 Shearing Resistance of Sands

4.2.1 Morphology and Undrained Shear Behavior

As the initial state of a sand which is represented by void ratio and effective confining stress determines to a large extent its behavior during undrained shear, to highlight the effects of particle morphology comparisons have to be made at the same state. In order to overcome some of the deficiencies inherent in the use of relative density and the state parameter ψ, Ishihara proposed a new state variable called the state index, I_s which may be defined as $I_s = (e_r - e)/(e_r - e_s)$. Here e is the initial value of the void ratio corresponding to a mean effective principal stress of $P' = (\sigma_1' + 2\sigma_3')/3$. e_r is a reference void ratio which is equal to e_0 for values of confining stress $P < P_{ms}$, where e_0 is a threshold void ratio differentiating conditions of zero and non-zero residual strengths and is determined as the void ratio on the steady state line (SSL) corresponding to $P = 0$. P'_{ms} is a critical stress obtained as the intersection of the $e = e_0$ line with the isotropic consolidation line (ICL) of the loosest possible sample obtained by moist placement. For $P' > P'_{ms}$ e_r is the void ratio on the ICL corresponding to the same value of P'. The other reference void ratio e_s is the void ratio at the steady state (quasi-steady state, QSS, used by Ishihara) at the same value of P'.

Here the effect of particle angularity on the undrained shear behavior of 3 sands having different morphological characteristics (see Table 5), will be reviewed (for details see Yudhbir and Thomas). To illustrate the effect of morphology, the ICL and SSL for subangular Toyoura sand (TS), subrounded Banding #6 sand (B_6) and angular Mine Tailings sand (MT), from tests performed (Castro et al. for B_6 and MT, Ishihara for TS) on specimens prepared by moist placement are shown in Fig. 19. It may be noted that the influence of grain morphology is reflected in the shapes of both ICL and SSL, with the change in void ratio for a given change in P' increasing as angularity increases. An examination of the extensive data for TS, MT and B_6 sands shows that, depending on the initial state of the sand in terms of e and P', four distinct modes of undrained shear behavior are possible. The influence of particle angularity on undrained shear behavior of saturated sands is most convincingly demonstrated by examining its influence on the range of initial states for which each type of behavior is indicated. For this purpose, using the

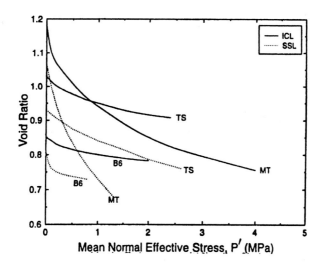

Fig. 19. ICL and SSL for Toyoura Sand (TS), Mine Tailings (MT) and Banding #6 Sand (B_6).

Table 5. Characteristics of sands studied

Sand	D_{50} mm	C_u	Fines, %	Grain shape	e_{max}	e_{min}
Toyoura	0.17	1.7	0	Subangular	0.977	0.597
Banding #6	0.157	1.7	0.2	Subrounded	0.82	0.52
Mine Tailings	0.256	2.5	7	Angular	1.08	0.68

data reported for 3 sands the I_s values were calculated (as indicated earlier, steady state instead of quasi-steady state was used for computing I_s). By comparing the initial states represented by I_s and P' with the behavior exhibited during undrained shear, the limiting values of I_s for each mode of behavior are determined.

The four types of behaviour — Type 1, Type 2, Type 3 and Type 4—are depicted in Figs. 20, 21, 22 for MT, TS and B_6 respectively. Each figure shows the limiting I_s values demarcating the range of initial states for which each type of behavior is exhibited. The salient features of each type of behavior in terms of stress-strain ($q = (\sigma_1 - \sigma_3)$ versus axial strain ε) and the effective stress path (q vs. P') are also illustrated. The ranges of initial states in terms of I_s and P' values for each type of behavior are summarized in Table 6.

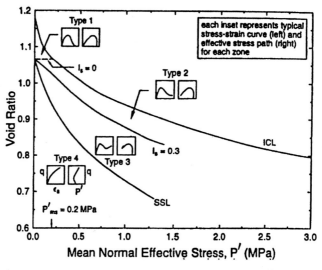

Fig. 20. Different types of undrained behavior and corresponding range of initial states of Mine Tailings

Table 6. Types of undrained shear behavior and corresponding initial states

Sand	e_0	P'_{ms}	Types of behavior and initial states				
			Type 1		Type 2	Type 3	Type 4
		MPa	I_s	P'	I_s	I_s	I_s
TS	0.93	1.5	≤ 0	$< P'_{ms}$	0-0.2	0.2-1.0	≥ 1.0
MT	1.065	0.2	≤ 0	$< P'_{ms}$	0-0.3	0.3-1.0	≥ 1.0
B_6	0.81	0.64	≤ 0	$< P'_{ms}$	0-0.55	0.55-1.0	≥ 1.0

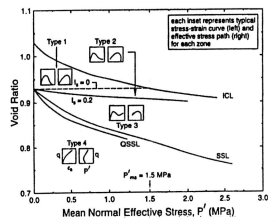

Fig. 21. Different types of undrained behavior and corresponding range of initial states of Toyoura Sand

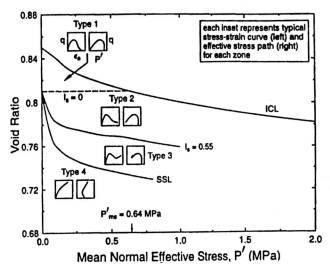

Fig. 22. Different types of undrained behavior and corresponding range of initial states of Banding #6 sand

The characteristic features of each type of behavior are:

Type 1. $(\sigma_1-\sigma_3)_{max}$ mobilized at small strain; zero residual undrained shear strength; mobilized effective angle of shearing resistance less than the steady state friction angle—characterizes meta-stable flow failure.

Type 2. After attaining $(\sigma_1-\sigma_3)_{max}$ shearing resistance decreases monotonically to reach steady state at large strains. Finite residual undrained shear strength.

Type 3. $(\sigma_1-\sigma_3)$ after attaining the peak decreases to a minimum and then increases again and approaches steady state at large strains—characterizes quasi-steady state (QSS) behavior.

Type 4. Dilatant response throughout shearing, except possibly at small strains
where behavior may be contractive.

From the foregoing discussion it will be noted that the range of I_s values for
which Type 1 and 2 modes are indicated, is significantly influenced by grain
morphology. The value of e_0 which together with the location of the ICL
determines the magnitude of critical stress P'_{ms} separating Type 1 and 2 modes,
increases as angularity increases (see Table 6). The range of initial e and P'
values for which Type 1 behavior is indicated is least for the angular sands,
whereas it is more for the subangular and subrounded sands. This is consistent
with the observation of Terzaghi and Peck that "the most unstable sands so far
encountered consist chiefly of rounded grains". It may also be seen that in the
case of angular and subangular sands, Type 2 behavior is indicated for I_s between
0 and (0.2 – 0.3), whereas subrounded sand exhibits the same behavior over a
wide range of I_s values (0 – 0.55) .

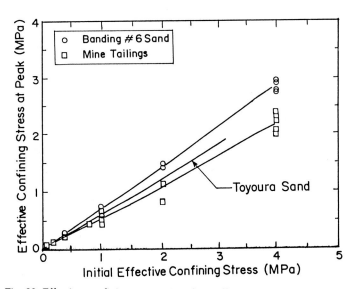

Fig. 23. Effective confining stress at peak vs effective confining stress
for B_6, TS and MT sands

Effect of grain morphology on some undrained shear parameters considered
by Ishihara is also illustrated in Figs. 23, 24, 25 and 26 in terms of initial confining
stress (P'_c) vs effective confining stress at peak (P'_p), initial stress ratio, r_c defined as
P'_c/P'_s, where P'_s is the value of P' at steady state, vs. I_s, peak undrained shear
strength $S_p = (\sigma_1 - \sigma_3)_{max}/2$, normalized with respect to initial confining stress σ'_0 vs
I_s and residual undrained shear strength $S_{us} = (q_{ss}/2) \cos \times \phi'_{ss}$, where $q_{ss} = (\sigma'_1 -
\sigma_3)$ at steady state and ϕ'_{ss} is ϕ' at steady state, normalized with respect to σ'_0 vs I_s
respectively. Fig. 23 shows that the slope of P'_p vs P'_c increases with angularity
from 0.54 for B_6 to 0.72 for MT; the value for TS being 0.61. These relationships are
applicable for Type 2 and 3 behavior described earlier. For TS in the case of Type 1

behavior, for which the results were kindly made available by Professor Ishihara, the slope increases to 0.74 from 0.61 for Type 2 and 3 behavior.

Figure 24 illustrates the extreme sensitivity of r_c to I_s. It may be noted that for a given value of I_s the subrounded sand shows the maximum value of r_c, which is indicative of very high pore water pressures at steady state and hence a high degree of instability. The angular sand exhibits the least value of r_c, for the same value of I_s, suggesting a lower degree of instability compared to subrounded and subangular sands.

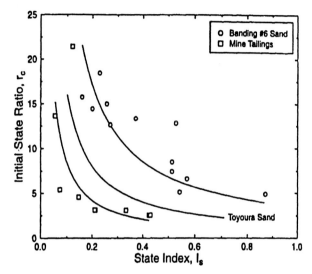

Fig. 24. Initial State Ratio vs State Index

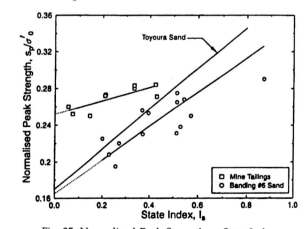

Fig. 25. Normalized Peak Strength vs State Index

The relationships of S_p/σ'_0 and S_{us}/σ'_0 with I_s in Figs. 25 and 26 respectively are linear and valid for Type 2 and 3 behavior only. The trends shown highlight the influence of angularity. The values of S_{us}/σ'_0 at $I_s = 0$ being 0.076 for angular (MT), 0.015 for subangualr (TS) and 0.0008 for subrounded (B_6) sands.

4.2.2 Morphology and Angle of Shearing Resistance

Prediction of the drained angle of shearing resistance (f') is an important pre-requisite for the satisfactory solution of design problems involving sands, where considerations of stability are important. Two approaches towards the prediction of ϕ' involving two state variables—the relative dilatancy index I_R, (Bolton) and the state parameter, ψ (Been and Jefferies) are currently available. While the relative dilatancy index does not consider various characteristics of sands including particle morphology, the state parameter approach reflects these. Here an approach similar to the state parameter approach is considered for the prediction of ϕ' and its components—constant volume friction angle, $\phi\square'_{cv}$, and a dilatancy component which is an emperical function of the state parameter, ψ (for details, see Thomas).

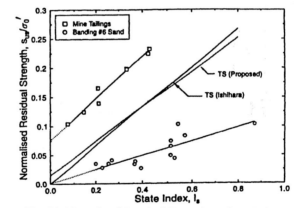

Fig. 26. Normalized Residual Strength vs State Index

Earlier work of Professor A. Casagrande for round and subangular sands, and Professor A. Vesic for angular sand recommends that variation of ϕ' with void ratio, within a sufficiently narrow range of confining stress, is such that their product, $e \tan \phi'$, may be taken as constant. Fig. 27, based on review of available data, shows variation of $e \tan \phi'$ with I_D for a variety of sands and it will be seen that $e \tan \phi'$ remains reasonably constant over a wide range of relative density. The value of this constant would in general be a function of the confining stress and the nature of the sand, especially its morphology. If it is assumed that for a sand with an initial void ratio which equal to the value of void ratio at steady state (e_{ss}) corresponding to the existing confining stress, ϕ' is equal to the steady state angle of shearing resistance (ϕ'_{ss}), then the value of the above constant can be expressed as $e_{ss} \tan \phi'_{ss}$. At extremely low levels of confining stress $e \tan \phi'$ may be taken equal to $e_{ss} \tan \phi'_{cv}$. Based on review of available data it is proposed that e_0 for most sands may be assumed approximately equal to void ratio corresponding to $I_D = 0.1$. The value of $e_0 \tan \times \phi'_{cv}$ was computed for poorly graded and predominantly quartz sands and is depicted in Fig. 28 as a function of roundness index R for the same value of $P = 0$. This clearly illustrates the effect of morphology on the constant $e \tan \phi'$ for a given stress level. Now for a sand at any value of initial void ratio e and initial confining stress, $e \tan \phi' = e_{ss} \tan \phi'_{ss} = e_{cr} \tan \phi'_{cv}$ (since steady state and critical state are considered practically identical, e_{cr} is void ratio at critical state for a given stress level).

Thus, it may be proposed that for $e \leq e_{cr}$, the angle of shearing resistance ϕ' may be estimated by using the following equation:

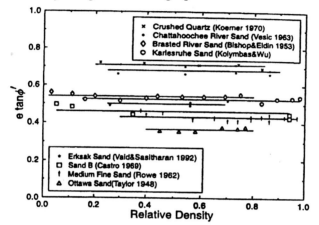

Fig. 27. $e \tan \phi'$ vs relative density I_D for different sands

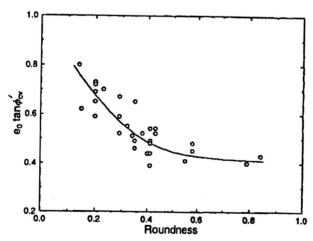

Fig. 28. $e_0 \tan \phi'_{cv}$ vs roundness index R for poorly graded, predominantly quartz sands

$$\phi' = \tan^{-1}\left(\frac{e_{ss}\tan\phi'_{ss}}{e}\right) = \tan^{-1}\left(\frac{e_{cr}\tan\phi'_{cv}}{e}\right) \tag{3}$$

For $e \geq e_{cr}$, it may be assumed that $\phi' = \phi'_{cv}$.

Figure 29 shows the effect of confining stress on ϕ' for MT, B_6 and TS at $I_D=$ 1.0 as predicted by equation (3). The angular MT exhibits maximum stress level dependency of ϕ' and subrounded B_6 shows the least.

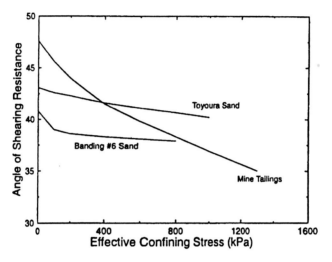

Fig. 29. Effect of confining stress on ϕ' for sands of different angularity and having $I_D=1.0$, as predicted by Equation #(3).

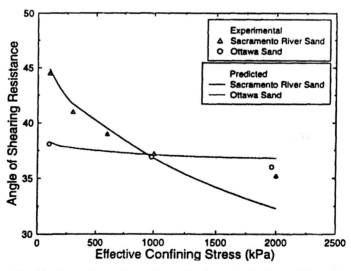

Fig. 30. Comparison of experimentally determined values of ϕ' with those predicted using Eq. # (3)

Figure 30 depicts comparison of experimentally determined values of ϕ' with those predicted by using equation (3) for Sacramento River and Ottawa sands at $I_d = 1.0$ (Lee and Seed). For rounded Ottawa sand the agreement is excellent whereas in case of Sacramento River sand at stress levels greater than 1000 kPa, equation (3) underpredicts ϕ'. This discrepancy is due to crushing of angular grains at stress levels greater than 1000 kPa. For details see Thomas.

As indicated earlier ϕ' comprises ϕ'_{cv} and dilatancy component. In order to be able to predict dilatancy component at different void ratios, we need an estimate of ϕ'_{cv}. Fig. 31 shows variation of ϕ'_{cv} with roundness index R for predominantly quartz sands. It will be seen that well graded sands, gravels and boulders have a larger ϕ'_{cv} in comparison to poorly graded sands. The effect of particle shape and size on the angle of repose (approximately equal to ϕ'_{cv}) is depicted in Fig. 32 (for details, see Stephenson, Cornforth, Thomas). Given ϕ' (equation (3)) and ϕ'_{cv} (Fig. 31) the dilatancy component $(\phi' - \phi'_{cv})$ can be estimated. Fig. 33 compares predicted values of $(\phi' - \phi'_{cv})$ with those from the relative density index model proposed by Bolton[5] (see Thomas for details).

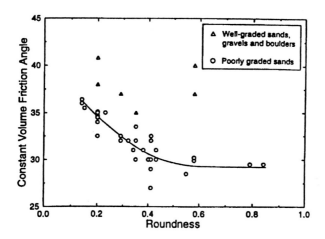

Fig. 31. ϕ'_{cv} vs roundness index R for predominantly quartz sands

4.3 1-D Deformation under Monotonic Loading and Cyclic Loading for Normally and Over-consolidated States

4.3.1 Morphology and Constrained Modulus

A quantitative evaluation of the effect of particle angularity on compressibility of sands is discussed in detail by Rahim. Most conveniently it can be evaluated in terms of certain parameters employed to model one-dimensional stress-strain behavior of sands as proposed by Janbu and Bellotti et al. Following Janbu, constrained modulus M for normally consolidated sand may be expressed as

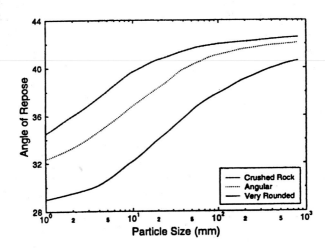

Fig. 32. Effect of particle size and angularity on angle of repose of rock fills

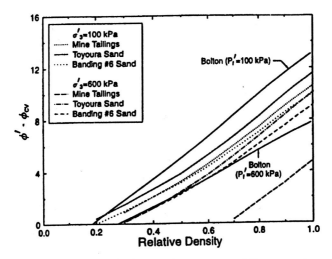

Fig. 33. Prediction of dilatational component of ϕ'—comparison of Equation #(3) with Bolton's method.

$$M = mP_a \left(\frac{\sigma'_v}{P'_a} \right)^{m_1}$$

(4)

where P_a is atmospheric pressure;

m is modulus number;

m_1 indicates rate of change of M with vertical effective stress σ'_v, usually taken as 0.5.

Based on the results of calibration chamber tests on two sands, Bellotti et al. suggest a more general relationship:

$$M = m_0 \left(\frac{\sigma_v^i}{P_a'} \right)^{m_1} \exp(m_2 I_D) P_a' \tag{5}$$

Effect of sand morphology on m (equation # 4) is brought out in Fig. 34 and the variation of m_0, m_1 and m_2 (equation # 5) with roundness index, R, is illustrated in Fig. 35. It will be seen (Fig. 35) that m_0 is very sensitive to grain morphology and m_1 is essentially independent of it. Coefficient m_2 suggests variation between 1 to 2 though more data, especially from 1-D tests using bender elements, is needed to clarify the picture.

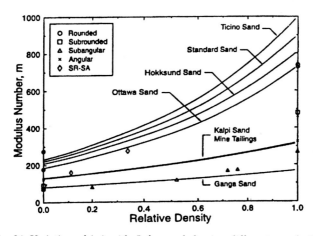

Fig. 34. Variation of (m) with I_D for sands having different angularity

4.3.2 Morphology and Secondary Compression

Though the magnitude of secondary compression (time dependent deformation under constant vertical effective stress) of sands is considerably less than that for clays, this effect is very significant for accurate prediction of settlement of foundations on sands (see Schmertmann and Burland and Burbidge). Mesri et al. and Schmertmann consider secondary compression as the most important mechanism of ageing of sands and therefore for a better understanding of the againg phenomena and development of methods for prediction of resulting improvements in behavior, a thorough understanding of the nature of secondary compression of sands is essential. Most of the results presented here are at

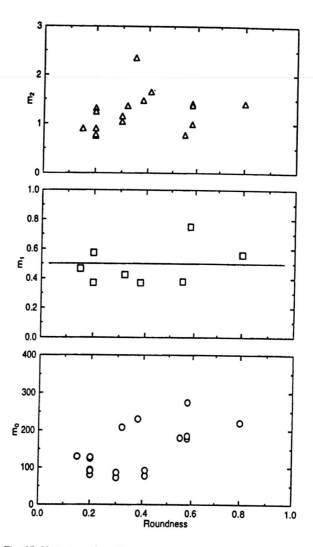

Fig. 35. Variation of coefficients m_0, m_1, m_2 with roundness index R

relatively elevated vertical effective stress range—1100 to 1600 kPa—primarily to ensure significant deformation measurements (given the fact that at low stress the side friction may affect very small deformation values (in conventional 1-D tests) for dry sands.

Figures 36, 37 and 38 depict time dependent increase in vertical strain for the materials tested and Fig. 39 compares these results. Rounded standard sand shows much less strain increment compared to angular Ganga sand for comparable I_D value. Steel balls show the least and very angular crushed rock (undergoing particle crushing during the test) shows the maximum strain

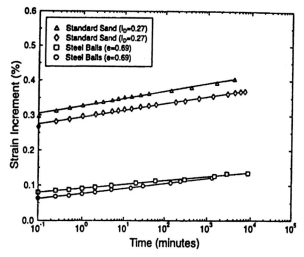

Fig. 36. Time dependent increase in vertical strain, stress increment 1100-1600 kPa

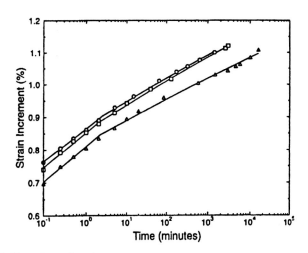

Fig. 37. Time dependent increase in vertical strain—Ganga sand, $I_D = 0.25$,
stress increment 1100-1600 kPa

increment. These results are presented in C_α vs. C_c format in Fig. 40 and are compared with Mesri's range of values for C_α / C_c for sands in general. While C_α and C_c clearly show increase with increasing grain angularity, the basic effort in this study (see Thomas) was to see the effect of grain morphology on C_α / C_c. The data in Fig. 40 does not clearly answer this enquiry though it may be argued that effect of side friction will be least for steel walls and rounded sand, and maximum for very angular rough grains. Adjustments for this error (if measured

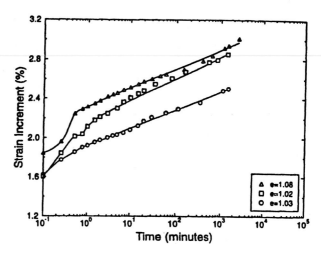

Fig. 38. Time dependent increase in vertical strain. Crushed rock,
stress increment 1100-1600 kPa

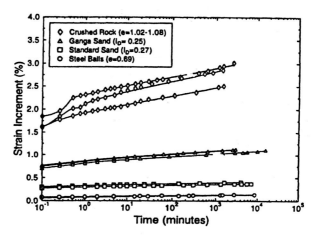

Fig. 39. Comparison of time dependent increase in vertical strain for materials
with different angularity, stress increment 1100-1600 kPa

accurately) would tend to push Ganga sand data on the $C_\alpha / C_c = 0.015$ line and
crushed rock data to 0.02 lines leaving steel balls and standard sand between
0.01 and 0.015 lines. It is felt that careful 1-D tests with bender elements and
minimization of side friction effects would help clarify the picture. The authors
feel that wide range of $C_\alpha / C_c = 0.01$ to 0.02 needs to be further explained before
values of this ratio can be assigned to natural sands as suggested by Terzaghi et
al.

4.3.3 Morphology and Behavior in 1-D Cyclic Loading

It is well known that over-consolidation and secondary compression result in significant improvements in the stress-deformation behavior of sands, due to the effects of strain hardening and increase in horizontal effective stress. Cyclic straining which in nature can be induced by seismic activity and wave loading, could also lead to similar changes. Sands with different morphology were subjected to repeated loading-unloading cycles over the stress ranges: $400 \leftrightarrow 100$ and $1600 \leftrightarrow 400$, in the oedometer. Results in Fig. 41 show strain increment

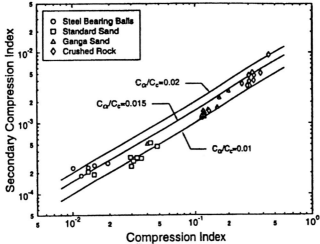

Fig. 40. C_α vs C_c for materials with different morphology

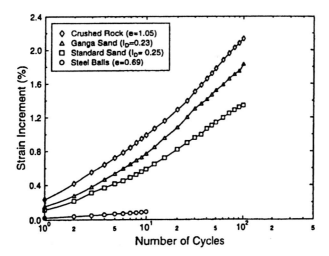

Fig. 41. Increase in vertical strain induced by cyclic variation of vertical stress between 400 and 1600 kPa, different materials

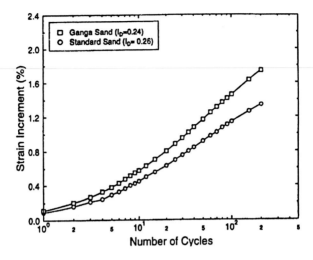

Fig. 42. Increase in vertical strain induced by cyclic variation of vertical stress between 400 and 1600 kPa, Ganga vs Standard Sand.

in different materials due to cyclic loading. While steel balls exhibit asymptotic response after 10 cycles, and standard sand requires 100 cycles to reach that state; Ganga and Crushed rock (angular sands) are still exhibiting increasing trend with number of cycles. Also for a given number of cycles the strain increment increases with increasing grain angularity. Fig. 42 compared response of rounded standard sand with angular Ganga sand for comparable initial relative density.

The nature of hysteresis loop in metals is usually attributed to crystal imperfections and in case of sands dissimilar morphological characteristics are expected to affect the nature of hysteresis loop. Figs. 43 and 44 depict results for rounded standard sand and angular Ganga sand. The hysteresis loop closes at the end of 100 cycles (though Ganga sand shows slight changes with additional cycles). Two differences are worth observing: firstly, the size of the final loop for rounded sand is considerably smaller than that of angular Ganga sand; and secondly the average slope (reciprocal of slope is a measure of an average elastic modulus) of the stabilized loop is steeper for angular Ganga sand (small average elastic modulus) compared to rounded standard sand. A higher value of modulus for rounded sand is consistent with the expected trend. These findings suggest that, under comparable conditions, strain increment vs number of cycles response during 1-D cyclic loading and nature of stabilized hysteresis loop can be used as relatively simple indices of differences in grain morphology of natural sands.

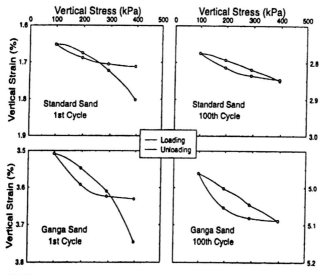

Fig. 43. Hysteresis loops for cyclic loading between 100 and 400 kPa for Standard and Ganga Sands with same I_D

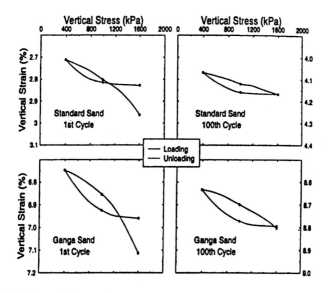

Fig. 44. Hysteresis loops for cyclic loading between 400 and 1600 kPa for Standard and Ganga Sands with same I_D

5. AGEING AND ENGINEERING BEHAVIOR OF SANDS—SOME ASPECTS ONLY

As indicated earlier, secondary compression following deposition is considered to be an important mechanism of ageing of sands in nature. As a result of over-consolidation resulting from removal of overburden by erosion, and cyclic pre-straining due to seismic loading during their geologic history, sand deposits undergo changes which are analogous to those reproduced by secondary compression. These three mechanisms of ageing are briefly reviewed in this section.

5.1 Effect of Secondary Compression, Cyclic Pre-straining and Over-consolidation on Stress-Strain Response

Results of oedometer tests under different mode of deformation for Ganga sand samples at initial $I_D = 0.25$ are depicted in Fig. 45. For a better comparison of the increase in stiffness, compared to normal compression, resulting from different mechanisms of ageing, data in Fig. 45 is replotted in Fig. 46 after subtracting the additional strain induced by each mechanism.

In general all three mechanisms produce increased stiffness up to a certain stress level after which the behavior of all samples becomes practically identical to that of the normally compressed specimens. It may also be noted that the response after secondary compression and cyclic pre-straining is almost identical, whereas over-consolidated specimens exhibit a stiffer response. Here for each sequence of loading a correction for the apparatus compliance was

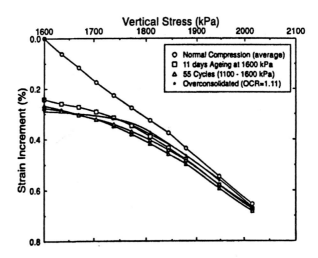

Fig. 45. Effects of secondary compression, cyclic-prestraining and over-consolidation on the stress-strain response of Ganga Sand during subsequent loading

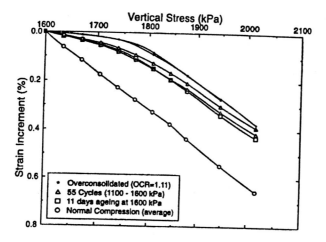

Fig. 46. Comparison of stress-strain curves of aged, cyclically pre-strained and over-consolidated specimens of Ganga Sand

applied using the appropriate calibration curve. The observed difference in behavior for over-consolidated samples is thus unlikely to be due to some error and this mechanism, unlike in case of clays, does not adequately simulate the ageing process due to secondary compression. It is suggested that cyclic pre-straining can be used to simulate ageing phenomenon in the laboratory. That this is so will be further reinforced in the following section by the strain-time response of sands subjected to all these three mechanisms producing additional strain at a given effective stress level.

5.2 Strain-Time Response of Normally Compressed, Cyclically Pre-strained and Over-consolidated Specimens

The effect of one day secondary compression at 1600 kPa, on the strain-time response on subsequent loading is illustrated in Fig. 47. Here the increase in vertical strain due to each vertical stress increment of 35 kPa is plotted against time. The shaded portion represents the response for normally compressed specimens for all increments of 35 kPa between 1600-1880 kPa, for which readings were available upto 5 minutes. Data for one normal compression test, where for the increment of 1845-1880 kPa strains were recorded for one day, is also shown in Fig. 47. For the specimens which have experienced one day of secondary compression at 1600 kPa, the strain-time curves for the first increments are relatively flat almost up to 100-200 minutes. As the stress level increases the curves steepen and approach that of the normally compressed ones.

The observed response is in accordance with Mesri's postulate that for a given sand C_α / C_s is constant. When the aged specimens are loaded beyond 1600 kPa, initially they exhibit a stiff response and as the stress level increases

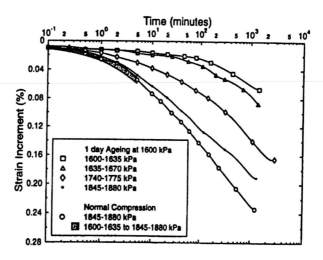

Fig. 47. Strain-time response of normally compressed and aged specimens
of Ganga sand for different stress increments

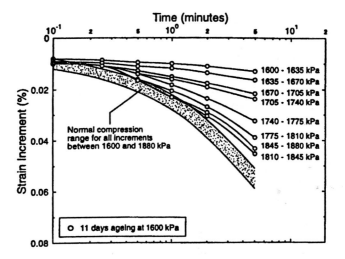

Fig. 48. Strain-time response of aged specimens of Ganga sand for
different stress increments

further, the behavior tends to that typical of normally compressed samples (see Figs. 45, 46). The increase in compressibility is being reflected in the increasing slopes of strain-time curves. This type of testing constitutes a simple and elegant way of demonstrating the validity of the concept of C_α / C_c being a constant for a given sand.

The effect of secondary compression, cyclic pre-straining and over-consolidation (with each mechanism resulting in approximately the same amount of additional strain at 1600 kPa) is illustrated in Figs. 48, 49, 50 respectively. The response of aged and cyclically pre-strained samples is practically identical (Figs. 48, 49) whereas in case of over-consolidated samples even though the broad pattern of behavior is similar in many ways to the aged and cyclically pre-strained specimens, for the first few stress increments the curves are flatter.

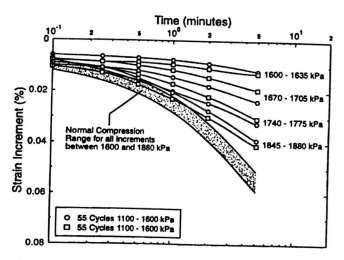

Fig. 49. Strain-time response of cyclically pre-strained specimens of Ganga sand for different stress increments

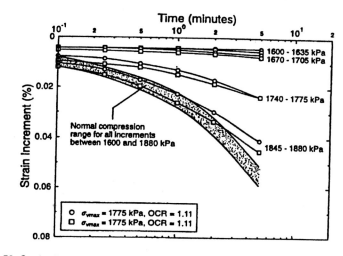

Fig. 50. Strain-time response of over-consolidated specimens of Ganga sand for different stress increments

Figure 51 shows compression response of Ganga sand specimens for different stress increments during loading following secondary compression, cyclic pre-straining and over-consolidation mechanisms. It will be seen that the effects of secondary compression and cyclic pre-straining on subsequent strain-time response are found to be practically identical at all stages. But until a stress level of 1705 kPa, the increase in strain with is suppressed to a much larger extent in the case of over-consolidated samples. Thus the similarity between strain-time response after secondary compression and cyclic pre-straining reinforces the observation made earlier that the consequences of these phenomena are practically identical with respect to stress-strain behavior. On the other hand over-consolidated specimens are again found to exhibit a stiffer response.

To recap, in this series of tests, after the samples were loaded to 1600 kPa, equal amounts of additional strains were induced by three mechanisms: Secondary compression, cyclic pre-straining and over-consolidation. On further loading the aged and cyclically pre-strained specimens were found to exhibit practically identical stress-strain and strain-time response, whereas the behavior of over-consolidated samples was significantly different. Thus even though the strain induced by the three mechanisms is the same, the consequences of over-consolidation are different form the other two. In the case of aged and cyclically pre-strained specimens yielding begins at a lower stress in comparison to the over-consolidated ones, as indicted by stress-strain and time-strain curves. Apparently the expansion of the yield locus caused by seconday compression and cyclic pre-straining is of a lower magnitude than that due to over-consolidation.

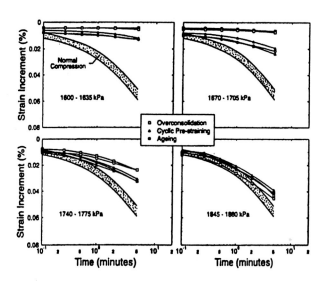

Fig. 51. Comparison of the effects of secondary compression, cyclically pre-straining and over-consolidation on the strain-time response of Ganga sand for different stress increments

The observed similarity between one-dimensional stress-strain and time-strain response of aged and cyclically pre-strained samples suggests that cyclic pre-straining may be successfully employed to simulate the effects of ageing in the laboratory. This is a very significant result since effects of much longer time of sustained loading can be easily investigated. It is important to note that the stress range in which cyclic loading is carried out, needs to be kept sufficiently small. It was found that when stress was cycled over a larger range (400-1600 kPa), the samples exhibited a much softer response. It is clear that the smaller this range, the closer would be the similarity between secondary compression and cyclic pre-straining.

5.3 Time-settlement of Building Foundations on Sand

These laboratory trends of time dependent deformation are well documented in the field through long term monitoring of settlement of foundations on sands. Figs. 52 and 53 reproduce the data discussed by Burland and Burbidge for building and chimney foundations on sands in Warsaw, Poland. These data show time dependent deformations over much longer period of time starting after a certain reference time which Burland Burbidge suggest three years. It will also be noted that in case of chimney foundation, where cyclic loading-unloading due to wind loads is also superposed from time to time, the ageing process is accelerated. It will also be seen form Fig. 53 that deformation due to secondary compression and that resulting from periodic wind loading follow the same time-deformation relationship. This further supports the suggestion made in the above discussion that cyclic pre-straining can be used to simulate ageing phenomenon in the laboratory.

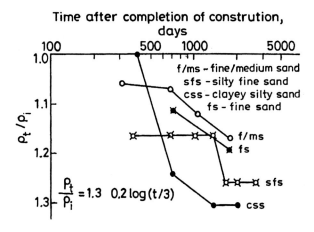

Fig. 52. Time-settlement characteristics of four buildings in Warsaw

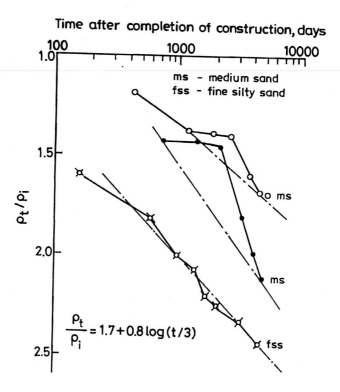

Fig. 53. Time-settlement characteristics of three chimneys in Warsaw

5.4 Shear Response of Aged Sands

To further explore the consequences of secondary compression and cyclic pre-straining, direct shear tests were carried out on loose dry samples ($I_D = 0.24$–0.27) of angular Ganga sand under a constant normal vertical stress of 800 kPa. In the first test the sample was sheared 5 minutes after the application of vertical stress (normally compressed sample). Two tests were performed on samples aged for 2 hours and 11 days before shearing and a cyclically pre-strained sample was also sheared (50 cycles of loading-unloading between 800-400 kPa). Because of some difficulties in measurement of equipment compliance, the exact magnitude of vertical strain induced by secondary compression and cyclic pre-straining could not be determined. However, it is thought that strain induced by 50 cycles is more than that resulting from 11-day ageing due to secondary compression.

Figure 54 presents the results of shear test. Some observations can be made. First the maximum value of τ/σ is found to be same in all the tests, indicting that the angle of shearing resistance (drained value) was not affected by secondary compression or cyclic pre-straining. This is consistent with other findings for uncemented sands. Daramola reports similar results for samples age by secondary compression (for 0, 10, 30 and 152 days) and sheared in the triaxial. Teachavorasinskon et al. report that cyclic pre-straining does not influence angle of shearing resistance, ϕ'.

Test results on undisturbed samples of natural aged sands show varying

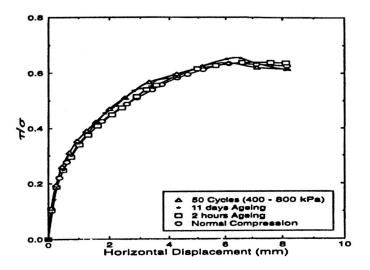

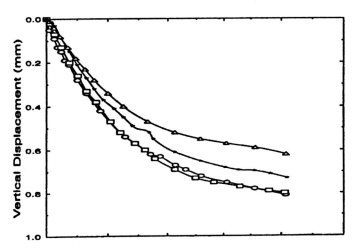

Fig. 54. Results of direct shear tests on normally compressed, aged and cyclically pre-strained specimens of Ganga sand having same initial I_D

increase in f which is primarily due to the enhanced magnitude of dilatational component of shearing resistance. Dusseault and Morgenstern report large values of ϕ' (55°-65°) for Canadian locked sands which are known to be cemented and had a complex stress history due to movement of very large glaciers over those sand deposits. However, in case of uncemented aged British sands undisturbed samples (Palmer and Barton) tested at normal stresses ranging form 50-900 kPa, showed increase in shearing resistance with increasing age.

Shearing resistance interpreted as c' and ϕ by assuming linear failure envelopes shows that ϕ_1 increases with grain contact index, I_c upto $I_c \approx 0.275$ after which the ϕ' remains constant even though I_c value increases upto 0.46 for samples as old as 170 million years before present. The increase in strength with ageing beyond $I_c = 0.275$ is primarily due to c' in the linear interpertation of failure envelope. Incidentally for the British aged sands relationship between c' and I_c, may be expressed as:

$$\ln c' = 12.5 \, I_c + 1 \qquad (6)$$

the implication of equation # 6 that for a theoretical value of $I_c = 0$ (100% tangential contacts c' is 2.71 kPa which is indeed a very samll value. In reality at very low normal stresses there is evidence to suggest that failure envelopes will be curvilinear but owing to the usual difficulty of obtaining reliable results in this stress range, testing is usually avoided at very low stress levels.

So in laboratory shear tests with very limited ageing periods, the nature of grain contacts is such that I_c or GC are so low (< 0.04) that c' is zero and ϕ does not change as compared to normally compressed samples (deviations may be within the experimental measurement error). However, in the field, ageing increases I_c which leads to more effective interlocking in case of angular sands and on shearing the dilatational component will contribute to the increased value of angle of shearing resistance.

The second significant observation is regarding the volume change during shearing of aged samples. Fig. 54 shows that sample aged for 11 days and the one cyclically pre-strained show reduction in contractancy during shear as compared to normally compressed samples. Daramola observed similar results for drained triaxial tests on aged samples. Teachavorasinskun et al. also observed that cyclic pre-straining reduces volume contraction during shearing.

Testing of undisturbed samples of locked Canadian sands (Dusseault and Morgensterro), because of their very high relative density ($I_D > 1.0$), cementation and highly developed interparticle fabric, show phenomenal increase in positive dilatancy with age. Barton et al. also show increased positive dilatancy during shearing with age for undisturbed samples of uncemented aged British sands.

Thomas further discusses the effects of ageing, cyclic pre-straining, over-consolidation and cementation on other mechanical characteristics of sands, such as: critical stress which demarcates the stress range up to which the aged/pre-strained/over-consolidated/cemented sand exhibits stiff response;

shear/deformation modulus; penetration resistance, and resistance to liquefaction.

5.5 Ageing, Interparticle Contacts and Mechanical Characteristics of Aged Sands

Based on the available information it is now generally agreed that due to secondary compression and cyclic pre-straining mechanisms (due to seismic, wind and wave loadings) results in changes in grain contact morphology (and some reduction in porosity) which is primarily responsible for increase in stiffness, shearing resistance, penetration resistance and resistance to liquefaction of in-situ saturated sands.

Here a brief discussion of the excellent data available for aged uncemented British sands will be presented and some quantitative relationships between grain contact index and unconfined compressive strength of sands and weak sandstones will be examined. Increase in c' and f' with increasing grain contact index and also changes in-situ porosity during ageing side by side with development of microfabric will be highlighted.

For the British aged sands the number of grain contacts (N) per grain (called contact index by Palmer and Barton) increases with age as shown in Fig. 55. As a consequence of increase in number of contacts per grain, there is a reduction in porosity with age as depicted in Fig. 56. It will be seen that in-situ porosity of most aged sands is less than the range of minimum remoulded porosities determined in the laboratory. This difference is a clear measure of well developed imcrofabric during ageing, especially of uncemented sands, which is well illustrated in Fig. 9. It will also be seen that a change in type of contacts occurs, principally as progressive reduction in the number of tangential contacts with the corresponding increase in straight and concave/convex contacts with age.

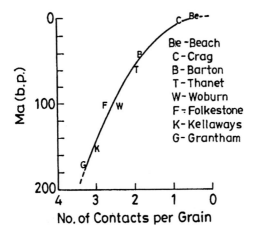

Fig. 55. Number of grain contacts per grain (contact index) vs chronological age.

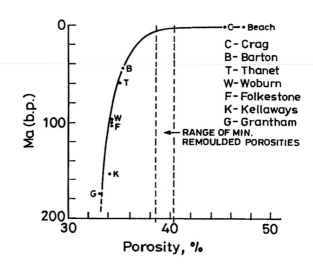

Fig. 56. Mean porosities *n*, of sands vs chronological age

5.5.1 Unconfined Strength and Microfabric Relationship

For the British aged sands Palmer and Barton had shown an increase in compressive strength, σ_c, with grain contact index (GC) and a decrease in σ_c with tangent index (TI) as reproduced in Fig. 57. Dovereiner had suggested similar trends for weak sandstones. Yudhbir and Rahim demonstrated that there is indeed a unique relationship between σ_c and measure of grain contact index, whether I_c as defined by Yuhdhbir and Rahim or GC as defined by Dobereiner. Such a relationship is depicted in Fig. 58 which can be expressed as:

$$\ln \sigma_c = 15.79 \, I_c + 1 \tag{7}$$

where σ_c is in kPa. Fig. 58 also lends support to the proposition that I_c = GC as assumed by Yudhbir and Rahim.

5.5.2 Development of Microfabric and Reduction in Porosity during Ageing

Ageing of sands in the field by secondary compression over very long periods of time (on geologic time scale) lelads to gradual changes in types of particle contacts from tangential to planar-type, which results in some reduction in porosity. In case of British aged uncemented fine sands, out of the total 0.120–0.13 decrease in porosity from the depositional value, roughly 0.09 (72%) is produced when the change in number of grain contacts per grain is practically

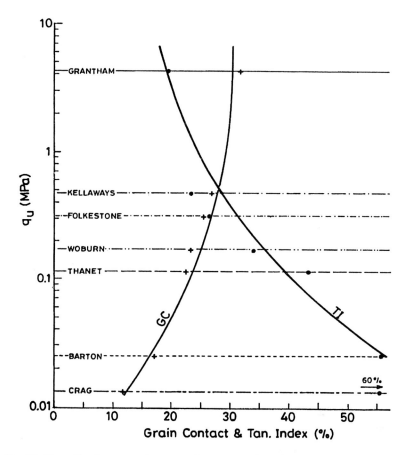

Fig. 57. Unconfined compressive strength, σ_c vs number of tangential contact (Tan. Index, TI) and Grain contact areas (Grain Contact, GC).

negligible (see Figs. 55, 56) and only 0.03-0.04 (28%) reduction in porosity occurs when the major increase in grain contacts takes place. It would seem that during the slow secondary compression phase when grain contacts are changing from predominantly tangential-type planar-type, very little change in porosity takes place. Most of the decrease in porosity is during the compaction under overburden when majority of the few grain contacts present are still of tangential type. Therefore, the increased stiffness, dilatational component of shear, penetration resistance and resistance to liquefaction exhibited by uncemented aged sands in the field is primarily due to very effective particle interlocking (highly developed microfabric) and there is very little contribution from decrease in porosity during ageing. It may be safely stated that the progressive development of microfabric by decrease of tangential contacts and increase in planar-type contacts with age (Fig. 9) takes place at a practically constant volume.

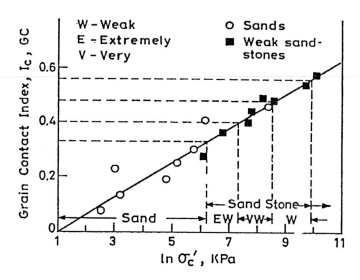

Fig. 58. Grain Contact Index, I_c, GC versus *ln* σ_c

Based on these observations, it would seem that Holocene and Pleistocene sands have developed a fabric with primarily tangential-type contacts with very small value of I_c or GC and for these sands the increase in stiffness, dilatational component of shearing resistance, penetration resistance and resistance to liquefaction due to to ageing is not as marked as that for those aged sands where the microfabric is dominantly controlled by the planar-type interlocked contacts. The Recent and Pleistocene sands may thus be classfied as lightly aged ones for which the field relative density $I_D \leq 1.0$ (as based on remoulded state e_{min} values) compared to those which have developed microfabric with predominantly planar-type interparticle contacts and with $I_D > 1.0$. These latter types may be termed highly aged sands. Canadian locked sands which are cemented and also possess a highly developed microfabric with planar-type interparticle contacts would certainly exhibit much more increase in stiffness and dilatational component of shearing resistance, penetration to resistance and resistance to liquefaction compared to highly aged British fine sands. For the same value of I_D (>1.0) it is suggesteed that highly aged uncemented sands develop relative soft planar-type interparticle contacts compared to stiff and brittle ones for "locked" aged sands. One logical inference from this discussion follows according to which laboratory studies to investigate development of microfabric and response of aged sands on the laboratory time scale (or even human-life span scale) are not likely to yield any significant measurable effects of ageing particularly if conventional laboratory test setup are used, and therefore much effort must be devoted to in-situ study of aged sands using surface wave measurements

(see Woods) or laboratory studies using bender elements and auto data logging system for tests on very high quality undisturbed samples of aged sands (for example those obtained by in-situ freezing). More data similar to that reported by Palmer and Barton would help better understand both the mechanisms of ageing and also the in-situ mechanical characteristics of these sands.

6. CONCLUSIONS

In this presentation it has been brought out that grain morphology and microfabric induced by ageing can be expressed in terms of indices which can then be correlated with mechanical characteristics of sands.

Influence of grain morphology on mechanical characteristics of sands has been highlighted. It is emphasized that correlation based on plenetration tests (SPT and/or SCPT), whether conducted in-situ in or in calibration chambers, obtained for a particular type of sand cannot be extrapolated for all types of granular deposits.

It is demonstrated that the influence of grain morphology is clearly reflected in the shaples of ICL and SSL, with the change in void ratio for a given change in P' increasing as grain angularity increases. Also the influence of particle angularity on undrained shear behavior of saturated sands is most convincingly demonstrated by examining its influence on the range of initial states for each of the four distinct modes of undrained shear behavior proposed by Ishihara.

Relationships between some undrained shear parameters, such as, $r_c = P_c/P'_s$, S_p/σ'_0, S_{us}/σ'_0 and state index, I_s have also been shown to be very significantly influenced by grain morphology. It is shown that for sands $e \tan \phi' = e_{ss} \tan \phi'_{ss} = e_{cr} \tan \phi'_{cv}$ is a constant and can be used to predict the effect of confining stress on ϕ'. Also making use of a well-defined relationship between ϕ'_{cv} and R, and $e \tan \phi' = $ constant, the dilatational component of shearing resistance, $(\phi' - \phi'_{cv})$ can be estimated at different values of relative density. The predicted patterns show reasonable agreement with the relationship proposed by Bolton.

Influence of grain morphology on stress-strain-time behavior of sands during 1-D deformation has been highlighted. It is shown that the secondary compression and cyclic pre-straining mechanisms of ageing of sands give identical results whereas over-consolidation indicates a much stiffer response. It is recommended that cyclic pre-straining can be used to investigate ageing of sands. During cyclic pre-straining it is observed that the size of the hysteresis loop (after 100 cycles) for rounded sand is much smaller than that of angular sand. Also the average slope (reciprocal of the slope is a measure of average elastic modulus) of the stabilized loop is steeper for angular sands (small average elastic modulus) compared to rounded sand. Characteristics of the stabilized hysteresis loop may thus be used as a reliable index of grain morphology of natural sands. Furthermore, the strain-time behavior is shown to be an elegant way to demonstrate the influence of grain morphology and also the constancy of C_α/C_c as proposed by Mesri.

Based on limited laboratory test results available, both triaxial and direct shear, for sands subjected to ageing by secondary consolidation and cyclic pre-straining, it is observed that ϕ' is practically constant (irrespective of ageing period on laboratory time scale) whereas the volume contraction during shearing reduces with increased ageing period and additional strain induced by cyclic pre-straining. Results based on tests on undistrubed aged sands, however, indicate increase in c' and ϕ' (for linearized interpretation of failure envelope) which is shown to be intimately related to the increase in grain contact index with time (on geologic scale). A unique relationship is demonstrated between grain contact index and unconfined compressive strength of aged uncemented sands and weak sand stones.

Based on data available for uncemented aged UK sands, it is seen that roughly 72% of change in porosity with reference to depositional value occurs when the grain contacts are few ($I_c \leq 0.04$) and are predominantly tangential type. The remaining 28% change in porosity occurs as I_c increases form 0.04 to 0.46 and majority of the contacts are planar-type (interlocked); these changes in contact types occur over very very long periods of time (on geologic time scale). It is thus inferred that laboratory investigations into development of microfabric during ageing and response of aged sands are not likely to yield any significant measurable effects, particularly if conventional laboratory test setup like oedometer are used. It is further recommended that much greater emphasis needs to be placed on in-situ investigations of aged sands using geophysical site characterization techniques; especially seismic method called SASW. Simultaneously sophisticated laboratory studies using bender elements and auto data logging system for test on high quality undisturbed samples of the type depicted in Fig. 9 are needed to improve our understanding of the mechanisms of ageing and their effect on mechanical characteristics of aged sands.

REFERENCES

Barton, M.E., et al. A geotechnical investigation of two Hampshire Tertiary Sand Beds: are they locked sands? Quart. Journal of Engineering Geology, 19, 399-412 (1986).

Been, K. and Jefferies, M.G. A state parameter for sands, Geotechnique, 35(2), 99-112 (1985).

Belloti, R. et al. Laboratory validation of in-situ tests, Geotechnical Engineering in Italy, an over-view published on the occasion of ISSMFE Golden Jubilee, 11th ICSMFE, San Francisco, U.S.A., (1985).

Bhatia, S.K., and Soliman, A.F. Frequency distribution of void ratio of granular materials determined by an image analyzer, Soils and Foundations, 30(1), 1-16 (1990).

Bolton, M.D.The strength and dilatancy of sands, Geotechnique, 36(1), 65-78(1986).

Burland, J.B., and Burbidge, M.C. Settlement of foundations on sands and gravel, Proc. of the Institution of Engineers, 78(1), 1325-1381 (1985).

Castro. G. et al. Liquefaction induced by cyclic loading, Report to National Science Foundation, Washington, D.C., No. NSF/CEE – 82018, (1982).

Cornforth, D.H. Prediction of drained shear strength of sands from relative density measurements, Evaluation of Relative Density and its Role in Geotechnical Engineering Projects Involving Cohesionless Soils, ASTM, STP, 523, 281-303 (1973).

Daramola, O. Effect of consolidation age on stiffness of sand, Geotechnicque, 30(2), 213-216 (1980).

De Jaeger, J. Influence de la morphologie des sables sur leur compartement mecanique, Ph.D. Thesis, Universite Catholique de Louvain (1991).

De Jaeger, J. Influence of grain size and shape on the dry sand behavior, Proc. 13th ICSMFE, New Delhi, India, 1, 13–16 (1994).

Dobereiner, L. Engineering geology of weak sandstones, Ph.D. Thesis, Imperial College, University of London (1984).

Dusseault, M.B., and Morgenstern, N.R. Locked sands, Quarterly Journal of Engineering Geology, 12, 117-131 (1979).

Grivas, D.A., and Harr, M.E. Particle contacts in discrete materials, Journal of the Geotechnical Engineering Div., ASCE, 106(5), 559-564 (1980).

Hansbo, S. A new approach to the determination of shear strength of clay of by the fall-cone test, Royal Swedish Geotechnical Institute Proc. No. 14 (1957).

Houlsby, G.T. Theoretical analysis of the fall cone test, Geotechnique, 32(2), 111-118 (1982).

Ishihara, K. Liquefaction and flow failure during earthquakes, Geotechnique, 43(3), 351-415 (1993).

Jamiolkowsky, M. et al. New correlations of penetration tests for design practice, Penetration Testing, ISOPT-1, De Ruiter (ed.), 1, 263-296 (1988).

Janbu, N. Soil models in offshore engineering, Geotechnique, 35(3), 241-281 (1985).

Kuo, C.Y., and Frost, J.D. Quantifying the fabric of granular materials—an image analysis approach, Report No. GIT-CEE/GEO-95-1, School of Civil and Environmental Engineering, The Geogia Institute of Technology, U.S.A. (1995).

Lee, K.L., and Seed, H.B. Drained strength characteristics of sands, Journal of SMFE Div., ASCE, 93(6), 117-141 (1967).

Mesri, G. et al. Post densification penetration resistance of clean sands, Journal of Geotechnical Engineering, ASCE, 116(7), 1095-1115 (1990).

Meyerhoff, G.G. The ultimate bearing capacity of wedge-shaped foundations, Proc. 5th ICSMFE, Paris, 2, 105-109 (1961).

Oda, M. Initial fabrics and their relations to mechnical properties of granular material, Soils and Foundations, 12(1), 17-36 (1972).

Oda, M. et al. Some experimentally based fundamental results on the mechanical bahavior of granular materials, Geotechnique, 30(4), 479-495 (1980).

Palmer, S.N., and Barton, M.E. Porosity reduction, microfabric and resultant lithification in UK uncemented sands, Diagenesis of Sedimentary Sequences, Marshal, J.D. (ed.) Geological Society Special Publication No. 36, 29-40 (1987).

Powers, M.C. A new roundness scale for sedimentary particles, Journal of Sedimentary Petrology, 23(2), 117-119 (1953).

Rahim, A. Effect of morphology and mineralogy on compressibility of sands, Ph.D. Thesis, Department of Civil engineering, IIT Kanpur, India (1989).

Schemertmann, J.H. Static cone to compute static settlement over sand, Journal of SMFE Div., ASCE, 96(3), 1011-1043, (1970).

Schmertmann, J.H. The mechanical ageing of soils, Journal of Geotechnical Engineering, ASCE, 117(9), 1288-1329 (1991).

Sahu, J.T., and Yudhbir. Model tests on sands with different angularity and mineralogy, Soils and Foundations, 38(4), 151-158 (1998).

Stephenson, D. Rockfill in Hydraulic Engineering, Elsevier Scientific Publishing Company (1979).

Teachavorasinskun, S. et al., Effects of the cyclic prestraining on dilatancy characteristics and liquefaction strength of sand, pre-failure Deformation of Geomaterials, Shibuya, Mitachi and Miura (eds.), Balkema, Rotterdam, 75-80 (1994).

Terzaghi, K., and Peck, R.B. Soil Mechanics in Engineering Practice, John Wiley and Sons, New York (1996)

Terzaghi, K. et al. Soil Mechanics in Engineering Practice, John Wiley and Sons, New York (1996)

Thomas, J. Morphology, ageing and engineering behavior of sands, Ph.D. Thesis, Department of Civil Engineering, I.I.T. Kanpur, India (1997).

Woods, R.D. Geophysical characterization of sites, edited vol. by ISSMFE Technical Committee # 10, for XIII ICSMFE, New Delhi, India (1994).

Youd, T.L. Factors Controlling Maximum and Minimum Densities of Sands, Evaluation of Relative Density and its Role in Geotechnical Projects Involving Cohesionless Soils, ASTM, STP, 523, 98-112 (1973).

Yudhbir and Rahim, A. Image analysis applied to Geotechnical engineering, Proc. ASCE Engineering Foundation Conference on Digital Image Processing: Techniques and Applications in Civil Engineering, Hawaii, U.S.A (1993).

Yudhbir and Thomas, J. Grain morphology, crushing and static cone penteration of natural sands, Proc. Bengt B. Broms Symposium in Geotechnical Engineering, 469-481 (1995).